INTRODUCTION TO
NUMERICAL COMPUTATIONS

This is a volume in
COMPUTER SCIENCE AND APPLIED MATHEMATICS
A Series of Monographs and Textbooks

Editor: WERNER RHEINBOLDT

A complete list of titles in this series appears at the end of this volume.

INTRODUCTION TO
NUMERICAL COMPUTATIONS

James S. Vandergraft

Computer Science Department
University of Maryland
College Park, Maryland

ACADEMIC PRESS   *New York   San Francisco   London*   *1978*
A Subsidiary of Harcourt Brace Jovanovich, Publishers

ACADEMIC PRESS, INC.
111 Fifth Avenue, New York, New York 10003

United Kingdom Edition published by
ACADEMIC PRESS, INC. (LONDON) LTD.
24/28 Oval Road, London NW1   7DX

Library of Congress Catalog Card Number: 78–18672

ISBN 0–12–711350–9

PRINTED IN THE UNITED STATES OF AMERICA

78 79 80 81 82   9 8 7 6 5 4 3 2 1

Dedicated to

Werner C. Rheinboldt

CONTENTS

PREFACE

The aim of this book is to introduce numerical algorithms as they are used in practice. In other words, it is assumed throughout that the algorithms will eventually be programmed and run on a modern digital computer. Accordingly, the effect of computer arithmetic with floating point numbers during the execution of such programs is always presented as an integral part of the study of the algorithms. Accuracy, efficiency, and reliability of the various methods are discussed in detail in order to show clearly their advantages and limitations. Whenever possible, practical and realistic techniques for estimating the accuracy of the computed solutions are given. At the same time, by analyzing the various algorithms completely and in a uniform manner, general techniques for algorithm analysis are illustrated. An understanding of these techniques provides a solid foundation for more advanced work in algorithm development.

The level of this book is suitable for the advanced undergraduate. More generally, the material included here is accessible to a rather wide audience because the required mathematical background has been kept at an elementary level. Most of the specific mathematical tools that are used are covered in a standard first-year calculus course. Some of the discussions, however, require mathematical maturity beyond this. For this reason, further practical experience with mathematics or additional courses in, say, linear algebra or differential equations are recommended. In addition, knowledge of some numerically oriented programming language is essential. Most methods are described by using an informal language that is based on structured programming ideas. It is expected that the serious student will program and run many of these methods in order to gain experience with their performance.

Most of the usual topics contained in introductory numerical analysis textbooks are included. All of the well-known and most frequently used algorithms for interpolation and approximation, numerical differentiation and integration, solution of linear systems and nonlinear equations, and for solving ordinary differential equations are discussed. Topics that were omitted are those that require more advanced mathematics (such as eigenvalue problems) or are not well suited for computer use (such as difference tables for interpolation). Some additional topics that are often not found in elementary textbooks are a complete discussion of computer arithmetic, problems that arise in the computer evaluation of functions, and a detailed treatment of cubic spline interpolation. Furthermore, by using a special notation devised by G. W. Stewart, we have been able to include simple but complete rounding error analyses of nearly every algorithm that is mentioned.

ACKNOWLEDGMENTS

The first draft of this text was written while on sabbatical at the Eidgenossische Technische Hochschule, Zürich. I would like to express my thanks to Professors P. Henrici, J. Marti, and B. Eckmann for making this visit possible. I am also very grateful to Mrs. Dawn Shifflett for the excellent typing of the manuscript and to various friends and associates for their encouragement. Finally a special thanks to my teacher, friend, and colleague, Werner C. Rheinboldt, for his careful reading and criticisms of the manuscript. Many of the best features of this text are due to him.

INTRODUCTION TO
NUMERICAL COMPUTATIONS

Chapter 1

BASIC ASPECTS OF NUMERICAL COMPUTATIONS

1.1 Numerical Algorithms

This section introduces some of the fundamental ideas involved in numerical computations. We stress especially those ideas that are peculiar to *numerical* computations as contrasted with nonnumerical aspects of computer applications.

1.1.1 Computing with Real Numbers

The term "numerical computations" refers to the use of computers to solve problems involving **real numbers**. Many real numbers can be expressed by a finite string of digits. In a certain mathematical sense, however, "most" real numbers require an infinite string of digits to represent them, even allowing for changes in the number base that is used. Some of the common real numbers that occur in applications and involve infinite representations are π, $\sqrt{2}$, e, etc. Computers, on the other hand, are finite machines. In fact, most scientific computers allow only a certain fixed quantity of digits to be used for the representation of a single number. Thus the actual set of numbers available for any numerical computation with a computer is comparatively small. Data, intermediate results, and the final answers, must all be approximated by these special computer numbers. It is important to realize that these approximations may have a serious effect on the overall computation. For example, most computer subroutines for evaluating trigonometric functions, e.g., sin θ, cos θ, first subtract (or add) multiples of 2π to the argument in order to obtain an argument between $-\pi$ and π. This change in the argument should not change the result because, for example, sin $\theta =$ sin$(\theta - k2\pi)$, $k = 1, 2, \ldots$. But the number π cannot be exactly represented in the computer, and hence this subtraction process *does* change the result. In fact, because of this, it is extremely difficult to compute sin θ accurately for any large value of θ.

More generally, *any* numerical process that involves real numbers that cannot be represented exactly in the computer must be used with care. Many simple " solutions" to problems are computationally infeasible because they rely too strongly on certain specific numbers that are not computable. Even when the problem and the solution involve only computer-representable numbers, the *computation* may introduce numbers that cannot be represented in the computer. In practice, one is rarely able to compute the *exact* solution to a problem, even though the solution may turn out to be numbers that can be represented exactly in the computer.

1.1.2   Computing with Functions

Most mathematical problems involve functions. There are only two ways to represent a function in a computer: by a table of values or by a subprogram that can compute values of the function at specific points. Because computers can perform only four mathematical operations—addition, subtraction, multiplication, and division—it follows that the only functions that can be evaluated by a computer are those that require just these four operations. Recall that functions that involve only the four basic arithmetic operations are called **rational functions**. Such functions can always be written in the form

$$r(x) = p(x)/q(x), \tag{1.1}$$

where $p(x)$ and $q(x)$ are polynomials in x; that is,

$$p(x) = a_0 + a_1 x + a_2 x^2 + \cdots + a_n x^n$$

$$q(x) = b_0 + b_1 x + b_2 x^2 + \cdots + b_m x^m.$$

A further restriction is imposed by the fact that, as just observed, the coefficients $a_0, a_1, \ldots, a_n$ and $b_0, b_1, \ldots, b_m$ must be among the limited set of numbers that can be stored in the computer. Also, of course, the values of x for which the function can be evaluated must be in this set. Thus, any function that occurs in a mathematical problem must first be approximated by a suitable rational function (1.1). This is a necessity created by the design of the computer and cannot be avoided by clever techniques or complicated algorithms. Rather, the algorithms that are used in numerical computations must be carefully chosen so that these various approximations do not destroy the accuracy of the computed results.

1.1.3   Discretizations

Most numerical problems are initially formulated in terms of mathematical objects such as functions, curves, and domains, or involve operations such as differentiation, integration, and limits. As just observed, functions must be represented in a computer by a table of values, or by a rational approximation. Similarly, geometrical domains can, at best, be represented by sets of points or by subprograms that compute points that are contained in them. Operations such as integration and differentiation must be replaced by summation and differencing. The process of replacing continuous-type mathematical concepts by computable objects is called *discretization*. Often it is a crucial step in solving a mathematical problem numerically. To illustrate this idea, consider the problem of computing the integral

$$\int_0^1 f(x)\, dx. \tag{1.2}$$

We assume that f is a continuous function that can be evaluated at any x in $[0, 1]$. **Numerical integration** methods for approximating (1.2) choose some points $x_0, x_1, \ldots, x_n$ in the interval $[0, 1]$. The integral is then replaced by a summation

$$\sum_{k=0}^n a_k f(x_k) \tag{1.3}$$

where the coefficients $a_0, a_1, \ldots, a_n$ are suitably chosen. To be more precise, suppose that the points $x_0, x_1, \ldots, x_n$ are evenly spaced in the interval $[0, 1]$; that is, $x_k = kh$, $k = 0, 1, \ldots, n$, with $h = 1/n$. Then the integral (1.2) can be written as

$$\int_0^1 f(x)\, dx = \sum_{k=0}^{n-1} \int_{x_k}^{x_{k+1}} f(x)\, dx.$$

If each of these integrals is approximated (see Fig. 1.1) by the area of the

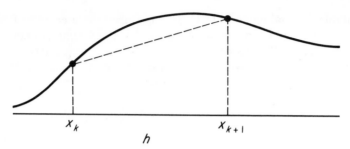

Fig. 1.1 Simple trapezoidal approximation.

trapezoid $\{x_k, x_{k+1}, f(x_{k+1}), f(x_k)\}$, i.e.,

$$\int_{x_k}^{x_{k+1}} f(x)\, dx \simeq \tfrac{1}{2}h[f(x_k) + f(x_{k+1})],$$

then the approximation

$$\int_a^b f(x)\, dx \simeq \sum_{k=0}^{n-1} \tfrac{1}{2}h[f(x_k) + f(x_{k+1})]$$

$$= \tfrac{1}{2}h[f(x_0) + 2f(x_1) + \cdots + 2f(x_{n-1}) + f(x_n)]$$

is just the well-known **trapezoidal rule**. It has the form (1.3) with $a_0 = a_n = \tfrac{1}{2}h$, and $a_k = h$, $k = 1, 2, \ldots, n-1$. The numerical algorithm simply computes this sum.

As a second example of discretization, consider the differential equation

| Example: Differential equation | $y'(x) = 2xy(x) - 2x^2 + 1, \quad 0 \le x \le 1$ $y(0) = 1.$ | (1.4) |

The exact solution is the function $\bar{y}(x) = e^{x^2} + x$, defined on the interval $[0, 1]$. Most numerical algorithms for solving (1.4) determine a set of values $y_0, y_1, \ldots, y_n$ that approximate the values of $\bar{y}(x_0), \bar{y}(x_1), \ldots, \bar{y}(x_n)$, where x_0, $x_1, \ldots, x_n$ are specified points between 0 and 1. That is, the computed solution will be a *table of values* for the function $\bar{y}(x)$. Once it is realized that the computed solution must be a *discretized* function, it is natural to discretize the differential equation itself. This may be done, for example, by replacing the derivative with a divided difference, that is, by introducing the approximation

$$y'(x_k) \approx (y_{k+1} - y_k)/(x_{k+1} - x_k). \tag{1.5}$$

If, for simplicity, the points $x_0, x_1, \ldots, x_n$ are chosen to be evenly spaced

$$x_k = kh, \quad k = 0, 1, \ldots, n, \quad h = 1/n,$$

so that $x_0 = 0$, $x_n = 1$, and $x_{k+1} - x_k = h$, $k = 0, 1, \ldots, n-1$, then (1.4) becomes

$$(y_{k+1} - y_k)/h = 2x_k y_k - 2x_k^2 + 1.$$

We can rewrite this as

Discretized equation	$y_{k+1} = y_k + h(2x_k y_k - 2x_k^2 + 1), \quad k = 0, 1, \ldots, n$
	$y_0 = 1.$

$$(1.6)$$

Thus, we obtain a discretized version of (1.4) that can be easily "solved" computationally for $y_1, y_2, \ldots, y_n$. The important point to observe here is that the numerical algorithm only attempts to solve the *discrete problems* (1.3) and (1.6) and knows nothing about the original continuous problems (1.2) and (1.4).

1.1.4 Iterations

Mathematical problems and their solutions are often formulated in terms of some kind of infinite process. Examples of such processes that have been discussed already are integration, differentiation, and, more generally, limits and infinite summations. Many such infinite processes can be formulated as *iterations*. That is, the desired object is the limit of a sequence $x_0, x_1, \ldots, x_k, x_{k+1}, \ldots$ of *iterates*. A particular iterate is determined according to some formula from the previous iterates.

For example, the constant e is defined by the infinite series

$$e = \sum_{k=0}^{\infty} \frac{1}{k!} \quad (0! = 1). \tag{1.7}$$

This can be written as the iteration

Example: Iteration for finding e	$x_k = x_{k-1} + (1/k!), \quad k = 1, 2, \ldots$
	$x_0 = 1.$

$$(1.8)$$

(In fact, an infinite series is *defined* to be the limit of its partial sums.) Similarly, the quantity $\sqrt{2}$ is defined to be the (positive) solution to the equation $x^2 = 2$. It can be shown (see Chapter 7) that the solution to this equation is the limit of the sequence defined by the iteration

| Example: Iteration for finding $\sqrt{2}$ | $x_k = \frac{1}{2}[x_{k-1} + (2/x_{k-1})],$ $k = 1, 2, \ldots$ $x_0 = 2.$ | (1.9) |

Iterative formulations of problems often suggest useful computational solutions. However, two problems arise here. The first one is due to the finiteness restriction imposed by limited time and computational resources. That is, only a finite number of iterates can be computed, and hence a decision must be made as to which iterate should be accepted as the desired approximation. This question of **stopping criteria** for iterative processes is a difficult problem that occurs often in numerical computations.

The second problem that arises in connection with iterative processes is the question of whether the process actually converges and, if so, does it converge to the desired quantity? For example, to solve the equation $x^2 = 2$, we might also consider the iteration

$$x_k = 2/x_{k-1}, \quad k = 1, 2 \cdots$$
$$x_0 = 2. \tag{1.10}$$

By taking limits as $k \to \infty$ in (1.10), it is clear that *if* the limit $x^* = \lim x_k$ exists, then x^* satisfies $x^* = 2/x^*$, i.e., $(x^*)^2 = 2$. But the sequence of values determined by (1.10) is $\{2, 1, 2, 1, 2, \ldots\}$, which obviously does not converge. In this example, the nonconvergence was discovered by actually computing the iterates. In more complicated iterations, it may be very expensive to compute even *one* iterate. For this reason, it is necessary to have a clear understanding of iterative processes so that one can predict in advance whether the iterates will be likely to converge.

1.1.5 Rounding and Truncation Errors

The previous discussions have pointed out several kinds of error that arise in the course of the numerical solution of mathematical problems. These errors can be separated into two groups—those caused by the finiteness of number representations in the computer and those caused by the finiteness of resources available for solving the problem. The former kind of error is called **rounding error** and occurs, for example, when data must be rounded to a certain number of digits before they can be accepted by the computer. More generally, rounding errors arise whenever numbers in the computer are combined, by means of one of the arithmetic operations, to produce a result that also must be approximated by a suitable computer number. In order to evaluate the effectiveness of any numerical algorithm, it is necessary to examine the effect of rounding errors. Indeed, many algorithms that would work quite well if there were no rounding errors fail completely in actual use because of them. Surprisingly, there are even situations in which the reverse is true; that is, there are algorithms that only work effectively because rounding errors perturb the results just enough to avoid certain anomalies.

The latter kind of error, which will be called **truncation error**, also occurs in several guises. The process of approximating functions by appropriate rational functions introduces this kind of error, as do the discretization processes discussed in Section 1.1.3. Also, the error caused by stopping an iteration after finitely many steps is of this type. In most cases, the truncation error is introduced *before* the numerical computation ever begins. For example, when the integral (1.2) is replaced by the summation (1.3), a certain truncation error is introduced, whereas the numerical process of evaluating the sum (1.3) only involves rounding error.

When truncation error is caused by a discretization process, it usually happens that the smaller this error is the more complicated is the resulting numerical process. Furthermore, generally the amount of rounding error is directly proportional to the complexity of the computation. Thus, in many cases, **reducing the truncation error causes the rounding error to increase**. Usually, these two errors will be analyzed separately. It must be realized, however, that there is often an indirect connection between them that must be taken into account.

EXERCISES

1   Use the sine function of your compiler to compute $\sin(k\pi)$ for $k = 1, 2, 3, \ldots, 100$. Observe the decrease in accuracy as k increases.

2   The following functions can all be evaluated using only the four basic arithmetic operations—addition, subtraction, multiplication, and division—but, as written, they do not have the form (1.1). Show how to rewrite them in this form.
   (a) $f(x) = x + (x^2 + 3x + 2)/(x + 3)$.
   (b) $g(x) = (x^2 + 2)/(x - 1) + (x - 1)/(x^2 + 2)$.
   (c) $h(x) = \{[(x + 1)(x - 1)]/(x + 2)\}/[(1/x) + 3]$.

3   Compute the values x_1, x_2, x_3, x_4 defined by (1.9) and compare them to the value $\sqrt{2} = 1.41421356 \ldots$.

4   Use (1.6), with $h = .1$, to solve (1.4). Compare your answers with the exact solution given by $\bar{y}(x) = e^{x^2} + 2$.

5   Which of the following errors are rounding and which are truncation errors?
   (a) Replace $\sin x$ by $x - (x^3/3!) + (x^5/5!)$.
   (b) Use 3.14159 for π.
   (c) Use the value x_{10} for $\sqrt{2}$ when $x_1, x_2, \ldots, x_{10}$ are given by (1.9).
   (d) Divide 1.0 by 3.0 and call the result .33333.

1.2   Unstable Problems (Ill-Conditioning)

Sometimes the main difficulty in solving a problem is caused by an extreme sensitivity of the solution to slight changes in the data. In this section we give several examples of such problems and examine the consequences of this sensitivity.

1.2.1   Solutions of Linear Equations

Consider the system of two linear equations

Example: Unstable linear system

$$x + 2y = 3$$
$$.499x + 1.001y = 1.5 \tag{1.11}$$

whose solution is $x = y = 1.0$. If the second equation is replaced by

$$.5x + 1.001y = 1.5,$$

then the solution becomes $x = 3$, $y = 0$ (and it is not hard to show that this is the *only* solution). Thus, the small change of the coefficient .499 to .500 has caused the solution to change drastically.

A mathematical problem in which the solution is very sensitive to changes in the data is said to be an **unstable** or **ill-conditioned** *problem*. Conversely, if the problem is such that small changes in the data cause equally small changes in the solution, then the problem is called **stable** or **well-conditioned**. Notice that we are not concerned here with a numerical technique for *finding* the solution. In the above example, the solutions were found by inspection. The important thing is to note how these exact solutions change when the data are changed. If a problem is unstable, then it may well happen that the various approximations mentioned in Section 1.1 cause drastic changes in the solution, even though they change the problem only slightly. Obviously, it is a waste of time to try to obtain a good approximation to this changed solution which is of no practical interest. The best that can be expected from a numerical algorithm in such a case is a warning that something may be amiss, and that perhaps the problem may be unstable. For example, good methods for solving linear equations, as described in Chapter 6, when applied to the system (1.11) can predict the fact that changes in the solution may be as much as 100 times larger than changes in the data.

1.2.2 Solutions of Nonlinear Equations

Let $f(x)$ be a nonlinear function that is continuous and can be differentiated. In order to examine the stability of the problem of solving the equation

$$f(x) = 0, \qquad (1.12)$$

consider small changes in f, which can be described by

$$\tilde{f}(x) = f(x) + \varepsilon g(x), \qquad (1.13)$$

where $g(x)$ is some other function, and ε is small. It can be shown, by using the implicit function theorem, that if x^* is a solution to (1.12), and $f'(x^*) \neq 0$, then for sufficiently small ε, $\tilde{f}(x) = 0$ has a solution $x^*(\varepsilon)$. Moreover, this solution satisfies

$$x^* - x^*(\varepsilon) \simeq \varepsilon g(x^*)/f'(x^*).$$

If $f'(x^*) = 0, f''(x^*) \neq 0$, then $x^*(\varepsilon)$ may still exist, but now

$$x^* - x^*(\varepsilon) \simeq (2\varepsilon g(x^*)/f''(x^*))^{1/2}.$$

Thus if $f'(x^*)$ is small or zero, then small changes in f may cause large changes in the solution to (1.12); that is, the problem (1.12) is unstable.

To illustrate this theoretical fact, consider the equation

Example: Unstable nonlinear equation

$$f(x) = x^3 + 96x^2 - 396x + 400 = 0, \qquad (1.14)$$

which has a solution $x^* = 2$. But for this function $f'(x^*) = 0$ and indeed the slightly modified equation

$$x^3 + 96x^2 - 396x + 399.9998 = 0 \qquad (1.15)$$

has a solution 2.001. Thus a change in the seventh digit of one coefficient of the equation has resulted in a change in the fourth digit of the solution. Changes in the seventh digit of input data are not uncommon, resulting often from conversion of decimal numbers to binary form. Because of the instability of the problem (1.14), it makes no sense even to look for a solution that is accurate to more than four figures. The real difficulty here is in detecting the instability of the problem and estimating its magnitude.

1.2.3 Subtraction

One of the most common problems that must be recognized as unstable is the simple operation of subtraction (or addition of numbers with opposite signs). Consider, for example, the numbers

Example:	$x = 12345678.0$
Two positive	
numbers	$y = 12345677.0$

whose difference is $z = x - y = 1.00000000$. But by slightly perturbing these numbers to

| Perturbed | $\tilde{x} = 12345678.1$ |
| numbers | $\tilde{y} = 12345676.9$ |

we obtain a difference $\tilde{z} = \tilde{x} - \tilde{y} = 1.20000000$. Thus, a change in the *ninth* figure of the data has caused a change in the *second* figure of the answer.

Of course, not all subtractions are unstable. The above phenomenon happens only when the two numbers are nearly equal. In this case we say that **subtractive cancellation** has occurred. This will be discussed in more detail in Chapter 3.

1.2.4   Differential Equations

As a final example of a typical mathematical problem that is unstable, consider the differential equation

Example:	$$y'(x) = (2/\pi)xy(y - \pi), \qquad 0 \le x \le 10 \qquad (1.16)$$
Differential	
equation	$$y(0) = y_0.$$

The exact solution to this equation is

| Unstable | $$y(x) = \pi y_0 / [y_0 + (\pi - y_0)e^{x^2}]. \qquad (1.17)$$ |
| solution | |

Hence, for $y_0 = \pi$, the solution is $y(x) = \pi$, for all x; whereas if $y_0 < \pi$, the solution tends to zero as $x \to \infty$, and for $y_0 > \pi$ the solution goes to infinity. Thus if the input data prescribed the value of π for y_0, then recalling that

$$\pi = 3.1415926536 \ldots$$

we see that the eight-digit (rounded) approximation 3.1415927 will lead to a solution that goes to infinity, while a nine-digit (rounded) approximation 3.14159265 will give a solution that goes to zero. Again, we stress the point that a numerical method for solving (1.16) can at best approximate the exact solution of the problem *as it appears in the computer*. But this exact solution changes drastically because of very small changes in the data. Notice also that parts of the "data" for the equation (1.16) are values of the right-hand side of the equation for various values of x and y. Since this also involves π, as well as the nonrational function e^{x^2}, there will also be approximations occurring here.

EXERCISES

1 Show that $x = y = 1.0$ is the *only* solution to (1.11) and $x = 3$, $y = 0$ is the only solution to the same system but with the second equation replaced by $.5x + 1.001y = 1.5$.

2 Solve the system

$$2x - 4y = 1$$

$$-2.998x + 6.001y = 2$$

using any method you know. Compare the solution with the solution to the system obtained by changing the second equation to $-2.998x + 6y = 2$. Is this problem stable?

3 Examine the stability of the equation

$$x^3 - 102x^2 + 201x - 100 = 0,$$

which has a solution $x^* = 1$. That is, change one of the coefficients (say, change 201 to 200), and show that $x^* = 1$ is no longer even close to a solution.

4 Verify that (1.17) is the solution to (1.16).

1.3   Unstable Methods

Many methods that have been proposed for solving certain problems would work quite well if exact arithmetic were used. The purpose of this section is to show how computer arithmetic and round-off errors can destroy the usefulness of an algorithm.

1.3.1   Subtractive Cancellation

It is well known from elementary algebra that the quadratic equation

Example:
Quadratic
formula

$$ax^2 + bx + c = 0, \qquad a \neq 0 \tag{1.18}$$

has at most two solutions, one of which is given by the formula

$$x = \frac{-b + \sqrt{b^2 - 4ac}}{2a}. \tag{1.19}$$

This is a mathematically correct statement that suggests a simple numerical method for solving (1.18). However, suppose that

$$a = 1, \qquad b = 1000.01, \qquad c = -2.5245315 \qquad (1.20)$$

so that the solution of (1.18), as given by (1.19), is $x_1 = .0025245 \ldots$ But in order to *compute* this solution using (1.19), it is necessary first to compute the square root of $(1000.01)^2 + 4 \cdot (2.5245315)$. If this square root is computed correctly to eight digits, the value of x given by (1.19) is .0025000000. This is a rather poor approximation since it provides only two-digit accuracy although eight digits were used for the square root. Obviously, there is something special about the data given here that causes this method to give an inaccurate result. In fact, it is not hard to see what is happening if the details are written out:

$$\tfrac{1}{2}(-b + \sqrt{b^2 - 4ac}) = \tfrac{1}{2}(-1000.0100 + 1000.0150) = .00250000.$$

Thus, the first six digits of the eight-digit square root 1000.0150 are cancelled out by the value of $-b$, so that the final result has only two remaining digits. Generally, the loss of digits through subtraction is called **subtractive cancellation** and is a major source of errors in numerical computation.

A numerical method that is basically good but that sometimes, either because of the data or because of the way in which the data are handled, produces extremely poor results, is said to be an **unstable method**. The term "unstable" is used to stress the nonuniformity of the accuracy; that is, sometimes the results are extremely accurate, sometimes extremely inaccurate. For example, if formula (1.19) is used to solve (1.18), but with the data

$$a = 1, \qquad b = -1000.01, \qquad c = -2.5245315,$$

then an accurate eight-figure result is obtained.

As another illustration of subtractive cancellation, consider the problem of evaluating the function

$$F(x) = \sin x - \cos x$$

for values of x near $\pi/4$. Since $\sin \pi/4 = \cos \pi/4 = \sqrt{2}/2$, subtractive cancellation may occur. Table 1.1 verifies this.

Subtractive cancellation is a frequent cause of instability in numerical methods, and great care must be taken to avoid it. We shall discuss some tchniques for this in Section 3.4.

1.3.2 Magnification of Errors—Recursions

Many numerical methods determine a sequence of values, each of which is obtained from previously computed values. If small errors in the earlier

TABLE 1.1 *Values of* $\sin((\pi/4) + 10^{-7}k)$ – $\cos((\pi/4) + 10^{-7}k)$

k	Computed values	Exact values, to eight figures
1	.18626451 × 10⁻⁶	.19340340 × 10⁻⁶
2	.32782555 × 10⁻⁶	.33482476 × 10⁻⁶
3	.46938658 × 10⁻⁶	.47624611 × 10⁻⁶
4	.61094761 × 10⁻⁶	.61766747 × 10⁻⁶
5	.75995922 × 10⁻⁶	.75908882 × 10⁻⁶

values are grossly magnified by later computations, then the final results may be very inaccurate. Consider, as an example, the problem of determining the value of the integral

$$\int_0^1 x^{20} e^{x-1} \, dx. \tag{1.21}$$

If we let $I_k = \int_0^1 x^k e^{x-1} \, dx$, then integration by parts gives

Example:
Unstable
recursion

$$I_k = x^k e^{x-1} \Big|_0^1 - k \int_0^1 x^{k-1} e^{x-1} \, dx = 1 - kI_{k-1}$$

and $I_0 = \int_0^1 e^{x-1} \, dx = e^{x-1} \big|_0^1 = 1 - (1/e)$. Thus we can find I_{20} by computing $I_1, I_2, \ldots, I_{20}$ *recursively* according to the formula

$$I_k = 1 - kI_{k-1}, \qquad k = 1, 2, \ldots, 20 \tag{1.22a}$$

with

$$I_0 = 1 - (1/e). \tag{1.22b}$$

TABLE 1.2 *Computed Values of (1.22).*

k	I_k
0	.63212056 × 10⁰
5	.14553225 × 10⁰
11	−.15222549 × 10⁰
15	−.75209072 × 10⁴
16	.12033551 × 10⁶
17	−.20457027 × 10⁷
18	.36822650 × 10⁸
19	−.69963034 × 10⁹
20	.13992607 × 10¹⁰

The UNIVAC 1108 produced the values for I_k shown in Table 1.2. It is not hard to see that this is nonsense for large k. For example, every I_k should be positive, which some of these are not. Furthermore, it can be shown that I_k actually decreases to zero as k increases, while the values in Table 1.2 increase in magnitude.

The large errors here are due to the fact that errors in I_{k-1} are *multiplied* by k and are then transmitted directly to I_k. Thus, even a small rounding error in I_0 is passed on to I_1 and then is multiplied by two when computing I_2. This doubled error is then multiplied by three when finding I_3, etc. By the time I_{20} is computed, the initial error in I_0 has been magnified by the factor $20! \simeq 2 \times 10^{18}$.

Many recursive computations exhibit this kind of instability. A technique for avoiding it, in certain cases, will be discussed in Section 2.5.

1.3.3   Magnification of Errors—Differential Equations

Another common example of a numerical computation in which errors made in early stages of a computation may have a strong effect on later stages is the numerical solution of differential equations. Consider, for example, the equation

$$y'(x) = y, \qquad 0 \le x \le 1$$
$$y(0) = 1 \tag{1.23}$$

whose exact solution is $y(x) = e^x$. If the derivative in (1.23) is replaced by the divided difference

$$\frac{-\tfrac{1}{2}y(x + 2h) + 2y(x + h) - \tfrac{3}{2}y(x)}{h}, \tag{1.24}$$

which will be derived in Section 5.1, then just as in Section 1.1.3, we obtain a discrete equation

Example:
Unstable
method for
differential
equations

$$y_{n+2} = 4y_{n+1} - 3y_n - 2hy_n, \qquad n = 0, 1, 2, \ldots, N$$
$$y_0 = 1 \tag{1.25}$$
$$y_1 = e^h.$$

However, if this is used with $h = .1$, then the value obtained for y_{10}, which should approximate $y(1) = e$, is $y_{10} = -6.55860$. The difficulty here is that the error in y_1 is multiplied by four when y_2 is computed. Similarly, the total error in y_2 is multiplied by four when computing y_3, etc. The stability of numerical methods for solving differential equations is an important and complicated topic that will be studied thoroughly in Chapter 8.

1.3.4   Stability of Algorithms

In Section 1.2 we gave several examples of unstable problems and in this section we have considered some unstable methods. A frequent mistake that is made when testing a new method is to apply it to an unstable problem. If the method introduces small rounding or truncation errors into the problem, then the instability of the *problem* will result in large errors. These errors should not be blamed on the method. In fact, the method may do quite well when applied to stable problems.

In order to properly judge numerical algorithms, the stability of the algorithm must not be confused with the stability of the problems to which the algorithm is applied. A naive definition of "stable algorithm" would probably involve a comparison of the "exact solution" with the "computed solution." As observed above, however, the difference between these solutions might be large only because the problem is unstable, not because the method is unstable. A **forward error analysis** of an algorithm does this kind of comparison. The important point is that such an analysis may be misleading.

Another kind of stability analysis for algorithms is the so-called **backward error analysis**. This type of analysis is independent of the stability of the *problem*, and hence is a true indication of the stability of the *method*. The basis for this analysis is to consider the *computed solution* as being the *exact solution* to a slightly perturbed problem. The method is stable if this perturbation is sufficiently small. Figure 1.2 illustrates the case when a stable

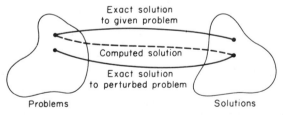

Fig. 1.2   A stable algorithm applied to stable problems.

algorithm is applied to a stable problem. Note that the computed solution is close to the exact solution. In Fig. 1.3, however, a stable algorithm is applied to an unstable problem. Here the computed solution is not close to the exact

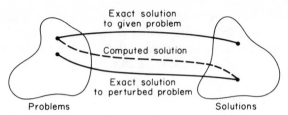

Fig. 1.3   A stable algorithm applied to unstable problems.

solution. The error, however, is caused by the problem, not the algorithm. Finally, in Fig. 1.4 we illustrate an unstable method applied to a problem. Here the instability of the method is reflected as a large perturbation in the problem.

We shall not attempt to give a precise definition of "stable algorithm." However, we shall analyze algorithms, sometimes in both the forward and backward senses, and use this analysis to recommend certain algorithms.

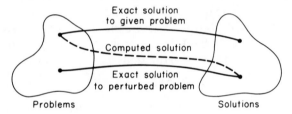

Fig. 1.4   An unstable algorithm.

EXERCISES

1   Show that the solution given by (1.19) can be rewritten as

$$x = -2c/(b^2 + \sqrt{b^2 - 4ac}).$$

That is, show that these two expressions for x are equivalent.

2   Write a computer program to evaluate (1.19) for several values of a, b, c, with b large and positive and a, c of moderate size. Compare the results with values given by the formula in Exercise 1 above.

3   If your compiler language has double-precision capability, evaluate the function $f(x) = (1 - \cos x)/x^2$ for $x = 1, .1, .01, .001, \ldots$ in both single and double precision. Compare the results.

4   Try to compute the value of the integral

$$I_{20} = \int_0^1 x^{20} \sin \pi x \, dx$$

using the recursion

$$I_k = (1/\pi) - [k(k-1)/\pi^2]I_{k-2}, \qquad k = 2, 4, 6, \ldots, 20.$$

Discuss the stability of this recursion. (*Note:* This recursion is obtained by integrating by parts *twice*.)

5 Use (1.25) with $h = .01$ to solve (1.23). Compare y_{100} with $y(1) = e$.

Suggestions for Further Reading

There are a great many texts that cover the basic ideas of numerical computations. Among those that extend or complement the material presented here are Isaacson and Keller (1966), Dorn and McCracken (1972), Hildebrand (1974), Conte and deBoor (1972), Dahlquist, et al. (1974), Henrici (1964), and Ralston (1965). In addition, there are several journals that are published at regular intervals and that contain up to date articles on numerical computations. These include *ACM Transactions on Mathematical Software, BIT* (a Scandinavian journal, but most of the articles are in English), *Communications of the ACM* (Association of Computing Machinery), *The Computer Journal* (published by the British Computer Society), *Mathematics of Computation* (published by the American Mathematical Society), *Numerische Mathematik* (published in Germany, but most articles are in English), and *SIAM Journal of Numerical Analysis* (published by the Society for Industrial and Applied Mathematics).

Chapter 2

COMPUTER ARITHMETIC

2.1 Number Representation

The decimal number system is convenient for use by people but very inconvenient for use by computers. In this section we review some basic ideas about number bases and number representations.

2.1.1 Base β Numbers

In order to understand how computers operate with numbers, it is necessary first to understand how numbers are represented inside the computer. Recall that any integer can be used as a base for a number system. That is, if β is some positive integer, then any number can be written in the form

$$\pm d_1 d_2 \ldots d_n \cdot d_{n+1} d_{n+2} \ldots d_{n+m} \ldots, \tag{2.1}$$

where $d_1, d_2, d_3, \ldots$ are digits between 0 and $\beta - 1$. The *value* of such a number is given by the formula

$$d_1 \beta^{n-1} + d_2 \beta^{n-2} + \cdots + d_{n-1} \beta + d_n$$
$$+ d_{n+1} \beta^{-1} + d_{n+2} \beta^{-2} + \cdots + d_{n+m} \beta^{-m} + \cdots. \tag{2.2}$$

The most commonly used number bases are 10, 2, 8, and recently 16. Numbers expressed with these bases are called, respectively, decimal, binary, octal, and hexadecimal numbers.

Most computers use the **binary number system** because of the fact that in this system only two digits are needed for representing numbers. That is, (2.1) becomes a string of zeros and ones. Hence, if a closed switch (or a positive polarity) denotes the digit 1, and an open switch (or negative polarity) denotes the digit 0, then any binary number can be represented by a series of opened and closed switches, or positive and negative polarities. A major drawback, however, with the binary number system is that it requires a great many binary digits to represent even rather small numbers. For example, the decimal number 45 requires six binary digits, 101101, and an eight-decimal-digit number may require as many as 27 binary digits. In most cases it is undesirable to print out numbers in their binary form because of the excessive number of zeros and ones that would result. To avoid this, the binary numbers are converted to octal (a process that is extremely simple) and then the octal form is printed.

Thus people use the decimal number system, whereas most computers make use of two additional number bases—binary and octal. Furthermore, certain modern computers (e.g., the IBM 360 series) use a base 16 number system. In order to make the following discussion as general as possible, we shall consider an arbitrary number base β and only use a specific base for examples. When there is a possibility for confusion, the notation $(\cdot)_\beta$ will be used to denote a base β number. For example, $(1011.10)_2$ and $(1101.1)_2$ denote binary numbers, while $(101.1)_{10}$ and $(287.56)_{10}$ mean that the numbers in parentheses are decimal numbers.

2.1.2 Conversion Difficulties

A difficulty with using a number base other than ten is that data are usually given in decimal form, and computed results, if left in, say, binary form, would have little meaning to most people. Thus, input and output numbers must be converted from one number system to another. Several aspects of this conversion problem will be discussed in later sections; however, a general remark is appropriate here. Specifically, as observed earlier, the *number of digits* needed to represent a certain value depends on the number base. In fact, the extreme case of this is the situation in which a certain quantity requires only a few digits with some number base but requires an *infinite* number of digits in another base. For example, the quantity " one third " is represented in the decimal system by .3333 ..., where the three dots indicate that the string of threes never ends. On the other hand, in the *base 3* number system, this quantity is written as .1, i.e.,

$$(.1)_3 = (.3333 \ldots)_{10}.$$

Other examples of this phenomena are

Number of digits versus number base	

$$(.27)_{10} = (.01000101000111101011 \ldots)_2$$
$$(.1)_{10} = (.0001100110011 \ldots)_2,$$

which show that certain simple decimal fractions may not have a finite binary representation. Such numbers cannot be represented exactly inside a computer no matter how many binary digits are available.

On the other hand, it can be shown (Exercise 2) that any number with finitely many binary digits also has finitely many decimal digits. In fact, any number with t binary digits has at most t decimal digits. For example,

$$(.1)_2 = (.5)_{10}$$
$$(.11)_2 = (.75)_{10}$$
$$(.1011)_2 = (.6875)_{10}.$$

However, most binary computers work with a rather large number of digits, at least 8 and as many as 64. To print out the result a 27-binary-digit computation as a 27-decimal-digit number would be expensive and extremely misleading. As we shall see in Section 2.4.2 a *computed* 27-binary-digit result may be accurate to only *eight* decimal digits.

Thus the data for a problem given as decimal numbers must be converted into suitable binary numbers for use by the computer. Conversely, computed results, as binary numbers in the computer, must be converted to *meaningful* decimal numbers.

2.1.3   Scientific Notation

Most scientists and engineers are accustomed to writing very large or very small numbers as medium-sized numbers multiplied by some power of 10. For example

Use of exponents	

.00001234   is written as $.1234 \times 10^{-4}$

12340000.   is written as $.1234 \times 10^{8}$.

This idea easily generalizes to base β numbers and leads to the floating point systems used by computers. The basis for this generalization is illustrated in the binary system by the equations:

$$(10)_2 = 2^1$$
$$(100)_2 = 2^2$$
$$(1000)_2 = 2^3$$
$$(.1)_2 = 2^{-1} \tag{2.3}$$
$$(.01)_2 = 2^{-2}$$
$$(.001)_2 = 2^{-3}$$
$$\text{etc.}$$

In the octal system we have

$$(10)_8 = 8^1$$
$$(100)_8 = 8^2$$
$$(.1)_8 = 8^{-1} \tag{2.4}$$
$$(.01)_8 = 8^{-2}$$
$$\text{etc.}$$

Thus the following equalities are valid:

$$(17.46)_8 = (.1746 \times 100)_8 = (.1746)_8 \times 8^2$$
$$(.001746)_8 = (.1746 \times .01)_8 = (.1746)_8 \times 8^{-2}$$
$$(101000.0)_2 = (.101 \times 1000000)_2 = (.101)_2 \times 2^6 \tag{2.5}$$
$$(.00101)_2 = (.101 \times .01)_2 = (101)_2 \times 2^{-2}.$$

In general, any base β number can be written in the form

Base β number with exponent

$$\pm .d_1 d_2 d_3 \cdots \times \beta^e, \tag{2.6}$$

where $d_1, d_2, d_3, \ldots$ are base β digits and e is a decimal integer (positive, negative, or zero) that determines the actual location of the decimal point.

EXERCISES

1 Use (2.2) to express the following base β numbers in decimal form:

$$(10.0)_8, \quad (1101)_2, \quad (10.0)_5, \quad (.1)_2,$$
$$(.11)_2, \quad (.401)_8, \quad (.17)_5, \quad (.775)_8,$$
$$(35.27)_8, \quad (101.101)_2, \quad (44.4)_5, \quad (7.77)_8.$$

2 Show, by examples or by a formal proof, that any number with t binary digits has at most t decimal digits. (Consider whole numbers, then fractions.)

3 Write the following base β numbers in the form (2.6):

$$(-12.345)_{10}, \quad (.0012345)_{10}, \quad (1.2)_8,$$
$$(.0123)_8, \quad (10101.001)_2, \quad (.000101)_2.$$

2.2 Floating Point Numbers

Computers represent numbers by using a system that is based on the scientific notation mentioned in the previous section. In this section we define this system precisely and consider some of the consequences of using it in numerical computations.

2.2.1 Representing Numbers in a Computer

All scientific computers work with certain fixed quantities of information as a single unit. These "units of information" are called *words* and consist of a fixed number of digits that can be used to represent program instructions, alphabetic symbols, numbers, etc. The number of digits in a word is called the **wordlength**. Our concern here is with the way in which *noninteger* numbers (i.e., real numbers) are represented by such a word of digits. It is important to understand this representation because of the strong influence it has on all numerical computations.

The standard way of representing real numbers in a computer is based on the scientific notation described in the previous subsection. That is, the representation has two parts—the *mantissa* or **fractional part**, which is the $\pm.d_1 d_2 d_3 \cdots$ of (2.6), and the **exponent** of the number base, which is the quantity e in (2.6). We shall always assume that the number is written with the decimal (or binary) point to the extreme left.

2.2.2 t-Digit Floating Point Numbers

Because of the limitation on the number of digits in a computer word, it is necessary to restrict the number of digits in the fractional part and to limit the size of the exponent. A t-**digit, base β floating point number** is any number that can be written as

$$\pm.d_1 d_2 \cdots d_t \times \beta^e,$$

where $.d_1 d_2 \cdots d_t$ is a base β fraction, i.e., $0 \leq d_i < \beta$, $i = 1, 2, \ldots, t$. Some examples of eight-digit base 10 floating point numbers are

$$.12345678 \times 10^0$$

$$-.87760000 \times 10^0$$

$$-.65476666 \times 10^{-3}$$

$$.10000000 \times 10^1.$$

Some six-digit base 16 numbers are

$$.89A0B1 \times 16^2$$

$$-.100000 \times 16^{-1}$$

$$.166666 \times 16^0$$

$$.ABC289 \times 16^1,$$

where A, B, C, ... denote the "digits" 10, 11, 12, ..., which are used in the hexadecimal system.

If the left most digit of the fraction part is nonzero, then the floating point number is said to be **normalized**. Notice that any nonzero floating point number can be normalized by simply adjusting the exponent. For example, in the eight-digit base 10 system

$$.00123456 \times 10^3 = .12345600 \times 10^1.$$

The fact that the fraction part of a normalized nonzero number cannot be less than β^{-1} will be important later.

2.2.3   Floating Point Number Systems

The limitation on the size of the exponent can be expressed by

Exponent range	$m \leq e \leq M,$

where m is a negative integer and M is some positive integer. Thus, the set of normalized floating point numbers that can be represented by a computer word is completely defined by specifying

(a)   the number base (10, 2, or 16, etc.),
(b)   the number of digits in the fractional part (t),
(c)   the range of the exponent (m, M).

It is convenient to use the notation $F(\beta, t, m, M)$ to denote the set of numbers consisting of zero and the normalized t-digit, base β floating point numbers with exponent between m and M. Every nonzero number x in this set can be written as

Normalized	
t-digit	$x = .d_1 d_2 \cdots d_t \times \beta^e, \quad 0 < d_1 < \beta, \quad 0 \le d_i < \beta,$
base β	
floating	$i = 2, \ldots, t,$
point	
numbers	$m \le e \le M.$

The number 0 will be represented as $+.000 \cdots 0 \times \beta^0$. Associated with any scientific computer is some basic set $F(\beta, t, m, M)$. Table 2.1 lists several popular computers, together with the values of β, t, m, M associated with each.

TABLE 2.1 *Floating Point Number Sets*

	β	t	m	M
UNIVAC 1108/1106	2	27	-128	127
IBM 360 and 370	16	6	-64	63
PDP 11	2	24	-128	127
CDC 6600	2	48	-975	1071
Burroughs B5500	8	13	-51	77

The values of t, m, M are essentially determined by the number of digits in a computer word. In order to be able to work with a larger set of floating point numbers, many computers have provisions for using two words to represent a single number. These **double-precision** (or **long precision**) **numbers** are characterized by values of $t, m,$ and M that are roughly twice as large as for single-precision numbers. Double-precision numbers will be used only for a few special applications in this book.

2.2.4 Distribution of Floating Point Numbers

Once a computer has been selected, the numbers that can be used in any computation with that computer are *only* those numbers that are in the corresponding set $F(\beta, t, m, M)$ (ignoring the possibility of double-precision numbers). Any number *not* in this set must be approximated by a floating point number in the set. One of the most serious objections to using such floating point numbers is the uneven way in which they are distributed. Since this unevenness also causes much of the complications that arise when analyzing numerical computations, it is worthwhile to examine this phenomenon carefully.

To be precise, consider the floating point system defined by $F(2, 3, -1, 2)$; that is, all numbers that can be written in the form $\pm .d_1 d_2 d_3 \times 2^e$, where $.d_1 d_2 d_3$ is a *binary* number and $-1 \leq e \leq 2$. These numbers (positive only) in increasing magnitude are

$$(.000)_2 \times 2^0 = 0$$
$$(.100)_2 \times 2^{-1} = \tfrac{1}{2} \times \tfrac{1}{2} = \tfrac{1}{4}$$
$$(.101)_2 \times 2^{-1} = (\tfrac{1}{2} + \tfrac{1}{8}) \times \tfrac{1}{2} = \tfrac{5}{16}$$
$$(.110)_2 \times 2^{-1} = (\tfrac{1}{2} + \tfrac{1}{4}) \times \tfrac{1}{2} = \tfrac{3}{8}$$
$$(.111)_2 \times 2^{-1} = (\tfrac{1}{2} + \tfrac{1}{4} + \tfrac{1}{8}) \times \tfrac{1}{2} = \tfrac{7}{16}$$
$$(.100)_2 \times 2^0 = (\tfrac{1}{2}) \times 1 = \tfrac{1}{2}$$
$$(.101)_2 \times 2^0 = (\tfrac{1}{2} + \tfrac{1}{8}) \times 1 = \tfrac{5}{8}$$
$$(.110)_2 \times 2^0 = (\tfrac{1}{2} + \tfrac{1}{4}) \times 1 = \tfrac{3}{4}$$
$$(.111)_2 \times 2^0 = (\tfrac{1}{2} + \tfrac{1}{4} + \tfrac{1}{8}) \times 1 = \tfrac{7}{8}$$
$$(.100)_2 \times 2^1 = \tfrac{1}{2} \times 2 = 1$$
$$(.101)_2 \times 2^1 = (\tfrac{1}{2} + \tfrac{1}{8}) \times 2 = \tfrac{5}{4}$$
$$(.110)_2 \times 2^1 = (\tfrac{1}{2} + \tfrac{1}{4}) \times 2 = \tfrac{3}{2}$$
$$(.111)_2 \times 2^1 = (\tfrac{1}{2} + \tfrac{1}{4} + \tfrac{1}{8}) \times 2 = \tfrac{7}{4}$$

etc.

If these numbers are represented as points along a line, they appear as shown in Fig. 2.1. That is, the points are clustered more closely together between $\tfrac{1}{4}$

Fig. 2.1   Positive numbers in the set $F(2, 3, -1, 2)$.

and $\tfrac{1}{2}$ than they are between 2 and 3. Thus if arbitrary numbers are to be approximated by these floating point numbers, it is clear that this approximation is more accurate for some numbers than for others. For example, numbers between $\tfrac{1}{4}$ and $\tfrac{1}{2}$ can be approximated rather well, whereas numbers between 2 and 3 may have an approximation error as large as $\tfrac{1}{2}$.

2.2.5   Rounded and Chopped Numbers

Since only relatively few numbers are contained in any set of floating point numbers, most real numbers will have to be approximated by one of these numbers. Consider a fixed set $F(\beta, t, m, M)$ of floating point numbers, and let x be any other number. The **rounded value of** x is defined to be that floating point number x_R that is closest to x, with the provision that if x lies exactly midway between two floating point numbers, then x_R will be the one of larger magnitude. The quantity

$$E_R = |x - x_R|$$

is called the **absolute rounding error**. The previous discussion shows that the size of the absolute rounding error may depend strongly on the number x. To illustrate this, consider again the example used above: $F(2, 3, -1, 2)$. Several rounded numbers, together with the corresponding absolute rounding errors are given here:

$$\left(\tfrac{1}{8}\right)_R = \tfrac{1}{4}, \quad E_R = \tfrac{1}{8} \qquad \left(2\tfrac{1}{4}\right)_R = 2\tfrac{1}{2}, \quad E_R = \tfrac{1}{4}$$

$$\left(\tfrac{9}{32}\right)_R = \tfrac{5}{16}, \quad E_R = \tfrac{1}{32} \qquad \left(2\tfrac{3}{4}\right)_R = 3, \quad E_R = \tfrac{1}{4}$$

$$\left(\tfrac{11}{16}\right)_R = \tfrac{3}{4}, \quad E_R = \tfrac{1}{16} \qquad \left(\tfrac{1}{16}\right)_R = 0, \quad E_R = \tfrac{1}{16}.$$

Note that, as expected from Fig. 2.1, numbers near $\tfrac{1}{2}$ generally have smaller absolute rounding errors than do numbers near 2.

A somewhat simpler way of obtaining a floating point approximation to a number x is to choose always the first floating point number that is smaller in magnitude than x. This number, denoted by x_C, is called the **chopped value** of x, and $E_C = |x - x_C|$ is called the **absolute chopping error**. Some examples, again with $F(2, 3, -1, 2)$ are

$$\left(\tfrac{1}{8}\right)_C = 0, \quad E_C = \tfrac{1}{8} \qquad \left(\tfrac{9}{32}\right)_C = \tfrac{1}{4}, \quad E_C = \tfrac{1}{32}$$

$$\left(\tfrac{1}{5}\right)_C = 0, \quad E_C = \tfrac{1}{5} \qquad (2.499)_C = 2, \quad E_C = .499.$$

2.2.6 Errors Caused by Rounding and Chopping

In order to analyze numerical computations, taking into account the effect of rounding error, it is necessary to obtain simple estimates for this error. For this purpose observe that it is always possible to write any nonzero decimal number x in the form

$$x = u \times 10^e + v \times 10^{e-t}, \qquad \tfrac{1}{10} \leq |u| < 1, \quad 0 \leq |v| < 1 \qquad (2.7)$$

when u is a normalized t-digit number. In fact, to obtain such a form, simply let u be the first t digits of x, with decimal point to the left; then v is given by the remaining digits of x. For example, with $t = 4$ we have

$$12.3456 = .1234 \times 10^2 + .56 \times 10^{-2}$$

Examples of (2.7)

$$-.0123456 = -.1234 \times 10^{-1} + (-.56) \times 10^{-5}$$

$$3.1415927\ldots = .3141 \times 10^1 + .5927\ldots \times 10^{-3}.$$

In exactly the same way any nonzero base β number can be written as

$$x = u \times \beta^e + v \times \beta^{e-t}, \qquad 1/\beta \le |u| < 1, \quad 0 \le |v| < 1, \qquad (2.8)$$

where u is a normalized t-digit base β number. Note that the restriction $|u| \ge 1/\beta$ simply says that the leftmost digit of u must be nonzero.

The first fact to be deduced from (2.8) is that if $m \le e \le M$, then $u \times \beta^e$ is just the *chopped* value of x in $F(\beta, t, m, M)$. This is verified by noting that $u \times \beta^e$ is indeed in $F(\beta, t, m, M)$, and any larger (in magnitude) such number must be larger than x in magnitude. Thus, to obtain the **chopped value** x_C, simply "chop off" those digits beyond the tth one and attach the proper exponent. More important than this procedure for finding x_C is the error formula

$$E_C = |x - x_C| = |v| \times \beta^{e-t} < \beta^{e-t}. \qquad (2.9)$$

Unfortunately, the right side of this inequality depends on e, which is a measure of the size of x. However, it also follows from (2.8) that $|x| \ge |u| \times \beta^e \ge \beta^{-1} \times \beta^e = \beta^{e-1}$. If $x \ne 0$, then this inequality combines with (2.9) to give

Bound on the relative chopping error	$$\|x - x_C\|/\|x\| < \beta^{e-t}/\beta^{e-1} = \beta^{1-t}. \qquad (2.10)$$

The quantity on the left of (2.10) is called the **relative chopping error** and will be denoted by ε_C; that is,

$$\varepsilon_C = \frac{|x - x_C|}{|x|}, \qquad \text{where} \quad x \ne 0. \qquad (2.11)$$

The quantity on the right of (2.10) is called the **unit chopping error** and is denoted by μ_C:

$$\mu_C = \beta^{1-t}. \qquad (2.12)$$

The **rounded value** of a number x can also be established from (2.8). The proper formula, assuming $m \le e \le M$, is

$$x_R = \begin{cases} u \times \beta^e & \text{if } |v| < \tfrac{1}{2} \\ u \times \beta^e + \beta^{e-t} & \text{if } u > 0 \text{ and } |v| \ge \tfrac{1}{2} \\ u \times \beta^e - \beta^{e-t} & \text{if } u < 0 \text{ and } |v| \ge \tfrac{1}{2}. \end{cases} \qquad (2.13)$$

Some examples, with $\beta = 10$ and $t = 4$, will help to justify this:

$$12.3456 = .1234 \times 10^2 + .56 \times 10^{-2}, \qquad |v| = |.56| \ge \tfrac{1}{2},$$

so

$$[12.3456]_R = .1235 \times 10^2$$
$$-123.456 = -.1234 \times 10^3 + (-.56) \times 10^{-1}, \qquad |v| = |.56| \geq \tfrac{1}{2}$$

so

$$[-123.456]_R = -.1235 \times 10^3$$
$$.3333333 = .3333 \times 10^0 + .333\ldots \times 10^{-4}, \qquad |v| = |.333\ldots| < \tfrac{1}{2}$$

so

$$[.333\ldots]_R \doteq .3333.$$

The importance of (2.13) is that it provides a bound on the rounding error. In fact, if $|v| < \tfrac{1}{2}$, then

$$|x - x_R| = |v| \times \beta^{e-t} < \tfrac{1}{2}\beta^{e-t},$$

whereas if $|v| \geq \tfrac{1}{2}$, then

$$|x - x_R| = |v - 1| \times \beta^{e-t} \leq \tfrac{1}{2} \times \beta^{e-t}.$$

Hence, in either case

$$|E_R| \leq \tfrac{1}{2} \times \beta^{e-t}.$$

Again, with the fact that $|x| \geq \beta^{e-1}$, we can write this as

Bound on the relative rounding error

$$|x - x_R|/|x| \leq \tfrac{1}{2}\beta^{e-t}/\beta^{e-1} = \tfrac{1}{2}\beta^{1-t}. \qquad (2.14)$$

The quantity

$$\varepsilon_R = |x - x_R|/|x|, \qquad x \neq 0, \qquad (2.15)$$

is called the **relative rounding error**, and the value

$$\mu = \tfrac{1}{2}\beta^{1-t}$$

is called the **unit rounding error**. Note that $\mu_C = 2\mu$.

Both of the estimates (2.10) and (2.14) are valid only if $m \leq e \leq M$, where e is the exponent in (2.7) and m, M are the bounds on the exponents of the floating point numbers. If the exponent e in (2.7) does not satisfy these bounds, then x is said to be *outside the range of the floating point number system*; i.e., x is too big or too small to be reasonably approximated by a floating point number. For example, in the number system represented by

Fig. 2.1, numbers between 0 and $\frac{1}{4}$ and numbers larger than $3\frac{1}{2}$ are outside the range of the number system. If, however, x is in the range of the floating point number system, then the estimates (2.10) and (2.14) can be summarized by

$$\boxed{\begin{aligned} \varepsilon_C &= |x - x_C|/|x| < \beta^{1-t} = \mu_C = 2\mu, \\ \varepsilon_R &= |x - x_R|/|x| < \tfrac{1}{2}\beta^{1-t} = \mu. \end{aligned} \qquad x \neq 0} \qquad (2.16)$$

The importance of the above inequalities is that the right-hand sides depend *only* on the floating point number system and are independent of the size of the number. This fact is also the reason for introducing *relative* errors in addition to the absolute errors defined earlier. Since (2.16) will be the key to most of the error analysis that follows, it is worthwhile to examine how much is "lost" in applying these inequalities; that is, how much bigger generally is the right side than the left side? To answer this, consider again the case $F(10, 4, -5, 4)$. Then $\mu_C = 10^{-3}$, $\mu = .5 \times 10^{-3}$, and

$$(765.4567)_C = .7654 \times 10^3,$$

$$\varepsilon_C = \frac{.0567}{765.4567} = .7407 \times 10^{-4} < \mu_C$$

$$(765.4567)_R = .7655 \times 10^3,$$

| Examples: |
| Chopping |
| and |
| rounding |
| errors |

$$\varepsilon_R = \frac{.0433}{765.4567} = .5657 \times 10^{-4} < \mu$$

$$(765.4999)_R = .7655 \times 10^3,$$

$$\varepsilon_R = \frac{.0001}{765.4999} = .1306 \times 10^{-6} < \mu$$

$$(100.05)_R = .1001 \times 10^3,$$

$$\varepsilon_R = \frac{.05}{100.05} = .4998 \times 10^{-3} < \mu.$$

The last example shows that for some numbers the relative rounding error may be nearly as large as μ. In fact, μ, as defined above, is the simplest quantity that depends only on the floating point number system and that can be used as a close upper bound on the relative rounding error of any number within the range of the floating point system. Indeed, if t is *very* large, then the number

$$1 + \tfrac{1}{2}\beta^{1-t}$$

has relative rounding error *very* near to μ (see Exercise 10).

EXERCISES

1   Which of the following numbers are *not* in the set $F(2, 4, -2, 2)$? Why? $-.1011 \times 2^1$, $-.11011 \times 2^{-2}$, 1.101×2^2, $.2121 \times 2^2$, $.01101 \times 2^{-2}$, $-.7615 \times 2^2$

2   Describe the set $F(\beta, t, m, M)$ for your computer. If your computer has double-precision capability, describe the set of double-precision floating point numbers.

3   Draw a diagram such as Fig. 2.1 for the positive numbers in the set $F(2, 3, -2, 1)$.

4   What is the biggest gap between two numbers in the set $F(10, 8, -50, 49)$? What is the smallest gap?

5   Using the floating point number system $F(10, 4, -2, 3)$, determine the following values:
     (a) $\left(\frac{2}{3}\right)_R$ (b) $\left(-\frac{2}{3}\right)_R$ (c) $(9.99999)_R$ (d) $(9.99999)_C$ (e) $(\pi)_C$.

6   Determine the absolute and relative rounding or chopping errors for each of the quantities in Exercise 5.

7   What is the unit rounding error for your computer?

8   Give a detailed proof that (2.13) is a valid formula.

9   Show that

$$|x - x_R|/|x_R| \le \mu$$

if $x_R \ne 0$ and x is within the range of the floating point numbers.

10   Assuming a floating point number system $F(\beta, t, m, M)$ show that the relative rounding error for the number $1 + \frac{1}{2}\beta^{1-t}$ is

$$\tfrac{1}{2}\beta^{1-t}/(1 + \tfrac{1}{2}\beta^{1-t}).$$

For $\beta = 2$, $t = 27$, evaluate this error and compare it with the unit rounding error $\frac{1}{2}\beta^{1-t}$.

11   Consider the set $F(2, 3, -1, 2)$. Show that for all x between 0 and $\frac{1}{4}$, the bound (2.10) does *not* hold. For which numbers does (2.14) not hold?

2.3   Arithmetic Operations

Many of the complications that arise in numerical calculations are caused by the fact that computers cannot do arithmetic in the exact mathematical sense. In order to understand the kind of arithmetic that is done by computers, we shall analyze in detail an idealized type of computer arithmetic that is similar to the actual arithmetic carried out by several of the common computers.

2.3.1   Computer Hardware

When any scientific computer does arithmetic, it takes two floating point numbers that are stored in its memory unit or in special registers, performs an arithmetic operation (addition, subtraction, multiplication, or division) and produces a new floating point number. In general, the number that is computed is *not* the exact mathematical result of this arithmetic operation. Different computer manufacturers have made different decisions as to how these computed values should be obtained. Usually these decision involve not only mathematical but also economic factors, and hence it turns out that different computers produce different results. In order to determine *exactly* what values are produced by a particular computer, it is necessary to study the descriptions of the arithmetic operations in the manuals supplied by the manufacturer. Often these descriptions are quite long and complicated due to special design characteristics. The complete analysis of the arithmetic properties of even one large-scale modern computer is beyond the scope of this book. However, it is possible to describe in some detail a hypothetical type of computer arithmetic that is very close to the actual arithmetic performed by several of the commonly used machines. This description will also show why different computers produce quite different results when doing certain calculations.

All computers have special registers into which numbers must be placed in order to perform the arithmetic. These **arithmetic registers** must be capable of being shifted right and left in order to align the decimal point during addition and subtraction and to normalize the result when necessary. One of the basic differences between computers is the length of the arithmetic registers relative to the word length of the computer. The most desirable but most expensive case is when these registers are *much* longer than the word length. More specifically, if the single-precision numbers have t digits in their fractional part, then the arithmetic registers should contain $t + s$ digits, where s is comparable to t. If $s = t$, the registers are said to be *double-length*, and if $s = 0$, the registers are called *single-length*. Motivated by economic considerations, several computer manufacturers have reached a "compromise" between these two extremes by letting $s = 1$. That is, the arithmetic registers have a single extra digit, called a *guard digit*. Surprisingly, this one extra digit has a noticeable effect on the accuracy of the computed results.

2.3.2 Computer Arithmetic—Ideal Case

We first consider the ideal situation in which the arithmetic registers are *very* long. That is, we assume that the arithmetic registers have $t + s$ digits, where s is at least as large as t.

In order to *add* (or *subtract*) two normalized floating point numbers x and y, where

$$x = .d_1 d_2 \cdots d_t \times \beta^e$$
$$y = .c_1 c_2 \cdots c_t \times \beta^f,$$

both numbers are loaded into special registers, which we call x *registers* and y *registers* for short. The sum will be accumulated in a register called the *sum register*. This latter register will have an extra digit at the left to hold a possible "carry-over" digit. That is, if, for example, $x = .9876 \times 10^0$ and $y = .7654 \times 10^0$, then the sum register has to contain 1.7530. The digit "1" will be held in the special "carry-over" position. Finally, for simplicity, we assume that x has the larger exponent; that is, $e \geq f$. The various steps of an addition process are described in Fig. 2.2. The examples that are used in this description are in $F(10, 4, -2, 3)$.

It is not hard to convince oneself that these steps will always produce the correctly *chopped* value of $x + y$, provided that overflow or underflow does not occur. The correctly *rounded* value can be obtained by rounding in step 6 to obtain a t-digit number. Notice, however, that rounding here may result in $b_0 \neq 0$, so that step 4 may have to be repeated. Finally, observe that the arithmetic registers need to have only $2t$ digits for this process to work correctly. That is, for addition, double-length registers are adequate.

Subtraction proceeds exactly as addition except that the sign of the subtrahend is changed before the process begins. *Multiplication* is considerably simpler because the decimal points do not have to be aligned. More precisely, with the same notation used above, we can compute the product xy by the process described in Fig. 2.3.

Several elementary but important properties of multiplication have been used in this description. First of all, in step 2 we have used the fact that the product of two t-digit numbers always contains at most $2t$ digits (see Exercise 4). Thus the exact product can be stored in a double-length product register. Secondly, in step 4 we have normalized the product by shifting at most one position. This is justified by the fact that the fraction parts of the normalized numbers x and y are both greater than or equal to β^{-1}. Hence their product will be greater than or equal to β^{-2}. Thus in step 4 if $b_1 = 0$, then $b_2 \neq 0$. Again, it can be verified that if overflow or underflow does not occur, then this process computes the correctly chopped value of the product. If the product is rounded to t digits in step 5, then the correctly rounded product is obtained.

The process of *division* is somewhat more complicated since the quotient of two t-digit numbers may contain an infinite number of digits. However, by using a process of repeated subtraction and shifting, it is always possible to compute, say, the first $t + s$ digits of the quotient. More precisely, we can describe the process by Fig. 2.4.

Steps 4 and 5 are justified by the fact that the *normalized* fractions, which are stored in the x and y registers, lie between β^{-1} and 1. Hence, the quotient that is computed in step 2 must lie between β^{-1} and β. If the quotient is close to β, then there will be one digit to the left of the decimal point, that is, the

1	$x \to x$ register	x register $\boxed{d_1 d_2 \cdots d_t \mid 0 \cdots 0}$
	$y \to y$ register	y register $\boxed{c_1 c_2 \cdots c_t \mid 0 \cdots 0}$
2	Align decimal points by shifting the y register $e-f$ positions to the right. If $e-f \geq t$, return x as the sum.	y register $\boxed{0 \cdots 0 c_1 c_2 \mid c_t 0 \cdots 0}$ *Example:* $\quad x = .1234 \times 10^3$ $y = .2387 \times 10^{-1}$ $e-f = 4 \geq t, \quad \text{sum} = .1234 \times 10^3$ Exact sum $= .12342387 \times 10^3$
3	Add the x and y registers. Set the sum exponent to e.	sum register $\boxed{b_0 \mid b_1 \cdots b_t \mid b_{t+1} \cdots}$ sum exponent $g = e$
4	If $b_0 \neq 0$: (i) Shift sum-register one position to the right. (ii) Increase the exponent by 1.	sum register $\boxed{0 \mid b_1 b_2 \cdots b_t \mid b_{t+1} \cdots}$ sum exponent $g = e + 1$ *Example:* $\quad x = .9876 \times 10^1$ $y = .8765 \times 10^1$ $\text{sum} = 1.8641 \times 10^1$ $= .18641 \times 10^2$
5	If $b_1 = 0$, *normalize:* (i) Shift left until $b_1 \neq 0$. (ii) Decrease exponent by 1 for each position shifted.	sum register $\boxed{0 \mid b_1 \cdots b_t \mid b_{t+1} \cdots} \quad b_1 \neq 0$ sum exponent $g = e \pm h$ *Example:* $\quad x = .1012 \times 10^1$ $y = -.9876 \times 10^{-1}$ $\text{sum} = .091324 \times 10^1$ $= .91324 \times 10^0$
6	Chop the result to t digits.	sum register $\boxed{0 \mid b_1 \cdots b_t 00 \cdots 0}$ *Example:* $\quad x = .1234 \times 10^2$ $y = .4567 \times 10^0$ $\text{sum} = .127967 \times 10^2$ $= .1279 \times 10^2$
7	Is the sum exponent too big? Yes: *Overflow* error sum not defined.	$g > M$? *Example:* $\quad x = .9876 \times 10^3$ $y = .4543 \times 10^3$ comp. sum $= .1441 \times 10^4$ Exponent too big
8	Is the sum exponent too small? Yes: *Underflow* error sum not defined.	$g < m$? *Example:* $\quad x = .4387 \times 10^{-2}$ $y = -.4325 \times 10^{-2}$ $\text{sum} = .0062 \times 10^{-2}$ Exponent too small
9	Attach sum exponent to sum.	sum $= .b_1 b_2 \cdots b_t \times \beta^g$.

Fig. 2.2 Computer addition.

1	Load x into x register. Load y into y register.	x register	$\boxed{d_1 d_2 \cdots d_t \mid 0 \cdots 0}$
	Assume neither is zero.	y register	$\boxed{c_1 c_2 \cdots c_t \mid 0 \cdots 0}$
2	Multiply x and y; put the product in product register.	product register	$\boxed{b_1 b_2 \cdots b_t \mid b_{t+1} \cdots}$
3	Compute product exponent by adding the exponents of x and y.	product exponent: $g = e + f$	

4	If $b_1 = 0$, normalize:
	
	(i) Shift left one position.
	(ii) Decrease product expon-
	      ent by 1.
		product register $\boxed{b_1 b_2 \cdots b_t b_{t+1} \cdots}$, $\quad b_1 \neq 0$
		
		product exponent $g = e + f - 1$
		
		Example: $x = .1234 \times 10^1$
		                        $y = .2468 \times 10^2$
		                        product $= .03045512 \times 10^3$
		                                   $= .3045512 \times 10^2$		

5	Chop the product to t-digits.	product register	$\boxed{b_1 b_2 \cdots b_t \mid 00 \cdots 0}$
6	If product exponent too big, *Overflow error* product not defined.	$g > M$	
7	If product exponent too small, *Underflow error* product not defined.	$g < m$	
8	Otherwise, affix exponent to fraction part.	product $= .b_1 b_2 \cdots b_t \times \beta^g$	

Fig. 2.3   Computer multiplication.

1	Load x into x register. Load y into y register. Assume $y \neq 0$.	x register $\boxed{d_1 \cdots d_t \mid 0 \cdots 0}$ y register $\boxed{c_1 \cdots c_2 \mid 0 \cdots 0}$
2	Determine the first $t + s - 1$ digits of the quotient. Place results in quotient register.	quotient register: $\boxed{b_0 \mid b_1 \cdots b_t \mid b_{t+1} \cdots b_{t+s-1} \; 0}$ (Quotient register contains an extra digit on the left. See example in step 4.)
3	Compute the quotient exponent by subtracting the exponents of x and y.	$g = e - f$
4	If $b_0 \neq 0$ (i) Shift right one position. (ii) Increase exponent by 1.	quotient register: $\boxed{0 \mid b_1 b_2 \cdots b_t \mid b_{t+1} \cdots b_{t+s}}$ $g = e - f + 1$ *Example*: $x = .5678 \times 10^1$ $y = .1234 \times 10^2$ $x/y = 4.601296596\ldots \times 10^{-1}$ quotient $= 4.6012965 \times 10^{-1}$ $= .46012965 \times 10^0$
5	If $b_1 = 0$, normalize: (i) Shift left one position. (ii) Decrease exponent by 1.	quotient register: $\boxed{0 \mid b_1 \cdots b_t \mid b_{t+1} \cdots}$ $\quad b_1 \neq 0$ quotient exponent $\quad g = e - f - 1$
6	Test for overflow or underflow. Result not defined.	$g > M$ or $g < m$?
7	Chop the quotient to t-digits; attach exponent.	quotient $= .b_1 b_2 \cdots b_t \times \beta^g$.

Fig. 2.4 Computer division.

"carry-over" digit b_0 will be nonzero. If the quotient is near β^{-1}, then the digit b_1 is zero, but $b_2 \neq 0$. Again, we observe that the above steps always determine the correctly chopped quotient, unless overflow or underflow occurs. The correctly rounded value is obtained by rounding in step 7. It is important that x and y are *normalized* floating point numbers.

2.3.3 Computer Arithmetic—Real Cases

None of the well-known computers currently in use does arithmetic exactly as described above. The main reason for this is speed and cost. One way to reduce the cost of the arithmetic unit is to reduce the value of s. Note that in the multiplication and division processes described above, at most a shift by one position can occur. Hence, $s = 1$, that is, the presence of a **guard digit**, is sufficient to guarantee that the correctly chopped product and quotient can be produced. The addition of two numbers with the same sign (or subtraction of numbers with opposite signs) can also be done accurately with $s = 1$. In this case the right shifting that is needed for decimal point alignment is equivalent to chopping off some digits. However, the sum (or difference) will be the correctly chopped t-digit value. The presence of a guard digit even makes it possible to obtain the correctly *rounded* value. The situation is not quite so nice when numbers of opposite sign are added (or numbers of the same sign are subtracted). In this case a right shift to align the decimal point followed by a left shift to normalize the result may introduce a slight error. The following example, with $\beta = 10$, $t = 4$, $s = 1$, illustrates this fact:

$$x = .1004 \times 10^0, \qquad y = -.9958 \times 10^{-2}$$

x register $\boxed{1004 \mid 0}$

y register $\boxed{9952 \mid 0}$

shifted y register $\boxed{0099 \mid 5}$

difference $\boxed{0904 \mid 5}$

normalized value $\boxed{9045 \mid 0}$

computed value $.9045 \times 10^{-1}$

exact value $\quad .90442 \times 10^{-1}$

correctly chopped value $.9044 \times 10^{-1}$.

As in this example, it can be shown generally that when $s = 1$, the difference between the computed value and the correctly chopped value is never more than 1 in the tth digit.

Finally, consider the case when $s = 0$. Several bad things can happen now. During multiplication and division, for example, if a t-digit value is produced, then a left shift introduces a spurious zero in the tth position. To illustrate this, consider again the case $\beta = 10$, $t = 4$, and $s = 0$:

$$x = .1234 \times 10^0, \qquad y = .1000 \times 10^1$$

x register $\boxed{1 \ 2 \ 3 \ 4}$

y register $\boxed{1 \ 0 \ 0 \ 0}$

product register $\boxed{0 \ 1 \ 2 \ 3}$ $(.1234 \times .1000 = .01234)$

normalized product $\boxed{1 \ 2 \ 3 \ 0}$

computed product $.1230 \times 10^0$
exact value $.1234 \times 10^0$.

Thus in a computer with single-length registers, multiplication by 1.0 may change the value of the multiplicand, that is, $1.0 \times x \neq x$. Even more serious is the error which results when two numbers of nearly equal value are subtracted. The following example (still with $\beta = 10$, $t = 4$, $s = 0$) shows this:

$$x = .1004 \times 10^2, \qquad y = .9952 \times 10^1$$

x register $\boxed{1004}$

y register $\boxed{9952}$

shifted y register $\boxed{0995}$

difference $\boxed{0009}$

computed difference $.9000 \times 10^{-1}$
exact difference $.8800 \times 10^{-1}$

The error here is caused by the loss of rightmost digits, during the right shifting, followed by the cancellation of some leftmost digits. The left shifting, which is necessary for normalization, causes the lost right digits to become important. This cancellation of high-order (leftmost) digits is often called *subtractive cancellation* since it can only occur during subtraction (or during the addition of numbers with opposite signs). However, this same term is also used to describe the purely numerical phenomenon discussed in Section 1.3.1, which occurs even on computers that have double length registers. Hence, we shall refer to this hardware-induced cancellation as **single-length register cancellation**.

Most modern computers, even those with single-length registers, have provisions for doing arithmetic with double-precision numbers. The operations are basically the same, except that often a computer with double-length registers does the double-precision arithmetic in those double-length registers. In this case cancellation may occur just as in single-precision arithmetic with single-length registers.

Finally, it should be pointed out that some computers fit into none of the categories described above. For example, the UNIVAC 1108 computer does addition and subtraction as though $s = 0$, whereas multiplication is done with $s = 1$. Also, some computers round certain intermediate values and chop others. In general, a careful study of the computer manual is necessary for a complete understanding of the arithmetic properties of a particular computer.

2.3.4   Computational Efficiency

The time required for completing a floating point arithmetic operation is usually many times longer than the time it takes to do other computer operations, such as integer arithmetic, fetching, storing, testing, etc. Furthermore, most numerical algorithms do not involve a great many nonarithmetic steps. Thus a reasonable estimate for the time required by a particular numerical algorithm can be obtained by counting the number of floating point arithmetic operations. On nearly all computers, addition and subtraction take approximately the same amount of time, so these two operations will be counted together. On many computers, however, multiplication and division times differ considerably (see Table 2.2). For this reason, we shall count the multiplications and divisions separately. We shall use the notation

$$aA + bM + cD \qquad (2.17)$$

to denote a computation that requires a floating point additions and subtractions, b multiplications, and c divisions. The number of floating point operations involved in a particular algorithm serves as a useful measure of the efficiency of that algorithm. Thus, an appropriate expression of the form (2.17) will be called the **efficiency measure** for that algorithm. If two algorithms for solving the same problem are available, then, in general, the one with the smallest efficiency measure is to be preferred. Examples of how (2.17) is determined will be given in Chapter 3.

TABLE 2.2 *Average Times for Floating Point Operations in μsec $(10^{-6}\ sec)$*

	Addition	Multiplication	Division
UNIVAC 1100/40	.76	1.65	5.3
UNIVAC 1108	2.2	3.0	8.63
IBM 360/50	6.13	20.75	21.25
IBM 360/85	.38	1.36	1.64
CDC 6600	.4	1.0	2.9

EXERCISES

1 Consult the manuals for your computer and write out a brief description of the addition operation. Indicate whether your computer is similar to the double- or single-length register type or has a guard digit.

2 Assume that a computer with $\beta = 10$, $t = 4$ performs addition exactly as described in Fig. 2.2. Find the computed sum for the following values of x and y:

(a) $x = .1234 \times 10^1$, $y = .1234 \times 10^2$
(b) $x = .1234 \times 10^1$, $y = .4321 \times 10^{-3}$
(c) $x = .1234 \times 10^2$, $y = .1234 \times 10^{-3}$
(d) $x = .1234 \times 10^2$, $y = -.1235 \times 10^2$
(e) $x = .1004 \times 10^2$, $y = -.9987 \times 10^1$.

3 Repeat Exercise 2, assuming single-length arithmetic registers, and that chopping occurs whenever digits are dropped.

4 Prove that the product of two t-digit numbers always contains no more than $2t$ digits.

5 Determine the exact add, multiply, and divide times for your computer. (Usually these times vary over a certain range. If so, give this range, and explain why the times vary.)

6 The following computation has been proposed as an indication of the rounding error in a computer:

$$H = 1.0/2.0 \qquad E = (X + X + X) - H$$
$$X = 2.0/3.0 - H \qquad F = (Y + Y + Y + Y + Y) - H$$
$$Y = 3.0/5.0 - H \qquad Q = (2.0)F/E.$$

The value of E and F in exact arithmetic is zero; however, due to rounding errors, the computed values will be nonzero. What is the value for Q computed by your computer? [W. Kahan, "A Problem," SIGNUM Newsletter, Vol. 6 (1971).]

7 Prove that the presence of a guard digit guarantees that a computed sum differs from the correctly chopped value by no more than 1 in the tth digit.

2.4 Errors in Computer Arithmetic

In this section we derive some estimates for the errors introduced by the computer arithmetic discussed in the previous section. These estimates are essential for analyzing the effectiveness of computational methods.

2.4.1 Basic Error Estimates

Once a number has been converted into the proper number base and then approximated by a floating point number by rounding or chopping, no further errors are introduced into this number. However, as was seen in Section 2.3, if this number is used as part of an arithmetic operation, the results will contain additional errors due to the limitations of the arithmetic unit of the computer. To simplify the discussion let $*$ denote any of the operations $+$, $-$, $\times$, $\div$. Then we shall use the notation $\text{fl}[x * y]$ to denote the **computed value** of $x * y$. It is assumed, of course, that x and y represent normalized floating point numbers from some given set $F(\beta, t, m, M)$ and that when $*$ represents $\div$, then $y \neq 0$. The difference between the exact value of $x * y$ and the computed value will be called the *absolute arithmetic error* for the operation $*$. The purpose of this section is to examine these errors.

If the processes described in Fig. 2.2–2.4 are used to compute $\text{fl}[x * y]$, then the correctly chopped values of $x * y$ are obtained. That is, with the notation of Section 2.2 we have

$$\text{fl}[x * y] = (x * y)_{\text{c}}. \tag{2.18}$$

When a guard digit is included in the arithmetic registers, then (2.18) remains valid when $*$ denotes $\times$ or $\div$. For addition or subtraction, a guard digit can only guarantee that (2.18) is correct to within one unit in the last digit. In the single-length register case (2.18) may not be even approximately true.

Whenever (2.18) holds, we can apply the relative error estimate (2.16) to obtain

Relative arithmetic error: Ideal case	$|(x * y) - \text{fl}(x * y)|/|x * y| \leq \mu_{\text{c}}, \qquad |x * y| \neq 0. \quad (2.19)$

Thus we have a simple bound on the **relative arithmetic error** *under the* *assumption that* (2.18) *is valid.* In case the arithmetic registers have a guard digit, as noted above, (2.18) and hence (2.19) will not hold for addition or subtraction. However, since the difference between $(x \pm y)_c$ and $fl(x \pm y)$ in this case is at most one in the tth digit, it follows that

<table>
<tr><td>Relative arithmetic error with guard digit</td><td>$|(x \pm y) - fl(x \pm y)|/|x \pm y| \le 2\mu_C, \qquad x \pm y \ne 0. \quad (2.20)$</td></tr>
</table>

In the case of single-length registers, the relative arithmetic error may be much larger than μ_C. Consider again the example discussed earlier in which $\beta = 10$, $t = 4$, $x = -.9952 \times 10^1$, and $y = .1004 \times 10^2$. Then with single-length registers, $fl[x + y] = .9000 \times 10^{-1}$ whereas the exact value is $x + y = .8800 \times 10^{-1}$. Thus, the relative arithmetic error is

$$|.8800 \times 10^{-1} - .9000 \times 10^{-1}|/.8800 \times 10^{-1} \approx .2273 \times 10^{-1}.$$

This is nearly 23 times larger than the unit chopping error $\mu_C = 10^{-3}$.

Thus in the ideal case discussed in Section 2.3.2, the error bound (2.19) is valid; but for real computers this must be modified to reflect the type of arithmetic that is actually performed by the computer. One way to summarize these observations is to write

<table>
<tr><td>Relative arithmetic error: Summary</td><td>$|(x * y) - fl(x * y)|/|x * y| \le r\mu, \qquad x * y \ne 0 \quad (2.21)$</td></tr>
</table>

where

$r = 1$ in ideal rounded arithmetic,
$r = 2$ in ideal chopped arithmetic or during multiplication and division with a guard digit,
$r = 4$ during addition or subtraction with a guard digit, or single-length register multiplication and division,
$r \ge 4$ in single-length register addition and subtraction.

We assume here that neither overflow nor underflow occurs. In many of the error analyses presented in this book, we shall use (2.21) with $r = 1$. It must be clearly understood that for some computers, this will result in error estimates that are much too conservative. For computers that have single-length registers, it may even be necessary to do some computations (especially additions and subtractions) in double precision. This is one way to avoid a large value of r in (2.21).

2.4.2   Computer Accuracy

The above estimates show that the unit chopping or rounding error serves two purposes: It is a measure of how accurately numbers can be approximated by the computer's floating point number system, and it measures the accuracy of the arithmetic done by the computer. Thus, the accuracy of two computers can be compared by comparing their respective unit chopping or rounding errors. In order to compare a binary computer with a decimal computer or, more importantly, to compare a computation done with a binary computer to a decimal computation done by hand or with the help of a hand calculator, it is necessary to compare 10^{1-t} with 2^{1-s}. In particular, the question How many decimal digits are equivalent to s binary digits? is answered by solving the equation

$$10^{1-t} = 2^{1-s}. \tag{2.22}$$

If we take logarithms of both sides, we find that

$$\log_{10} 10^{1-t} = \log_{10} 2^{1-s}$$

$$(1-t)\log_{10} 10 = (1-s)\log_{10} 2$$

$$1 - t = (1-s)\log_{10} 2$$

$$t = 1 + (s-1)\log_{10} 2.$$

The value of $\log_{10} 2$ is approximately .30103, so that

$$t \approx 1 + (s-1)(.30103). \tag{2.23}$$

To illustrate this result, we let $s = 27$, so that

$$t \approx 1 + 26(.30103) \approx 8.8.$$

This says that a 27-digit binary computer is more accurate, in this sense, than an eight-digit decimal computer but is less accurate than a nine-digit decimal computer.

Another application of (2.23) is to determine how many decimal digits should be used to represent a certain *computed* binary number. That is, suppose the result of a certain computation inside a binary computer has produced a value whose fractional part has, say, 20 correct binary digits. (That is, the relative error in this computed value is less than 2^{-20}.) The question is How many *decimal* digits should be printed out to represent this computed result? The answer to this question is obtained by using (2.23) with $s = 20$. The result is $t \approx 6.720$, which says that no more than six decimal digits are correct.

2.4.3   Another Kind of Error Estimate

The relative error estimates (2.10), (2.14), and (2.21) are cumbersome and impractical to use in complicated analyses. More useful estimates are derived as follows. Consider (2.10); that is,

$$\boxed{\text{Basic error inequality}} \qquad |x - x_C|/|x| \leq \mu_C. \qquad (2.10)$$

If we let $\varepsilon = (x_C - x)/x$, then $|\varepsilon| \leq \mu_C$, and the definition of ε can be rewritten as

$$x_C = x(1 + \varepsilon).$$

That is, the error bound (2.10) is equivalent to the formula

$$\boxed{\substack{\text{Relation} \\ \text{between} \\ \text{exact and} \\ \text{chopped} \\ \text{values}}} \qquad x_C = x(1 + \varepsilon) \qquad \text{for some} \quad \varepsilon \quad \text{with } |\varepsilon| \leq \mu_C. \qquad (2.24)$$

The advantage of (2.24) is that the chopped value x_C is expressed *precisely* in terms of the exact value x. The only place where an inequality occurs is in the estimate for the chopping error ε.

In the same manner, (2.14) can be written as

$$x_R = x(1 + \varepsilon) \qquad \text{for some} \quad \varepsilon \quad \text{with } |\varepsilon| \leq \mu, \qquad (2.25)$$

and (2.21) becomes

$$\text{fl}[x * y] = (x * y)(1 + \varepsilon) \qquad \text{where} \quad |\varepsilon| \leq r\mu. \qquad (2.26)$$

Formula (2.26) is especially useful for analyzing the effect of computer arithmetic on numerical algorithms. Recall, however, that if $*$ represents addition or subtraction with *single-length registers*, then the value of r may be quite large. A careful study of single-length register arithmetic, however, leads to a formula very similar to (2.26) but with a smaller bound for the error. Specifically, it can be shown (Exercise 8) that for all of the addition processes considered in Section 2.3 we can write

| More useful form of (2.26) | $$\mathrm{fl}[x \pm y] = x(1 + \varepsilon_1) \pm y(1 + \varepsilon_2),$$ | (2.27a) |

for certain values of ε_1 and ε_2 that satisfy

$$|\varepsilon_1| \leq r\mu, \qquad |\varepsilon_2| \leq r\mu, \qquad r \leq 4. \tag{2.27b}$$

By comparison note that (2.26) with $*$ replaced by $+$ gives

$$\mathrm{fl}[x + y] = x(1 + \varepsilon) + y(1 + \varepsilon), \qquad |\varepsilon| \leq r\mu. \tag{2.28}$$

Now, however, the value of r may be quite large; that is, ε may be large. To illustrate the difference between (2.27) and (2.28), consider again the example $x = -.9952 \times 10^1$, $y = .1004 \times 10^2$, with $\beta = 10$, $t = 4$, and single-length register addition. Then

$$\mathrm{fl}[x + y] = .9000 \times 10^{-1}$$

$$x + y = .8800 \times 10^{-1}$$

so that

$$\varepsilon = \{\mathrm{fl}[x + y] - (x + y)\}/(x + y) = .022727\ldots.$$

Thus for the formula (2.28) to hold, the value of ε must be

$$\varepsilon = .022727\ldots = 22.727\ldots \times \mu_C \cong 23\mu_C.$$

On the other hand, if we use the form (2.27), we have

$$\mathrm{fl}[x + y] = (-.9952 \times 10^1)(1 - .000201\ldots) + (.1004 \times 10^2)$$

so that $|\varepsilon_1| \leq .2\mu_C$ and $\varepsilon_2 = 0$.

For simplicity, we summarize these estimates as follows:

| For any number x in the range of the floating number system |
| $$x_R = x(1 + \varepsilon) \qquad \text{where} \quad |\varepsilon| \leq \mu$$ $$x_C = x(1 + \varepsilon) \qquad \text{where} \quad |\varepsilon| \leq 2\mu. \tag{2.29a}$$ |

> For any floating point numbers x, y:
>
> $$\text{fl}[x \pm y] = x(1 + \varepsilon_1) \pm y(1 + \varepsilon_2), \qquad |\varepsilon_i| \leq r\mu$$
> $$\text{fl}[x \times y] = (x \times y)(1 + \varepsilon), \qquad |\varepsilon| \leq r\mu \qquad\qquad (2.29\text{b})$$
> $$\text{fl}[x/y] = (x/y)(1 + \varepsilon) \qquad \text{if } y \neq 0, \quad |\varepsilon| \leq r\mu,$$
>
> where r is 1, 2, or 4, depending on the operation, the length of the arithmetic registers, and whether rounding or chopping is used.

As a final comment on these error estimates, consider the equation $\text{fl}[x \times y] = x(1 + \varepsilon) \times y = \hat{x} \times y$. Suppose now that x is a rounded (or chopped) approximation to some exact value $\bar{x}$, and that the quantity that is *really* wanted is $\bar{x} \times y$. Then (2.29a) says that x may differ from $\bar{x}$ by as much as μ, and (2.29b) says that x differs from $\hat{x}$ by less than μ. It *could* just happen that $\bar{x} = \hat{x}$; that is, our *computed* product $\text{fl}[x \times y]$ could be equal to the *exact* product $\bar{x} \times y$. Certainly this would be an unusual coincidence, but this argument does show that it is not possible to *guarantee* a more accurate result by using, say, double-precision multiplication. This reasoning applies to any arithmetic operation and is often used to justify statements to the effect that "computer arithmetic is as accurate as the (rounded) data warrants."

2.4.4   Relative versus Absolute Errors

In all of the above discussion, *relative* errors rather than *absolute* errors were used. There are several reasons for preferring to use relative rather than absolute errors, aside from their natural occurrence in the various error estimates. First of all, it makes sense to speak of small or large relative errors, whereas the size of an absolute error depends on the size of the number itself. For example, if $x = 9999.9$ is an approximation to $\bar{x} = 10000.0$, and $y = .99999$ is an approximation to $\bar{y} = 1.0$, then the absolute error in x is considerably larger than the absolute error in y, although x and y are both accurate approximations to $\bar{x}$ and $\bar{y}$, respectively. (Note that the *relative* errors in x and y are 10^{-5}.) On the other hand, a relative error near 1.0 means that the absolute error is nearly the same size as the number being approximated, and any reasonable interpretation would call this a large error. Thus a relative error is large or small depending on its size relative to 1.0. Another justification for emphasizing relative errors is that for floating point numbers it only makes sense to examine the accuracy of the mantissa—if the exponent is wrong, then something really bad has happened. The mantissas of x and $\bar{x}$ can be compared by computing $|x - \bar{x}|\beta^{-e}$, where e is the (common) exponent of x and $\bar{x}$. Unfortunately,

β^{-e} is troublesome to compute, and anyway $|x - \bar{x}|/|\bar{x}|$ is a good enough approximation to this error estimate. Finally note that all relative errors used so far were defined as the absolute error divided by the exact value, i.e.,

$$|x - \bar{x}|/|\bar{x}|,$$

where $\bar{x}$ is the exact value, x the approximate value. In many instances it will be more convenient to use the quantity

$$|x - \bar{x}|/|x|,$$

that is, to divide by the approximate value. Generally, it should be clearly understood that the main reason for dividing by $|\bar{x}|$ is to eliminate the dependence of the error measurement on the size of the number; and if x is a reasonably good approximation to $\bar{x}$, then this dependence can just as well be eliminated by dividing by $|x|$.

EXERCISES

1 Show that the unit rounding error μ can also be defined to be the largest number so that $fl[x + 1.0] = 1.0$ for every x with $0 \le x < \mu$.

2 Write a program for your computer that will determine the unit rounding error by using the definition given in Exercise 1. Take care that binary-to-decimal conversion does not affect your result.

3 Use the method described in this section to compare the accuracy of the computers listed in Table 2.1.

4 How many decimal digits should be printed out to represent the following quantities:
   (a) A binary number that has a relative error of 2^{-25}.
   (b) An octal number that has an error in the sixth digit.
   (c) A binary number that has 15 correct digits.

5 With $\beta = 10$, $t = 4$, and single-length registers, what is the *largest* relative arithmetic error that can result from subtracting two positive floating point numbers?

6 Find the relative arithmetic errors for the following expressions, assuming $\beta = 10$, $t = 4$, single-length arithmetic registers, and chopping.
   (a) $fl[(.1234 \times 10^0)/(.2000 \times 10^1)]$ (b) $fl[(.9876 \times 10^0)(.9900 \times 10^2)]$
   (c) $fl[(.9968 \times 10^2) - (.1006 \times 10^3)]$ (d) $fl[(.3333 \times 10^1) + (.6789 \times 10^0)]$.

7 Use Exercise 9 of Section 2.2 to show that $x_R = x/(1 + \varepsilon)$, where $|\varepsilon| \le \mu$, and hence that instead of (2.26) we can write

$$fl[x * y] = (x * y)/(1 + \varepsilon), \qquad |\varepsilon| \le r\mu.$$

8 Trace through the process of addition in single-length registers, as was done in Fig. 2.2, and show that

$$fl[x + y] = x(1 + \varepsilon_1) + y(1 + \varepsilon_2),$$

where $|\varepsilon_1|$ and $|\varepsilon_2|$ are bounded by $r\mu$ with $r \leq 4$. Justify the following special cases:
(a) $\varepsilon_2 = 0$ if x has the larger exponent, and both numbers have the same sign.
(b) $r \leq 2$ if the carry-over digit was nonzero after the addition.

9 Write the computations indicated in Exercise 6 in the form (2.29b). In each case give an approximate (one- or two-digit) value for the relative error ε (or ε_1, ε_2).

2.5 Analysis of Errors

In this section we show how the estimates that were derived in the previous section can be used to analyze rounding error. Two important applications of such an analysis are also discussed.

2.5.1   Accumulation of Errors

The estimates (2.21) and (2.26) can be thought of as error analyses for very simple computational processes—namely, the computation of a single sum, difference, product, or quotient. In order to analyze complicated algorithms, it is necessary to see how errors accumulate during more extensive computations. To illustrate how this is done, consider the computation of the sum of five positive floating point numbers:

$$s = x_1 + x_2 + x_3 + x_4 + x_5.$$

Because all of these numbers are positive, cancellation is not possible, and hence (2.27) is valid with small r; for simplicity we shall use $r = 1$. If the numbers are added together from left to right, then the computed sum is

$$\hat{s} = \text{fl}[\text{fl}[\text{fl}[\text{fl}[x_1 + x_2] + x_3] + x_4] + x_5]$$

Now (2.27) implies that

$$\text{fl}[x_1 + x_2] = x_1(1 + \varepsilon_1) + x_2(1 + \varepsilon_2), \qquad |\varepsilon_i| \leq \mu$$

so

$$\begin{aligned}
\text{fl}[\text{fl}[x_1 + x_2] + x_3] &= \text{fl}[(x_1(1 + \varepsilon_1) + x_2(1 + \varepsilon_2)) + x_3] \\
&= (x_1(1 + \varepsilon_1) + x_2(1 + \varepsilon_2))(1 + \varepsilon_3) + x_3(1 + \varepsilon_4) \\
&= x_1(1 + \varepsilon_1)(1 + \varepsilon_3) + x_2(1 + \varepsilon_2)(1 + \varepsilon_3) \\
&\quad + x_3(1 + \varepsilon_4).
\end{aligned}$$

Similarly we can write

$$\begin{aligned}
\text{fl}[\text{fl}[\text{fl}[x_1 + x_2] + x_3] + x_4] &= x_1(1 + \varepsilon_1)(1 + \varepsilon_3)(1 + \varepsilon_5) \\
&\quad + x_2(1 + \varepsilon_2)(1 + \varepsilon_3)(1 + \varepsilon_5) \\
&\quad + x_3(1 + \varepsilon_4)(1 + \varepsilon_5) \\
&\quad + x_4(1 + \varepsilon_6),
\end{aligned}$$

where all $|\varepsilon_i|$ are bounded by μ. The pattern should now be clear, and the final result can be written as

| Computed sum |

$$\hat{s} = x_1(1 + \delta_1) + x_2(1 + \delta_2) + x_3(1 + \delta_3)$$
$$+ x_4(1 + \delta_4) + x_5(1 + \delta_5) \tag{2.30a}$$

where

| Accumulated errors |

$$1 + \delta_1 = (1 + \varepsilon_1)(1 + \varepsilon_3)(1 + \varepsilon_5)(1 + \varepsilon_7)$$
$$1 + \delta_2 = (1 + \varepsilon_2)(1 + \varepsilon_3)(1 + \varepsilon_5)(1 + \varepsilon_7)$$
$$1 + \delta_3 = (1 + \varepsilon_4)(1 + \varepsilon_5)(1 + \varepsilon_7) \tag{2.30b}$$
$$1 + \delta_4 = (1 + \varepsilon_6)(1 + \varepsilon_7)$$
$$1 + \delta_5 = (1 + \varepsilon_8).$$

2.5.2   Backward and Forward Error Estimates

There are two useful interpretations of (2.30). The first results from rewriting (2.30a) in the form

$$\hat{s} = (x_1 + x_2 + x_3 + x_4 + x_5) + x_1\delta_1 + x_2\delta_2 + x_3\delta_3 + x_4\delta_4 + x_5\delta_5$$

so that

$$\hat{s} - s = x_1\delta_1 + x_2\delta_2 + \cdots + x_5\delta_5.$$

Thus, a bound on the **error in the computed sum** is

$$|\hat{s} - s| \leq x_1|\delta_1| + x_2|\delta_2| + x_3|\delta_3| + x_4|\delta_4| + x_5|\delta_5|. \tag{2.31}$$

The second interpretation of (2.30) is that

$$\hat{s} = \hat{x}_1 + \hat{x}_2 + \hat{x}_3 + \hat{x}_4 + \hat{x}_5 \tag{2.32}$$

where

$$\hat{x}_i = x_i(1 + \delta_i), \qquad i = 1, 2, 3, 4, 5 \tag{2.33a}$$

or

$$(\hat{x}_i - x_i)/x_i = \delta_i, \qquad i = 1, 2, 3, 4, 5. \tag{2.33b}$$

That is, $\hat{s}$ is the *exact* sum of five numbers $\hat{x}_1, \ldots, \hat{x}_5$ that are related to the original numbers according to (2.33).

The first interpretation, in which the computed result is compared to the exact result, is called a **forward error analysis**, as described in Section 1.3.4. The second interpretation (2.32) and (2.33) was also mentioned in Section

1.3.4 and was called a **backward error analysis.** Here we think of the computed result s as being the exact result of a computation with slightly changed data (2.33a). That is, the computational errors are "thrown back" onto the data.

A forward error analysis, such as (2.31), clearly is a reasonable way to examine the effect of computer arithmetic on a computational process. Often, however, such an analysis confuses the inaccuracy of the algorithm with the instability of the problem. A backward error analysis does not generally give quite so direct a measurement of the accuracy of the computed answers. Such an analysis, however, is often quite simple and can be extremely useful in determining the stability of an algorithm. Examples of this will be given later.

The error analyses discussed here are called **a priori analyses** because they are carried out before the computations are actually done. In later chapters we shall consider **a posteriori estimates** in which the computed answer is used to estimate its own accuracy.

2.5.3 Bounds on the Accumulated Errors

Before any of the above formulas can provide information about the accuracy of the computational process, the quantities δ_i, $i = 1, 2, \ldots, 5$ must be examined more closely. From (2.30b) we see that δ_4, for example, can be written as

Accumulated error	$$\delta_4 = \varepsilon_6 + \varepsilon_7 + \varepsilon_6\varepsilon_7,$$

where ε_6 and ε_7 are the relative errors in certain computed additions. The only thing that is really known about these errors is that they are bounded by some small multiple of the unit rounding error μ.

Most rounding error analyses lead eventually to products of the form (2.30b). More generally, consider the expression

Accumulated errors: General form	$$\frac{(1 + \varepsilon_1)(1 + \varepsilon_2) \cdots (1 + \varepsilon_k)}{(1 + \varepsilon_{k+1})(1 + \varepsilon_{k+2}) \cdots (1 + \varepsilon_n)}.$$	(2.34)

Such expressions occur when quantities are computed by a combination of n arithmetic operations. To show why we consider a *quotient* of products in

(2.34), consider the following example:

Example: Quotient of products	

$$\text{fl}\left[\frac{(xy)}{(uv)}\right] = \frac{\text{fl}(xy)}{\text{fl}(uv)}(1 + \varepsilon_1) = \frac{(xy)(1 + \varepsilon_2)(1 + \varepsilon_1)}{(uv)(1 + \varepsilon_3)}$$

$$= \frac{xy}{uv}\frac{(1 + \varepsilon_1)(1 + \varepsilon_2)}{(1 + \varepsilon_3)}, \qquad (2.35)$$

where ε_1 denotes the relative error in the division, while ε_2 and ε_3 are the relative errors in the two multiplications.

In order to simplify the analysis of rounding errors, we shall introduce a special notation for expressions of the form (2.34). To motivate this notation, observe that in analyses such as (2.30) and (2.35) the important thing is *how many* rounding errors are involved. That is, the *values* of the errors ε_1, ε_2, $\ldots$ are unimportant since we can never know them anyway. Thus we shall use the symbol[†]

Rounding error counter	$\langle n \rangle$

to denote any expression of the form (2.34). As we shall see, it is immaterial how many terms appear in the numerator and how many in the denominator; only the total number is important.

With this notation (2.30a) can be written as

$$\hat{s} = x_1\langle 4 \rangle + x_2\langle 4 \rangle + x_3\langle 3 \rangle + x_4\langle 2 \rangle + x_5\langle 1 \rangle$$

and (2.35) becomes

$$\text{fl}[xy/uv] = xy/uv\langle 3 \rangle.$$

To obtain error bounds from these analyses, it is necessary to write (2.34) in the form

$$\frac{(1 + \varepsilon_1) \cdots (1 + \varepsilon_k)}{(1 + \varepsilon_{k+1}) \cdots (1 + \varepsilon_n)} = 1 + \delta_n \qquad (2.36)$$

and then estimate δ_n. This is precisely what happened in (2.33b), and it is easily seen that the absolute rounding error in (2.35) is given by

$$\text{fl}\left[\frac{xy}{uv}\right] - \left(\frac{xy}{uv}\right) = \left(\frac{xy}{uv}\right)\delta_3, \qquad \text{where} \quad 1 + \delta_3 = \frac{(1 + \varepsilon_1)(1 + \varepsilon_2)}{1 + \varepsilon_3}.$$

We shall use the following theorem due to Wilkinson (1963) [see also Forsythe and Moler (1967)]. The proof is left as an exercise (Exercise 2).

[†] This symbol was first used by G. W. Stewart (1973) to analyze rounding error in the solution of linear systems.

Theorem 2.1   Let $\varepsilon_1, \varepsilon_2, \ldots, \varepsilon_n$ be rounding errors that satisfy $|\varepsilon_k| \leq r\mu$, $k = 1, 2, \ldots, n$. If $nr\mu \leq .1$, then there is a δ_n that satisfies (2.36) and

$$|\delta_n| \leq n(1.06r\mu).$$

The condition $nr\mu \leq .1$ simply says that n, the number of rounding errors, cannot be too large, but "too large" here depends on the accuracy of the computer. If $r\mu = 10^{-7}$, for example, then n must be less than 10^6, whereas if $r\mu = 10^{-15}$ then $n = 10^{14}$ is acceptable.

To make use of this result, we introduce the symbol

$$\langle n \rangle_0$$

to denote a value δ_n that satisfies an expression of the form (2.36). Thus we shall write

$$\langle n \rangle = 1 + \langle n \rangle_0 .$$

Theorem 2.1 says that, if $nr\mu \leq .1$, then

Bound on the accumulated error

$$|\langle n \rangle_0| \leq n\mu', \qquad\qquad\qquad (2.37a)$$

where

$$\mu' = (1.06)r\mu. \qquad\qquad\qquad (2.37b)$$

Some relationships between these quantities are

$$\langle n \rangle \langle m \rangle = \langle n + m \rangle \qquad\qquad\qquad (2.38a)$$

$$\langle n \rangle / \langle m \rangle = \langle n + m \rangle \qquad\qquad\qquad (2.38b)$$

$$|\langle n \rangle_0 + \langle m \rangle_0| \leq |\langle n + m \rangle_0|. \qquad\qquad\qquad (2.38c)$$

Notice that, on the basis of (2.38c) it is tempting to write $\langle n \rangle_0 + \langle m \rangle_0 = \langle n + m \rangle_0$. This, unfortunately, is not true; however, in deriving **bounds for accumulated errors**, it is permissible to replace the sum $\langle n \rangle_0 + \langle m \rangle_0$ by $\langle n + m \rangle_0$ because this can only increase the error bound.

To illustrate the use of the above notation, we shall complete the error analysis begun in Section 2.5.2. That is, (2.30a) is

$$\hat{s} = x_1 \langle 4 \rangle + x_2 \langle 4 \rangle + x_3 \langle 3 \rangle + x_4 \langle 2 \rangle + x_5 \langle 1 \rangle,$$

which can be written,

$$\hat{s} = x_1(1 + \langle 4 \rangle_0) + x_2(1 + \langle 4 \rangle_0) + x_3(1 + \langle 3 \rangle_0) + x_4(1 + \langle 2 \rangle_0)$$
$$+ x_5(1 + \langle 1 \rangle_0).$$

Hence,

$$\hat{s} - s = x_1\langle 4\rangle_0 + x_2\langle 4\rangle_0 + x_3\langle 3\rangle_0 + x_4\langle 2\rangle_0 + x_5\langle 1\rangle_0.$$

The estimate (2.37a) then gives the **forward error bound**

$$|\hat{s} - s| \le (4x_1 + 4x_2 + 3x_3 + 2x_4 + x_5)\mu'. \tag{2.39}$$

The **backward error analysis** is completed by observing, from (2.33b), that

$$(\hat{x} - x_1)/x_1 = \langle 4\rangle_0, \qquad (\hat{x}_i - x_i)/x_i = \langle 6 - i\rangle_0, \qquad i = 2, 3, 4, 5.$$

Thus, we can apply (2.37) again to obtain

$$|\hat{x}_1 - x_1|/|x_1| \le 4\mu', \qquad |\hat{x}_i - x_i|/|x_i| \le (6 - i)\mu', \qquad i = 2, 3, 4, 5. \tag{2.40}$$

Throughout the remainder of this book, we shall use the notation introduced above. The hypothesis $nr\mu \le .1$ will always be assumed, even if not so stated explicitly. The value of r here and in the definition of μ', (2.37b), will often be taken as 1, only for the sake of simplicity.

In much of the discussion of rounding errors, the terms *error* **bound** and *error* **estimate** have been used. Clearly, an error bound should be a number that is larger than the error; however, without anything further being said, such a quantity could be quite useless. For example, 10^{10} is an upper bound on the rounding error in a seven-digit floating point system but serves no useful purpose in analyzing the rounding error. In fact, any useful bound should also be an estimate in the sense that, at least for some data, the bound must be nearly attained. The unit rounding error $\mu = \frac{1}{2}\beta^{1-t}$ is such an estimate. In fact, for $x = 1 + \frac{1}{2}\beta^{1-t}$, we have

$$|x - x_R|/|x| = \frac{1}{2}\beta^{1-t}/(1 + \frac{1}{2}\beta^{1-t}),$$

If t is of reasonable size, then this quantity is *very* close to $\frac{1}{2}\beta^{1-t}$ since $1 + \frac{1}{2}\beta^{1-t}$ is *very* near to 1. Great care must be exercised when deriving error bounds. Too much should not be "thrown away" when deriving a bound; otherwise the result is a uselessly large bound.

2.5.4 Applications—Analyzing Algorithms

Error estimates such as (2.39) and (2.40) usually do not provide very reasonable estimates for the actual error in the computed answer. If, for example,

$$
\begin{aligned}
x_1 &= .1234 \times 10^1 \\
x_2 &= .3429 \times 10^0 \\
x_3 &= .1289 \times 10^{-1} \\
x_4 &= .9895 \times 10^{-3} \\
x_5 &= .9763 \times 10^{-5}
\end{aligned}
\tag{2.41}
$$

then, by using four-decimal-digit chopped floating point arithmetic, we compute the value $\hat{s} = .1588 \times 10^1$ whereas the exact sum is $s = .1590789\ldots \times 10^1$. Thus $|\hat{s} - s| = .2789\ldots \times 10^{-2}$. The error bound (2.39) simply says that $|\hat{s} - s| \leq 6.348\ldots \times 10^{-3} \approx .6348 \times 10^{-2}$, which is certainly true but not very accurate (by a factor of 2). On the other hand, (2.39) does give some interesting *qualitative* information; namely, that the numbers should be summed in increasing order of size. This follows from the fact that, in the bound (2.39), the number x_1 will be multiplied by four, while the number x_5 has a factor of 1 in the bound. Thus, x_1 should be as small as possible. Indeed, if the numbers in (2.41) are written in reverse order so that $x_1 = .9763 \times 10^{-5}$, $x_5 = .1234 \times 10^1$, then four-digit arithmetic gives $.1590 \times 10^1$ as the computed sum. This agrees with the exact sum to four figures. Hence, the error analysis has suggested a simple rearrangement of the algorithm for summing the five numbers in (2.41), and this rearrangement produces a noticeably more accurate result.

To illustrate the kind of conclusion that can be drawn from a **backward error analysis** such as (2.40), suppose that the numbers $x_1, \ldots, x_5$ are actually approximations to some *exact* numbers $\bar{x}_1, \ldots, \bar{x}_5$, with relative accuracy of the order of 100μ. Then it *could be* possible that the $\hat{x}_i$ defined by (2.33a) are, in fact, the exact values $\bar{x}_i$, and hence the computed sum is *equal* to the exact sum of the exact values. While such a coincidence is not very likely, this argument does show that one cannot guarantee a more accurate sum will be obtained by using, say, double-precision addition.

2.5.5   Applications—Detecting Numerical Instability

One important use of a priori error analyses is the detection of algorithms that are susceptible to numerical instability. By its very nature, a priori analysis cannot take into account particular sets of data. Hence the estimates must hold for *all* possible data—even those data that cause instability, if such exist. Thus very large a priori error estimates, especially those of the forward type, often suggest the possibility of numerical instability. (This observation is another reason for insisting that error bounds should always be as small as possible. Otherwise, stable algorithms may be suspected of being unstable merely because of a sloppy error analysis.)

Consider again the analysis of the summation of five numbers x_1, x_2, x_3, x_4, x_5 as given in Section 2.5.2. Suppose that some of the numbers are positive, some negative. Then (2.39) must be written as

$$|\hat{s} - s| \leq (4|x_1| + 4|x_2| + 3|x_3| + 2|x_4| + |x_5|)\mu'. \qquad (2.42)$$

If now we consider the situation where the sum s is approximately equal to μ', but all of the x_i have magnitude 1, then (2.42) gives the *relative* estimate

$$|\hat{s} - s|/|s| \leq 14\mu'/|s| \approx 14, \qquad (2.43)$$

which is obviously a very undesirable situation. This suggests then that there are situations where the process of summation can produce a very inaccurate result. Specifically, if some of the numbers are positive, and some are negative so that the resulting sum may be very much smaller than the individual terms, then the results might be quite bad. To illustrate this, consider the calculation of the sine function by means of the Taylor series expansion

| Example:
Instability
of summation | $\sin x = x - (x^3/3!) + (x^5/5!) - \cdots,$ | (2.44) |

which is valid for all values of x. It is well known that for $x > 0$, if the series is summed only up to the term $x^n/n!$, then the error is less than $|x^{n+2}/(n+2)!|$. Thus by summing this series until the terms are suitably small, one should be able to obtain a good approximation to $\sin x$ for any x. Indeed, for small values of x, this is certainly true; however, if $x = 25.65633$ is used in such a computation on the UNIVAC 1108, the computed result is -33.9347, which clearly is nonsense since $|\sin x| \leq 1$ for *all* x. Individual terms in the series (2.44) with this value of x and various subtotals are given in Table 2.3. An examination of this table shows that the difficulty is caused by the fact that some of the intermediate values ($.5503\ldots \times 10^{10}$, for example) are much larger than the final result (which should be 0.5). Notice that an a priori error bound such as (2.42) can predict this large error once it is known that some of the terms are so large.

It is worth stressing here that the simple computational process of summation is subject to instability even if all the numbers are positive and not tremendously large as in the above example. Indeed, the example (2.41) clearly illustrates that adding the numbers in descending order can cause the error to be unnecessarily large.

TABLE 2.3 *Computed Values of $s_n(x) = x - (x^3/3!) + (x^5/5!) - \cdots \pm (x^n/n!)$ for $x = 25.65633$*

n	s_n	$x^n/n!$
5	$.89849366 \times 10^5$	$.92639410 \times 10^5$
15	$-.78291331 \times 10^9$	$.10505516 \times 10^{10}$
25	$.55036908 \times 10^{10}$	$.10945006 \times 10^{11}$
45	$.51358077 \times 10^7$	$.21574005 \times 10^8$
65	$-.33308531 \times 10^2$	$.4800761 \times 10^1$
81	$-.33934695 \times 10^2$	$\sim 10^{-7}$

Another common type of summation calculation that is subject to this kind of instability is a summation of the form

> Example:
> Unstable
> summation
> problem

$$s = a_1 f(x_1) + a_2 f(x_2) + \cdots + a_n f(x_n), \qquad (2.45)$$

where $f(x)$ is some given continuous function, and the coefficients $a_1, a_2, \ldots,$ a_n sum to zero; i.e.,

$$a_1 + a_2 + \cdots + a_n = 0. \qquad (2.46)$$

If $x_1, x_2, \ldots, x_n$ are all nearly equal, then the value of the expression (2.45) will be nearly zero. But if $|f(x_i)|$, $i = 1, 2, \ldots, n$ are quite large, then the errors in the multiplications $a_k f(x_k)$ will be reflected as large errors in the computed results. To show this more precisely, ignore the errors caused by addition and write the computed value of (2.45) as

$$\hat{s} = a_1 f(x_1)\langle 1 \rangle + a_2 f(x_2)\langle 1 \rangle + \cdots + a_n f(x_n)\langle 1 \rangle$$

$$= [a_1 f(x_1) + a_2 f(x_2) + \cdots + a_n f(x_n)] + a_1 f(x_1)\langle 1 \rangle_0$$

$$+ \cdots + a_n f(x_n)\langle 1 \rangle_0$$

$$= s + a_1 f(x_1)\langle 1 \rangle_0 + \cdots + a_n f(x_n)\langle 1 \rangle_0. \qquad (2.47)$$

Here we have used the rounding error counters $\langle 1 \rangle$ to denote only the multiplication errors. Now, let $\bar{x}$ be the average of $x_1, x_2, \ldots, x_n$ and assume that $f(x_i) = f(\bar{x}) + \delta_i$, where $|\delta_i|$, $i = 1, \ldots, n$ are small. Then the exact sum can be written as

$$s = a_1(f(\bar{x}) + \delta_1) + \cdots + a_n(f(\bar{x}) + \delta_n)$$

$$= (a_1 + \cdots + a_n)f(\bar{x}) + a_1\delta_1 + \cdots + a_n\delta_n$$

$$= a_1\delta_1 + \cdots + a_n\delta_n$$

since (2.46) holds. Thus from (2.47) we have

$$\frac{\hat{s} - s}{s} = \frac{a_1 f(x_1)\langle 1 \rangle_0 + \cdots + a_n f(x_n)\langle 1 \rangle_0}{a_1\delta_1 + \cdots + a_n\delta_n}$$

$$= \frac{(a_1\langle 1 \rangle_0 + \cdots + a_n\langle 1 \rangle_0)f(\bar{x}) + a_1\delta_1\langle 1 \rangle_0 + \cdots + a_n\delta_n\langle 1 \rangle_0}{a_1\delta_1 + \cdots + a_n\delta_n}.$$

If $\delta_i \approx \langle 1 \rangle_0$, $i = 1, 2, \ldots, n$, then this implies that

$$\left| \frac{\hat{s} - s}{s} \right| \approx |f(\bar{x})|. \tag{2.48}$$

Thus the relative error in the computed value of s, caused only by the multiplication errors, will be as large as $|f(\bar{x})|$. If this value is greater than 1, then the computed value will be *totally* wrong. There is no general technique for eliminating the instability in a computation of the form (2.45).

The above examples are related to the problem of subtractive cancellation. In order to compute a small number by adding together large numbers, it must happen that at some point two numbers of nearly equal value are subtracted. To see this more clearly, suppose that $\hat{x}$ and $\hat{y}$ are computed approximations to certain values x and y, so that $\hat{x} = x \langle k \rangle$ and $\hat{y} = y \langle l \rangle$. That is, $\hat{x}$ and $\hat{y}$ contain k and l rounding errors, respectively. Suppose also that x and y are nearly equal in the sense that, say, $x = y \langle 1 \rangle$. Then we have

$$\frac{fl(\hat{x} - \hat{y}) - (x - y)}{x - y} = \frac{(\hat{x} - \hat{y})\langle 1 \rangle - (x - y)}{x - y}$$

$$= \frac{(x\langle k \rangle - y\langle l \rangle)\langle 1 \rangle - (x - y)}{(x - y)}$$

$$= \frac{(x\langle k + 1 \rangle - y\langle l + 1 \rangle) - (x - y)}{(x - y)}$$

$$= \frac{x\langle k + 1 \rangle_0 - y\langle l + 1 \rangle_0}{x - y}$$

$$= \frac{y\langle 1 \rangle_0 \langle k + 1 \rangle_0 - y\langle l + 1 \rangle_0}{y\langle 1 \rangle_0}.$$

Thus

$$\frac{fl(\hat{x} - \hat{y}) - (x - y)}{(x - y)} = \frac{\langle 1 \rangle_0 \langle k + 1 \rangle_0 - \langle l + 1 \rangle_0}{\langle 1 \rangle_0}.$$

This is an expression for the relative error in the computed difference between the computed values, as compared with the exact difference between the exact values. If, for example, we set $\langle 1 \rangle_0 = \mu'$, $\langle k + 1 \rangle_0 = (k + 1)\mu'$, and $\langle l + 1 \rangle_0 = -(l + 1)\mu'$, then this expression becomes $(k + 1)\mu' + (l + 1)$. This indicates a total loss of accuracy due to subtractive cancellation.

As a final example of using an error analysis to predict instability, consider the recursive computation

Example:
Unstable
recursion

$$I_k = 1 - kI_{k-1}, \qquad k = 1, 2, \ldots, 20 \qquad\qquad (2.49)$$

$$I_0 = 1 - (1/e).$$

This was described in Section 1.2 as a simple method for finding the value of the integral (1.21). Table 1.2 gives some computed values that show that **the method is unstable**. This instability was examined in Section 1.2 by considering how the error in I_0 is propagated through the computation. More precisely, we assume that the computed value $\hat{I}_0$ of I_0 has an error E_0 so that

$$\hat{I}_0 = I_0 + E_0.$$

Then, assuming that *no other errors* are committed, we have computed values of $\hat{I}_k$ given by

$$\hat{I}_1 = 1 - \hat{I}_0 = 1 - (I_0 + E_0) = (1 - I_0) - E_0 = I_1 - E_0$$

$$\hat{I}_2 = 1 - 2\hat{I}_1 = 1 - 2(I_1 - E_0) = (1 - 2I_1) + 2E_0 = I_2 + 2E_0$$

$$\hat{I}_3 = 1 - 3\hat{I}_2 = 1 - 3(I_2 + 2E_0) = (1 - 3I_2) - 6E_0 = I_3 - 6E_0$$

$$\vdots$$

$$\hat{I}_{20} = 1 - 20\hat{I}_{19} = 1 - 20(I_{19} - 19!\,E_0) = I_{20} + 20!\,E_0.$$

Thus, as predicted in Section 1.2, the initial error in I_0 is multiplied by 20! in $\hat{I}_{20}$. A study of this error propagation suggests a way to **stabilize this computation**. Suppose we *reverse* the recursion by writing it in the form

$$I_{k-1} = (1/k) - (1/k)I_k, \qquad k = K, K-1, \ldots, 20. \qquad\qquad (2.50)$$

Now we see that errors in early stages of the computation are multiplied by $1/k$, which will reduce them. The only difficulty with using (2.50) is in finding a starting value. But from the fact that

$$I_k = \int_0^1 x^k e^{x-1}\, dx \to 0 \quad \text{as} \quad k \to \infty,$$

it seems reasonable to set $I_K = 0$ for some (large) value of K and then use (2.50) to compute $I_{K-1}, I_{K-2}, \ldots, I_{20}$. To analyze this process and especially to examine the effect of setting $I_K = 0$, let $\hat{I}_k$ denote the computed values. Again, we shall ignore all errors except for the error in the starting value $\hat{I}_K$. Then we have

$$\hat{I}_K = I_K + E_0$$

$$\hat{I}_{K-1} = \frac{1}{K} - \frac{1}{K}\hat{I}_K = \frac{1}{K} - \frac{1}{K}(I_K + E_0) = I_{K-1} - \frac{1}{K}E_0$$

$$\hat{I}_{K-2} = \frac{1}{K-1} - \frac{1}{K-1}\hat{I}_{K-1} = \frac{1}{K-1} - \frac{1}{K-1}\left(I_{K-1} - \frac{1}{K}E_0\right)$$

$$= I_{K-2} + \frac{1}{K(K-1)}E_0$$

$$\vdots$$

$$\hat{I}_{20} = I_{20} + \frac{1}{K(K-1)(K-2)\cdots(K-(K-20))}E_0.$$

Hence even if E_0 is fairly large, after a few steps its effect will be unnoticeable. For example, taking $K = 30$ gives $K(K-1)\cdots(K-(K-20)) = 20 \cdot 21 \cdot 22 \cdots 30 > 10^{10}$, so that $\hat{I}_{20} \approx I_{20} \pm 10^{-10}E_0$. An interesting experiment is to compute I_{20} by using (2.50) with, say, $K = 30$ and several values of I_{30}, such as $I_{30} = 0$, $I_{30} = -1$, and $I_{30} = 1$. The results will be the same in all cases because the error in I_{30} is totally eliminated.

EXERCISES

1 Do a backward and forward error analysis on the computation of a product of five nonzero numbers: $p = x_1 x_2 x_3 x_4 x_5$.

2 Prove Theorem 2.1 by first proving the following lemmas:

Lemma 1 If $0 \le r\mu < 1$, then $1 - nr\mu \le (1 - r\mu)^n$.
[*Hint*: Expand $f(y) = (1 - y)^n$ in a Taylor series.]

Lemma 2 If $0 \le x \le .1$, then $1 + x \le e^x \le 1 + 1.06x$.
[*Hint*: Expand e^x in a Taylor series.]

Lemma 3 If $0 \le nr\mu \le .1$, then $(1 + r\mu)^n \le 1 + 1.06nr\mu$.
[*Hint*: Use Lemma 2.]

Lemma 4 If $|\varepsilon_i| \le r\mu$ for $i = 1, 2, \ldots, n$, and $nr\mu \le .1$, then
$$1 - nr\mu \le (1 + \varepsilon_1)(1 + \varepsilon_2)\cdots(1 + \varepsilon_n) \le 1 + 1.06nr\mu.$$

Lemma 5 If $0 \le x \le .1$, then $(1 - x)^{-1} \le 1 + 1.06x$.

Lemma 6 If $|\varepsilon_i| \le r\mu$ for $i = 1, 2, \ldots, n$ and $nr\mu \le .1$, then

$$1 - nr\mu \le \frac{1}{(1 + \varepsilon_1)\cdots(1 + \varepsilon_n)} \le 1 + 1.06nr\mu.$$

[*Hint*: Use Lemmas 4 and 5.]

Now combine Lemmas 4 and 6 to prove Theorem 2.1.

3 Prove the relations (2.38).

4 Do a forward and a backward error analysis for the following computations:

(a) $fl\{fl[x^2] + y\}$ (b) $fl\left\{\frac{[(x + y)(u + v)]}{(a + b)(c + d)}\right\}$

(c) $fl\left[\frac{(x + y)^2}{(u + v)}\right]$.

5 Find five positive numbers for which the bound (2.39) is nearly attained.

6 Consider the series $e^{-x} = 1 - x + (x^2/2) - (x^3/3!) + \cdots$. Write a computer program to evaluate e^{-x} for large positive x using this series. Use the program to show that if x is too large, rounding error causes instability. How can this instability be avoided?

7 The formula

$$f'(x) \sim \frac{-3f(x - 2h) - 10f(x - h) + 18f(x) - 6f(x + h) + f(x + 2h)}{12h}$$

is shown in Chapter 6 to give, theoretically, a good approximation to $f'(x)$, when $h > 0$ is small. Show that this is an expression of the type (2.45), (2.46). Using small h, and the function $f(x) = e^{10x}$, illustrate the instability by approximating $f'(0)$ with this formula.

8 Compute I_{20} using (2.50) with $K = 30$ and several values for I_{30}.

Suggestions for Further Reading

The basic reference for error analysis is Wilkinson (1963). A more modern discussion of floating point numbers and other numbers systems is given by Knuth (1969). The notation $\langle \; \rangle$, introduced in Section 2.4 is taken from Stewart [1973]. A complete analysis of floating point arithmetic, as done by IBM 360 model computers, is given by Sterbenz (1974). Forsythe and Moler (1967) give a nice discussion of rounding error during the solution of systems of linear equations.

Chapter 3

EVALUATION OF FUNCTIONS

3.1 Summation

As a first example of the analysis of a nontrivial numerical algorithm, we discuss the process of summing a set of numbers. Even this elementary computation is seen to have unsuspected complications. We also introduce an informal language for describing algorithms.

3.1.1 Nonassociativity and Subtractive Cancellation

The example (2.41) showed that because of the floating point arithmetic used by computers, the *order* in which numbers are added together affects the value of the computed sum. More precisely, if x, y, and z are any three floating point numbers, then it is *not* necessarily true that $fl(x + fl(y + z)) = fl(fl(x + y) + z)$. The mathematical term for the equation

$$x + (y + z) = (x + y) + z \qquad \text{all} \quad x, y, z$$

is *associativity*; hence we say that **computer addition is nonassociative.** The error bound (2.39) indicates that the accuracy of a computed sum may depend on the relative sizes of the numbers being added together. An even more serious cause of error in the computed sum is subtractive cancellation. To illustrate this phenomenon again, consider the four numbers

$$.1025 \times 10^4, \quad -.0123 \times 10^3, \quad -.9773 \times 10^2, \quad -.9315 \times 10^1. \quad (3.1)$$

If they are summed as written from left to right with four-digit arithmetic, the result is the exact sum $.6755 \times 10^1$. However, if they are summed in reverse order, i.e., from right to left, the result is $.7000 \times 10^1$, which is a rather poor approximation. This should clearly show that the simple mathematical problem of summing a set of numbers is *not* a simple computational problem. In particular, it may be necessary to arrange the numbers in a certain order or examine them as they are added together, in order to obtain a reasonably accurate value.

3.1.2 INFL Programs

Since even the order in which computations are carried out affects the final result, any description of a numerical algorithm must be extremely precise. Compiler languages such as ALGOL, FORTRAN, PL/1, etc., provide one means of describing algorithms with suitable precision; however, for the present discussion such descriptions are often *too* detailed, and many of the important ideas become lost. For this reason, we shall use a very informal kind of language, called INFL, related somewhat to ALGOL and FORTRAN but without any strict syntax rules.[†] Since this is an *informal* language, it will not be defined in a formal sense but rather through examples.

Summation Algorithm

$$\left[\begin{array}{l} \text{An INFL description of a simple algorithm for summing} \\ \text{together } N \text{ numbers } x_1, x_2, \ldots, x_N \text{ is now given.} \end{array}\right]$$

$$
\begin{array}{ll}
1 & \text{Input } \{N, x_1, x_2, \ldots, x_N\} \\
2 & S \leftarrow 0 \\
3 & \text{For } k = 1, 2, \ldots, N \\
  & 3.1 \quad S \leftarrow S + x_k \\
4 & \text{Output } \{S\} \text{ Halt}
\end{array}
\qquad (3.2)
$$

Anyone familiar with ALGOL, FORTRAN, or other similar languages should have no trouble in understanding this algorithm. Indeed it is a simple matter to convert this description into a computer program. Other "elements" of the INFL language will be introduced as needed.

3.1.3 Error Analysis

The error analysis for the addition of five numbers, which was described in Section 2.5, can easily be generalized to the addition of N numbers. As with many error analyses, the crucial step in such a generalization is to guess the general form of equations such as (2.30). A more formal and mechanical way

---

[†] INFL, as used here, is related to the language introduced by G. W. Stewart (1973).

to do this is to observe that computationally Step 3.1 in Algorithm (3.2) has the form

$$3.1 \quad S^{[k]} = \text{fl}[S^{[k-1]} + x_k].$$

We have used superscripts here to emphasize the fact that the value of S changes as the value of k changes. By applying (2.27) to this expression and using the notation introduced in Section 2.5.3, we have

$$S^{[k]} = S^{[k-1]}\langle 1 \rangle + x_k \langle 1 \rangle.$$

Thus,

$$\begin{aligned}
S^{[1]} &= S^{[0]}\langle 1 \rangle + x_1\langle 1 \rangle = x_1\langle 1 \rangle \\
S^{[2]} &= S^{[1]}\langle 1 \rangle + x_2\langle 1 \rangle = x_1\langle 1 \rangle\langle 1 \rangle + x_2\langle 1 \rangle = x_1\langle 2 \rangle + x_2\langle 1 \rangle \\
S^{[3]} &= S^{[2]}\langle 1 \rangle + x_3\langle 1 \rangle = (x_1\langle 2 \rangle + x_2\langle 1 \rangle)\langle 1 \rangle + x_3\langle 1 \rangle \\
&= x_1\langle 3 \rangle + x_2\langle 2 \rangle + x_3\langle 1 \rangle,
\end{aligned} \tag{3.3a}$$

where each of these expressions is obtained from the preceding one by a simple substitution. That is, the formula for $S^{[3]}$ is obtained by replacing $S^{[2]}$ with the formula just derived for $S^{[2]}$. It is easy to see that the **general formula** is

$$S^{[k]} = x_1\langle k \rangle + x_2\langle k-1 \rangle + \cdots + x_k\langle 1 \rangle. \tag{3.3b}$$

The computed sum, which is $S^{[N]}$, is then given by

$$S^{[N]} = x_1\langle N \rangle + x_2\langle N-1 \rangle + \cdots + x_{N-1}\langle 2 \rangle + x_N\langle 1 \rangle. \tag{3.3c}$$

Notice that, in fact, there is no error in $S^{[1]}$, which is just equal to x_1. However, it simplifies the notation to include an error factor here. If we denote the computed sum by $\hat{S}$, then we have

$$\hat{S} = \hat{x}_1 + \hat{x}_2 + \cdots + \hat{x}_N, \tag{3.3d}$$

where

$$\hat{x}_k = x_k\langle N - k + 1 \rangle, \qquad k = 1, 2, \ldots, N.$$

This can also be written as $\hat{x}_k = x_k + x_k\langle N - k + 1 \rangle_0$. Thus if we assume $nr\mu \le .1$, then (2.37) gives the expression

$$|\hat{x}_k - x_k|/|x_k| = |\langle N - k + 1 \rangle_0| \le (N - k + 1)\mu'.$$

We can summarize these results by writing the **backward error estimate**

$$\hat{S} = \hat{x}_1 + \hat{x}_2 + \cdots + \hat{x}_N$$
$$|\hat{x}_k - x_k|/|x_k| \le (N - k + 1)\mu', \qquad k = 1, 2, \ldots, N. \tag{3.4}$$

We can also derive a **forward error estimate** by writing (3.3d) as

$$\hat{S} = (x_1 + x_2 + \cdots + x_N) + x_1\langle N \rangle_0 + x_2\langle N - 1 \rangle_0 + \cdots + x_N\langle 1 \rangle_0,$$

so that
$$\hat{S} - S = x_1\langle N\rangle_0 + x_2\langle N - 1\rangle_0 + x_3\langle N - 2\rangle_0 + \cdots + x_N\langle 1\rangle_0.$$
Again we assume $nr\mu < .1$, so that

Forward error estimate	$|\hat{S} - S| \leq (N|x_1| + (N - 1)|x_2| + (N - 2)|x_3| + \cdots + |x_N|)\mu'.$

$$\tag{3.5}$$

This gives a forward error bound on the *absolute* error for this algorithm. Notice that by setting $N = 5$ in (3.5) we obtain the bound (2.39) except that x_1 is now multiplied by five instead of four because we did not use the fact that $S^{[1]} = x_1$ when deriving (3.5). To simplify (3.5) further, let
$$\bar{x} = \max\{|x_1|, |x_2|, \ldots, |x_N|\},$$
so that
$$|\hat{S} - S| \leq (N + (N - 1) + (N - 2) + \cdots + 1)\bar{x}\mu'.$$
Now we use the formula
$$N + (N - 1) + (N - 2) + \cdots + 3 + 2 + 1 = \tfrac{1}{2}N(N + 1) \tag{3.6}$$
to obtain
$$|\hat{S} - S| \leq \tfrac{1}{2}N(N + 1)\bar{x}\mu' \tag{3.7}$$
or, as a *relative* error bound,

Relative error estimate	$\dfrac{|\hat{S} - S|}{|S|} \leq \dfrac{N(N + 1)}{2}\left(\dfrac{\bar{x}}{|S|}\mu'\right).$

$$\tag{3.8}$$

A few remarks concerning the above error analysis are appropriate here. First of all, even the analysis of this simple computation tends to become somewhat tedious and, at first encounter, complicated. It is important to realize that most of the analysis is completely mechanical. In fact, the only place where any real cleverness is involved is in guessing the general form of the computed quantities [formula (3.3b)] from the specific form derived for the first few quantities $S^{[1]}$, $S^{[2]}$, $S^{[3]}$ in (3.3a). Once a certain proficiency is developed, much of the tedious detail in such an analysis can be omitted. In this chapter, however, we shall continue, for the sake of clarity, to include these details at the risk of making the error analysis appear to be overly complicated.

Just as in the simple case $N = 5$ discussed in Section 2.5, the estimates (3.5) and (3.8) do not give very accurate *quantitative* information. In deriving them, it is necessary to assume always that the worst possible error is being committed. That is, it must be assumed that in *each* addition, the rounding error is equal to μ, whereas in reality many of these errors will be considerably less than μ. For emphasis we repeat the conclusions that were drawn from the estimates with $N = 5$ in Section 2.5.

(i)   The numbers should be summed in increasing order of magnitude; i.e., the numbers should be arranged so that $|x_1| \leq |x_2| \leq \cdots \leq |x_N|$.

(ii)   The relative error may be quite large if some of the numbers are many times larger than the final sum S.

Conclusion (i) follows from the fact that in (3.5) $|x_1|$ is multiplied by N, whereas $|x_N|$ is only multiplied by one. This conclusion was "experimentally" verified by the numbers given in (2.41). Conclusion (ii) is a result of the factor $\bar{x}/S$ in the error estimate (3.8). The values given in Table 2.3 verify this conclusion.

3.1.4   Stable Algorithms

Algorithm (3.2) may produce rather bad results if the numbers are summed in the wrong order, or if some of them are many times larger than the final answer. Thus it is worthwhile considering other, perhaps more complicated, algorithms that give better results. Notice that if all of the numbers are positive (or all negative), then the final answer must be larger than any of the numbers, and hence the term $\bar{x}/S$ in (3.8) will be less than one. Furthermore, the possibility of adding the numbers in a bad order can be avoided by searching for the smallest number that has not yet been added. The following algorithm adds together a set of positive numbers in increasing order of magnitude. In this algorithm, xxx denotes the *largest* floating point number that can be represented in the computer.

A Stable Summation Algorithm

$$\left[\begin{array}{c} \text{Algorithm to sum a set of positive numbers} \\ \text{in order of increasing size.} \end{array} \right]$$

1   Input $\{N, x_1, x_2, \ldots, x_N\}$
2   $S \leftarrow 0$
3   For $k = 1, 2, \ldots, N$
   3.1   min $\leftarrow xxx$
   3.2   For $i = 1, 2, \ldots, N$ (3.9)
      3.2.1   If $x_i \neq 0$ and $x_i < $ min,
         then min $\leftarrow x_i$, i min $\leftarrow i$
   3.3 $S \leftarrow S + $ min
   3.4 $x_{i\,min} \leftarrow 0$
4   Output $\{S\}$ Halt

In this algorithm, once a number has been added into the sum, its value is set to zero in Step 3.4. This zeroed number is then never again considered as a summand because of the first part of Step 3.2.1. The statement 3.2 is a kind of *sorting* algorithm, called a *bubble sort*. It can generally be used to put a set of numbers into ascending or descending order, or can be used to alphabetize a set of names. More efficient sorting algorithms are available, but we use this one because of its simplicity.

If we want to have a general algorithm that will also accept negative numbers, then we can use the above algorithm as a subalgorithm to the following:

A Very General Stable Summation Algorithm

[Algorithm to sum any set of numbers in a stable manner.]

1 Input $\{N, x_1, x_2, \ldots, x_N\}$
2 $m_1 \leftarrow 0, m_2 \leftarrow 0$
3 For $k = 1, 2, \ldots, N$
   3.1 If $x_k \geq 0$, then $m_1 \leftarrow m_1 + 1$, $y_{m_1} \leftarrow x_k$
   3.2 If $x_k < 0$, then $m_2 \leftarrow m_2 + 1$, $z_{m_2} \leftarrow -x_k$ (3.10)
4 Input $\{m_1, y_1, \ldots, y_{m_1}\}$ to Algorithm (3.9), with output $\{PLUS\}$
5 Input $\{m_2, z_1, \ldots, z_{m_2}\}$ to Algorithm (3.9) with output $\{MINUS\}$
6 SUM = PLUS − MINUS
7 Output $\{PLUS, MINUS, SUM\}$, Halt

This algorithm sums all of the positive numbers, then all of the negative ones, and outputs these sums as well as the complete sum. If the positive and negative sums are quite different in magnitude, then the final sum will be very accurate; otherwise it may be necessary to compute these sums in double precision.

The testing involved in Algorithm (3.9) may substantially increase the time (i.e., cost) for this algorithm. To see this very clearly, and also to illustrate how the efficiency of an algorithm is determined, consider Step 3.2.1. In certain computers the comparison "$x_i < \min$" requires a floating point subtraction followed by a test of whether the result is positive or negative. Thus there will be a floating point subtraction in 3.2.1 every time $x_i \neq 0$, so that

when $k = 1$, $x_i \neq 0$ for all i, giving N subtractions in 3.2.1;
when $k = 2$, $x_i \neq 0$ for $(N - 1)$ i, giving $N - 1$ subtractions in 3.2.1;
when $k = 3$, $x_i \neq 0$ for $(N - 2)$ i, giving $N - 2$ subtractions in 3.2.1;
   $\vdots$
when $k = N - 1$, $x_i \neq 0$ for two i, giving two subtractions in 3.2.1;
when $k = N$, $x_i \neq 0$ for one i, giving one subtraction in 3.2.1.

This gives a total of $N + (N - 1) + (N - 2) + \cdots + 2 + 1 = \frac{1}{2}N(N + 1)$ subtractions in step 3.2. Finally, step 3.3 involves N additions, so the efficiency measure for this algorithm is $(\frac{1}{2}N(N + 1) + N)A + OM + OD$. This is to be compared with $(N)A + OM + OD$ for the simpler Algorithm (3.2). If $N = 100$, for example, (3.9) requires 5150 additions, while (3.2) takes only 100 additions. Some computers have a special instruction for comparing two numbers. On these computers Algorithm (3.9) is much more efficient than is indicated here.

3.1.5   Inner Products—Use of Double Precision

We next consider a simple computation that occurs frequently in various applications. The *inner product* of two sets of numbers $x_1, x_2, \ldots, x_N$ and $y_1, y_2, \ldots, y_N$, is defined by

$$I = x_1 y_1 + x_2 y_2 + \cdots + x_N y_N. \tag{3.11}$$

Clearly this computation can be done by a slight modification of the summation algorithm. Again, however, the possibility of large errors caused by summing in decreasing order or by a very small final result must be considered. In fact, many times the purpose for computing (3.11) is to see whether it is *zero*. In this case the final sum *will* be much smaller than the individual terms. Furthermore the individual terms must themselves be computed and hence will contain rounding errors. Thus the total error may be somewhat larger than is indicated by the error analyses for the summation algorithm. To avoid these difficulties, it is often recommended that the products be computed in *double precision*. These double-precision numbers are then added together and the final sum rounded to single-precision. Such a procedure is called *accumulated inner products* and is described by the following algorithm.

Accumulated Inner Product Algorithm

$$\left[\begin{array}{c} \text{Algorithm to compute accurate inner products. DI and DT are} \\ \text{double-precision numbers.} \end{array} \right.$$

$$\begin{array}{ll} 1 & \text{Input } \{N, x_1, \ldots, x_N, y_1, \ldots, y_N\} \\ 2 & \text{DI} = 0 \\ 3 & \text{For } k = 1, 2, \ldots, N \\ & \quad 3.1 \quad \text{DT} = x_k y_k \\ & \quad 3.2 \quad \text{DI} = \text{DI} + \text{DT} \\ 4 & I = \text{DI} \\ 5 & \text{Output } \{I\}. \text{ Halt} \end{array} \tag{3.12}$$

The error analysis for this algorithm is obtained from (3.3) by simply replacing each of the x_k by $\text{fl}(x_k y_k)$. But formula (2.27) implies that

$$\text{fl}(x_k y_k) = (x_k y_k)\langle 1 \rangle,$$

where $\langle 1 \rangle$ represents the relative rounding error in computing this product. Thus the computed inner product is

$$S^{[N]} = x_1 y_1 \langle N + 1 \rangle + x_2 y_2 \langle N \rangle + \cdots + x_N y_N \langle 2 \rangle. \tag{3.13}$$

This is just a minor modification of (3.3c) that shows that the various estimates will all have one additional rounding error in each term. Moreover, because of the double-precision arithmetic, the relative rounding errors will be much smaller, of the order β^{1-2t} instead of β^{1-t}. Thus the forward error estimate will be

Error analysis of algorithm (3.12)	$\begin{aligned} \lvert \hat{I} - I \rvert \leq &((N + 1)\lvert x_1 y_1 \rvert \\ &+ N\lvert x_2 y_2 \rvert + \cdots + 2\lvert x_N y_N \rvert)(1.06r\beta^{1-2t}) \end{aligned}$ (3.14)

and the corresponding **relative error estimate is**

$$\lvert \hat{I} - I \rvert / \lvert I \rvert \leq \tfrac{1}{2} N(N + 3)(T/\lvert I \rvert)1.06r\beta^{1-2t}, \tag{3.15}$$

where $T = \max\{\lvert x_1 y_1 \rvert, \lvert x_2 y_2 \rvert, \ldots, \lvert x_N y_N \rvert\}$.

These estimates have been derived only to illustrate how double-precision calculations can be used in an error estimate. No essentially new information has been gained from them, which is not surprising since the algorithm is so similar to the previously analyzed summation algorithm.

EXERCISES

1   Is computer *multiplication* associative? That is, does the equation $\text{fl}[\text{fl}[xy]z] = \text{fl}[x\text{fl}[yz]]$ always hold? Explain your answer.

2   Write Algorithms (3.2) and (3.9) as computer programs using any compiler language.

3   Describe, using INFL, an algorithm for computing the product of n numbers $x_1, x_2, \ldots, x_n$.

4   Describe, using INFL, an algorithm for finding the maximum of n numbers $x_1, \ldots, x_n$.

5 Do an error analysis for the computation of the *product* of n numbers. In particular, show that the accumulated error does *not* depend on the order in which the numbers are multiplied.

6 Referring to the add-time for your computer, estimate how much computer time it will take to add together 10,000 positive numbers. Use Algorithm 3.2. Then use Algorithm 3.9 and assume that each comparison requires a floating point subtraction.

3.2 Polynomial Evaluation

One of the most common numerical computations is the evaluation of polynomials. In this section we describe two algorithms for doing this and show how a careful analysis of accuracy and efficiency can be used to select the best algorithm.

3.2.1 Two Algorithms

The most complicated kind of function that can be evaluated using *only* addition, subtraction, and multiplication is a polynomial function:

$$p(x) = a_n x^n + a_{n-1} x^{n-1} + \cdots + a_1 x + a_0. \tag{3.16}$$

Recall that the numbers $a_0, a_1, \ldots, a_n$ are called the *coefficients* of the polynomial, and n is called the *degree* of the polynomial. The obvious way to compute the value of $p(x)$ for some floating point number x is given by the following algorithm.

An (Inefficient) Polynomial Evaluation Algorithm

[Algorithm to evaluate a polynomial.]

$$
\begin{aligned}
&1 \quad \text{Input } \{n, a_0, a_1, \ldots, a_n\} \\
&2 \quad \text{Input } \{x\} \\
&3 \quad p \leftarrow a_0 \\
&4 \quad z \leftarrow 1 \\
&5 \quad \text{For } k = 1, 2, \ldots, n \\
&\quad\quad 5.1 \quad z \leftarrow zx \\
&\quad\quad 5.2 \quad p \leftarrow p + a_k z \\
&6 \quad \text{Output } \{x, p\}. \text{ Halt}
\end{aligned}
\tag{3.17}
$$

The value of z in this algorithm is just x^k.

The above algorithm is *not* recommended because a much more efficient way to do this same computation is given by the following.

Horner's Method Algorithm

[Algorithm to evaluate a polynomial by Horner's method.]

$$
\begin{aligned}
&1 \quad \text{Input } \{n, a_0, a_1, \ldots, a_n\} \\
&2 \quad \text{Input } \{x\} \\
&3 \quad p \leftarrow a_n \\
&4 \quad \text{For } k = n - 1, n - 2, \ldots, 0 \\
&\quad 4.1 \quad p \leftarrow xp + a_k \\
&5 \quad \text{Output } \{x, p\}. \text{ Halt}
\end{aligned}
\tag{3.18}
$$

This method for evaluating a polynomial is called *Horner's method*, or *nested evaluation*. It is derived by writing the polynomial of degree 5, for example, in the form

$$p(x) = x(x(x(x(xa_5 + a_4) + a_3) + a_2) + a_1) + a_0. \tag{3.19}$$

Each of the expressions enclosed in parentheses in (3.19) has the form $xp + a_k$, where p is an expression of the same form except when $p = a_5$.

Many scientific and engineering problems involve the use of polynomials. A very common source of polynomials is the approximation of more complicated functions. Several such approximations will be derived in Chapter 4. However, to illustrate the above algorithms, consider the polynomial

$$
\begin{aligned}
p(x) = {}&1.0 + .99739581x + .50001602x^2 \\
&+ .17708332x^3 + .04165064x^4.
\end{aligned}
$$

It can be shown that this polynomial approximates the function e^x, for x between -1 and 1, with absolute error less than .005. To evaluate it, for $x = .5$, Algorithm (3.17) gives

$$p = 1.0 \quad \text{(step 3)}$$

$$p = 1.0 + .99739581(.5) = 1.4986980 \quad \text{(step 5.2)}$$

$$p = 1.4986980 + (.50001602)(.5)(.5) = 1.6237019 \quad \text{(step 5.2)}$$

$$p = 1.6237019 + (.17708332)(.25)(.5) = 1.645837325 \quad \text{(step 5.2)}$$

$$p = 1.645837325 + (.04165064)(.125)(.5) = 1.6484405 \quad \text{(step 5.2)}.$$

For this computation, we have used $4A + 7M$ operations. On the other hand, Algorithm (3.18) produces the values

$$p = .04165064 \quad \text{(step 3)}$$

$$p = (.5)(.04165064) + .17708332 = .19790864 \quad \text{(step 4.1)}$$

$$p = (.5)(.19790864) + .50001602 = .59897034 \quad \text{(step 4.1)}$$

$$p = (.5)(.59897034) + .99739581 = 1.296880980 \quad \text{(step 4.1)}$$

$$p = (.5)(1.296880980) + 1.0 = 1.6484405 \quad \text{(step 4.1)}.$$

Here we have used only $4A + 4M$ operations. Observe that these two algorithms produce the final value through completely different sets of computations. That is, the intermediate values of p that are determined by Algorithm (3.17) are quite different from those produced by (3.18). This suggests that the accumulated rounding error for the two algorithms will also differ—a conjecture that will be justified in the next subsection.

To show generally that (3.18) is more efficient than (3.17) consider the operation counts for the two algorithms:

Algorithm (3.17): n multiplications in step 5.1
n multiplications in step 5.2
n additions in step 4.2
Efficiency: $(n)A + (2n)M$.

(In the above example we did not count the first multiplication in (5.1), which is just $1 \cdot x$. Thus we had only seven multiplications instead of eight as derived here.)

Algorithm (3.18): n multiplications in step 4.1
n additions in step 4.1
Efficiency: $(n)A + (n)M$.

Thus (3.18) requires only half the number of multiplications. Often, however, a less efficient algorithm is more accurate. For this reason it is worthwhile to compare the accuracy of these two algorithms before rejecting (3.17) in favor of (3.18).

3.2.2   Error Analysis

We shall first examine the rounding error associated with the Horner method, i.e., Algorithm (3.18). In this algorithm floating point operations are contained only in step 4.1. We write this step as

$$p^{[k]} = \text{fl}[\text{fl}[xp^{[k+1]}] + a_k], \qquad k = n - 1, \quad n - 2, \ldots, 0, \qquad (3.20)$$

where $p^{[n]} = a_n$. The value $p^{[0]}$ is the computed value of $p(x)$, which we denote by $\hat{p}$. It follows then that

$$\hat{p} = p^{[0]} = \text{fl}[xp^{[1]}\langle 1 \rangle + a_0] = xp^{[1]}\langle 2 \rangle + a_0\langle 1 \rangle \qquad (3.21)$$

and similarly

$$p^{[1]} = xp^{[2]}\langle 2 \rangle + a_1\langle 1 \rangle. \qquad (3.22)$$

By combining these expressions we obtain

$$p^{[0]} = x\underbrace{(xp^{[2]}\langle 2 \rangle + a_1\langle 1 \rangle)}_{p^{[1]}}\langle 2 \rangle + a_0\langle 1 \rangle = x^2 p^{[2]}\langle 4 \rangle + a_1 x\langle 3 \rangle + a_0\langle 1 \rangle.$$

This process finally leads to

$$\hat{p} = p^{[0]} = x^n a_n \langle 2n \rangle + x^{n-1} a_{n-1} \langle 2n-1 \rangle + x^{n-2} a_{n-2} \langle 2n-3 \rangle$$
$$+ \cdots + x a_1 \langle 3 \rangle + a_0 \langle 1 \rangle. \tag{3.23}$$

From this we easily obtain the **backward error estimate for Algorithm (3.18)**

$$\hat{p} = \hat{a}_n x^n + \hat{a}_{n-1} x^{n-1} + \cdots + \hat{a}_1 x + \hat{a}_0 \tag{3.24}$$

where

$$\hat{a}_k = \begin{cases} a_k \langle 2n \rangle, & k = n \\ a_k \langle 2k+1 \rangle, & k = n-1, \ldots, 1, 0. \end{cases} \tag{3.25}$$

If we assume that $nr\mu \le .1$, then (2.37) can be applied to give the relative error bounds

$$\frac{|\hat{a}_k - a_k|}{|a_k|} \le \begin{cases} 2n\mu', & k = n \\ (2k+1)\mu', & k = 0, 1, \ldots, n-1. \end{cases} \tag{3.26}$$

A forward error estimate is obtained by writing (3.23) as

$$\hat{p} = x^n a_n (1 + \langle 2n \rangle_0) + x^{n-1} a_{n-1} (1 + \langle 2n-1 \rangle_0) + \cdots + a_0 (1 + \langle 1 \rangle_0)$$
$$= (a_n x^n + a_{n-1} x^{n-1} + \cdots + a_0) + x^n a_n \langle 2n \rangle_0 + x^{n-1} a_{n-1} \langle 2n-1 \rangle_0$$
$$+ \cdots + a_0 \langle 1 \rangle_0.$$

Thus,

$$\hat{p} - p(x) = a_n x^n \langle 2n \rangle_0 + a_{n-1} x^{n-1} \langle 2n-1 \rangle_0 + \cdots + a_0 \langle 1 \rangle_0. \tag{3.27}$$

Again, we assume that $nr\mu \le .1$, so (2.37) implies that

$$|\hat{p} - p(x)| \le [2n |a_n x^n| + (2n-1) |a_{n-1} x^{n-1}| + \cdots + 3 |a_1 x| + |a_0|]\mu'. \tag{3.28}$$

Formulas (3.24), (3.26), and (3.28) constitute a complete **rounding error analysis for Algorithm (3.18)**. A similar analysis of algorithm (3.17) leads to the results

Error analysis for Algorithm (3.17)	$\bar{p} = \bar{a}_n x^n + \bar{a}_{n-1} x^{n-1} + \cdots + \bar{a}_1 x + \bar{a}_0$ $\|\bar{a}_k - a_k\|/\|a_k\| \le (n+1)\mu', \quad k = 0, 1, \ldots, n$	(3.29)

and

$$|\bar{p} - p(x)| \le (n + 1)(|a_n x^n| + |a_{n-1} x^{n-1}| + \cdots + |a_1 x| + |a_0|)\mu'.$$
(3.30)

In (3.29) and (3.30) $\bar{p}$ denotes the value of $p(x)$ as computed by Algorithm (3.17). The derivation of these estimates is left as an exercise (Exercise 6). (This exercise is especially recommended as a test of the reader's understanding of rounding error analysis.)

As often happens when comparing two algorithms for solving the same problem, it is not possible to claim that one of these algorithms is always more accurate than the other. The forward error bounds (3.28) and (3.30) show clearly that the accumulated error depends on the sizes of the coefficients $a_0, a_1, \ldots, a_n$ as well as on the value of x. If, for example, the coefficients are all close to 1.0 in magnitude, then these error estimates have the form

$$|\hat{p} - p(x)| \le (2n|x^n| + |(2n - 1)|x^{n-1}| + \cdots + 3|x| + 1)\mu'$$
(Horner's Algorithm)   (3.31)

$$|\bar{p} - p(x)| \le (n + 1)(|x^n| + |x^{n-1}| + \cdots + |x| + 1)\mu'$$
(Algorithm 3.17).   (3.32)

When $|x|$ is very large, the dominant part of these bounds is

Estimates for large x	$|\hat{p} - p(x)| \approx 2n|x^n|\mu'$ (Horner Algorithm) $|\bar{p} - p(x)| \approx (n + 1)|x^n|\mu'$ (Algorithm 3.17).	(3.33a)

Thus (3.17) is more accurate. If x is very small, however, then the bounds are essentially

Estimates for small x	$|\hat{p} - p(x)| \approx \mu'$ (Horner Algorithm) $|\bar{p} - p(x)| \approx (n + 1)\mu'$ (Algorithm 3.17).	(3.33b)

Now, it seems that Horner's method is more accurate. Similarly, it can be seen that the comparative size of these error bounds also depends on the relative sizes of the coefficients.

It must always be kept in mind, however, that error bounds such as these are overly pessimistic. For values of n that are not too large, say, $n \le 20$, the bounds for the two algorithms are of the same order of magnitude. That is, the values given by (3.26) and (3.29) are roughly of the order $40\mu'$ and $21\mu'$, respectively. Thus for general use there is no strong reason to believe that either algorithm will be noticeably more accurate than the other. Only in specific situations as, for example, when x is very large or very small and high accuracy is desired, should the choice of algorithm be based on these error bounds. At the same time, these analyses dispel the notion that the algorithm with the fewer number of floating point operations is always more accurate. That is, it is *not* sufficient to compare only the efficiencies of algorithms. For some applications, the (slight) increase in accuracy obtained from Algorithm (3.17) may be worth the extra multiplications.

3.2.3 Applications

A special kind of polynomial that arises very frequently in applications is a *truncated power series*. Recall from calculus that if $f(x)$ is any function that can be differentiated sufficiently many times, then it can be expanded in a **Taylor series**. That is, f can be written as

$$f(x) = a_0 + a_1 x + a_2 x^2 + \cdots + a_n x^n + E_n(x).$$

Under certain conditions on f (see Appendix), the *remainder term* $E_n(x)$ has the property that it goes to zero as n becomes large. Thus to evaluate f at some point x, it suffices to choose n so large that $|E_n(x)|$ is very small and then evaluate the polynomial

$$p(x) = a_0 + a_1 x + \cdots + a_n x^n. \qquad (3.34)$$

We call such a polynomial a *truncated power series*. Examples of these that should be familiar from calculus are

Examples:
Truncated
power
series

$$e^x \approx 1 + x + \frac{x^2}{2!} + \frac{x^3}{3!} + \frac{x^4}{4!}$$

$$\sin x \approx x - \frac{x^3}{3!} + \frac{x^5}{5!} - \frac{x^7}{7!} \qquad (3.35)$$

$$\cos x \approx 1 - \frac{x^2}{2!} + \frac{x^4}{4!} - \frac{x^6}{6!}.$$

There are two characteristics of truncated power series that tend to make them more difficult to evaluate accurately and efficiently than other kinds of polynomials. The first is that, if x is large, then it may be necessary to take many terms in the series to get a reasonably accurate approximation to the value of the function; i.e., n must be large. But if n is large, then a_n is very small. In fact, it may be outside the range of the floating point number system of the computer even though the *product* $a_n x^n$ may be a reasonable number. Consider, for example, the series for e^x, given in (3.35). Here we have $a_{35} = (1/35!) \approx 10^{-40}$, which is outside the range of most floating point number systems. However, with $x = 10$, the product $a_{35} x^{35}$ is approximately 10^{-5}, which is well within the range of most computers. Thus, at least for large x, truncated power series cannot be evaluated using Horner's method since that algorithm requires the *coefficients* to be available as floating point numbers.

The second property of power series that affects the way in which they should be evaluated is that in most cases the coefficients a_k are rather simple functions of k. In fact, in many cases a_k is a very simple function of a_{k-1} or a_{k-2}. Thus, in the series for e^x we have $a_k = a_{k-1}/k$ and in the series for $\sin x$, $a_k = -a_{k-2}/[(k-1)k]$. An efficient algorithm should use these relations to reduce the amount of computation.

Finally it should be noted that any polynomial evaluation algorithm is closely related to a summation algorithm in which the terms to be summed are $a_k x^k$, $k = 0, 1, \ldots, n$. Thus for accuracy the terms should be summed in decreasing order. In the truncated power series case the smaller terms are $a_k x^k$, where k is large. Thus the order $(\cdots((a_n x^n + a_{n-1} x^{n-1}) + a_{n-2} x^{n-2}) + \cdots + a_0)$ should be used. This is in fact the order in which the terms are summed with Horner's method. We can combine all these observations to obtain the following "partial" algorithm for evaluating a truncated power series.

Truncated Power Series Algorithm

[Algorithm to evaluate truncated series.]

1   Input $\{x, n\}$
2   If n is large, go to 6
3   For $k = 0, 1, \ldots, n$
  3.1   Compute a_k
4   Input $\{n, a_0, a_1, \ldots, a_n, x\}$ to Algorithm (3.18) with output p
5   Output $\{x, p\}$. Halt
6   For $k = 0, 1, \ldots, n$ (3.36)
  6.1   Compute $b_k = a_k x^k$
7   $p \leftarrow b_n$
8   For $k = n - 1, n - 2, \ldots, 0$
  8.1 $p \leftarrow p + b_k$
9   Output $\{x, p\}$. Halt

The reason this is called a *partial* algorithm is that steps 2, 3.1, and 6.1 cannot be completely specified. They depend on the particular power series to be evaluated as well as on the computer being used. For example, in the series for e^x, using the UNIVAC 1108, the coefficients a_k for $k = 0, 1, 2, 3, 4, 5$ can be easily computed without danger of underflow. Thus statement 2 should be

$$2 \quad \text{If } n > 5 \text{ go to } 6. \qquad (3.37a)$$

Furthermore statement 3 for this series should be

$$3 \quad a_0 = 1$$
$$3.1 \quad \text{For } k = 1, 2, \ldots, n \qquad\qquad (3.37b)$$
$$3.1.1 \quad a_k = a_{k-1}/k$$

and statement 6 should be

$$6 \quad b_0 = 1$$
$$6.1 \quad \text{For } k = 1, 2, \ldots, n \qquad\qquad (3.37c)$$
$$6.1.1 \quad b_k = (b_{k-1}x)/k.$$

In this manner all of the coefficients are computed with a minimum of arithmetic and a minimum of rounding errors.

For another useful application of the polynomial evaluation algorithm, recall from Chapter 1 that a base β number

$$(d_t d_{t-1} \cdots d_2 d_1.)_\beta \qquad\qquad (3.38)$$

has the decimal (i.e., base 10) value given by

$$(d_t \cdots d_2 d_1)_\beta = (d_1 + d_2 \beta + d_3 \beta^2 + \cdots + d_t \beta^{t-1})_{10}.$$

Thus to convert (3.38) to a decimal number, the polynomial

$$p(x) = d_1 + d_2 x + d_3 x^2 + \cdots + d_t x^{t-1}$$

must be evaluated at $x = \beta$. For example, to convert $(1101011)_2$ to decimal, use Horner's method to evaluate the polynomial

$$p(x) = 1 + x + x^3 + x^5 + x^6$$

at $x = 2$. This gives in terms of the notation used in Algorithm (3.18)

$$p = 1$$
$$p = 2 + 1 = 3$$
$$p = 2 \cdot 3 + 0 = 6$$
$$p = 2 \cdot 6 + 1 = 13$$
$$p = 2 \cdot 13 + 0 = 26$$
$$p = 2 \cdot 26 + 1 = 53$$
$$p = 2 \cdot 53 + 1 = 107.$$

Thus Algorithm (3.18) with coefficients $d_1, d_2, \ldots, d_t$ and $x = \beta$ can be used to convert any base β integer to a decimal integer.

Similarly to convert a base β fraction

$$(.d_1 d_2 \cdots d_t)_\beta \tag{3.39}$$

to a decimal fraction, simply note that (3.39) has the decimal value

$$d_1 \beta^{-1} + d_2 \beta^{-2} + \cdots + d_t \beta^{-t}, \tag{3.40}$$

which is just the polynomial

$$p(x) = d_1 x + d_2 x^2 + \cdots + d_t x^t \tag{3.41}$$

evaluated at $x = 1/\beta$.

The converse problem—converting a decimal number to a base β number—is slightly more complicated and will not be considered here.

Further algorithms for evaluating polynomials and their derivatives are discussed in Section 7.5.

EXERCISES

1   Evaluate the polynomial $p(x) = x^5 + 2x^4 - x^3 + 3x^2 + x - 1$ at $x = 3$ by *hand*, using the method of (3.17) and then by Horner's method. Which is easier? Which one should *always* be used?

2   For the polynomial $p(x) = x^4 + 29x^3 - 1305x^2 + 6775x - 5500$ at $x = 5.1$, determine the bounds (3.29), (3.33) using the value of μ' for your computer. Which algorithm is more accurate for this polynomial?

3   Write an INFL program to evaluate the cos x series on your computer. That is, complete the algorithm description (3.36) with statements such as (3.37).

4   For the truncated power series $1 + x + (x^2/2!) + (x^3/3!) + (x^4/4!) + (x^5/5!)$, determine the bounds (3.29) and (3.33). Which algorithm, (3.17) or (3.18), would you expect always to be more accurate when applied to a truncated power series with small n?

5   Use Horner's method to convert the following numbers to decimal:
   (a) $(10111010)_2$ (b) $(34067)_8$ (c) $(.1101)_2$
   (d) $(.74201)_8$ (e) $(1011.1101)_2$.

6   Derive the rounding error estimates (3.29) and (3.30) for algorithm (3.17).
   [*Hints:* (a)   Write steps 5.1 and 5.2 of the algorithm as

   5.1 $z^{[k]} = \text{fl}[z^{[k-1]}x]$
   5.2 $p^{[k]} = \text{fl}[p^{[k-1]} + \text{fl}[a_k z^{[k]}]]$
   where $z^{[0]} = 1$, $p^{[0]} = a_0$, and $p^{[n]} = \bar{p}$. Then show that

   $$p^{[k]} = p^{[k-1]}\langle 1 \rangle + a_n z^{[k]}\langle 3 \rangle.$$

   (b)   Combine the above results to get

   $$\bar{p} = p^{[n]} = a_n x^n \langle n+1 \rangle + a_{n-1} x^{n-1}\langle n+1 \rangle + \cdots + a_1 x \langle n+1 \rangle + a_0 \langle n \rangle.$$

   (c)   Conclude from (b) that

   $$\bar{p} = \bar{a}_n x^n + \bar{a}_{n-1} x^{n-1} + \cdots + \bar{a}_1 x + \bar{a}_0$$

where

$$\bar{a}_k = \begin{cases} a_k \langle n + 1 \rangle, & k = 1, 2, \ldots, n \\ a_0 \langle n \rangle, & k = 0. \end{cases}$$

(d) Complete the forward error analysis by writing $\langle n + 1 \rangle = 1 + \langle n + 1 \rangle_0$ in part (b).]

3.3 Rational and Transcendental Functions

Rational functions are the most complicated kinds of functions that can be evaluated by using only the four basic arithmetic operations. Several methods for evaluating these functions are described and analyzed. The evaluation of more general functions is discussed briefly.

3.3.1 Rational Functions

As noted in Section 1.1, the most general class of functions that can be evaluated directly in a computer are the rational functions

$$r(x) = \frac{p_n(x)}{q_m(x)} = \frac{a_n x^n + a_{n-1} x^{n-1} + \cdots + a_1 x + a_0}{b_m x^m + b_{m-1} x^{m-1} + \cdots + b_1 x + b_0}. \tag{3.42}$$

Clearly, to evaluate such a function one need only use Algorithm (3.18) to evaluate the two polynomials $p_n(x)$, $q_m(x)$ and then divide these values. Specifically, we have

Evaluating Rational Functions

[Algorithm to evaluate rational functions.]

1 Input $\{n, a_0, a_1, \ldots, a_n\}$
2 Input $\{m, b_0, b_1, \ldots, b_m\}$
3 Input $\{x\}$
4 Input $\{n, a_0, \ldots, a_n, x\}$ to Algorithm (3.18) with output $\{p\}$
5 Input $\{m, b_0, b_1, \ldots, b_m x\}$ to Algorithm (3.18) with output $\{q\}$ (3.43)
6 If $q = 0$, output error message; halt
7 $r \leftarrow p/q$
8 Output $\{x, r\}$. Halt

Step 6 is to prevent attempted division by zero. This does not, however, preclude the possibility of q being so small as to cause exponent overflow in step 7. Such a possibility can and does cause problems when working with rational functions.

An important reason for considering rational functions is that some functions cannot be easily approximated by polynomials. For example, any function that goes to infinity in the interval of interest cannot be approximated by a polynomial. A rational function, however, may provide a very accurate approximation to such a function. As another indication of the importance of rational functions, consider the approximations

$$e^x \approx 1 + x + \frac{x^2}{2} + \frac{x^3}{6} + \frac{x^4}{24}$$

$$= 1 + x + .5x^2 + .16666\ldots x^3 + .0416666\ldots x^4 \qquad (3.44a)$$

$$e^x \approx (6 + 2x)/(6 - 4x + x^2). \qquad (3.44b)$$

The first approximation is just a truncated power series (3.35), while the second is a rational approximation of the form (3.42) with $n = 1$, $m = 2$. Both approximations are of the same accuracy; however, (3.44a) requires $4A + 4M$ operations, whereas (3.44b) takes only $3A + 3M + 1D$ operations to evaluate it. Thus if division is no more expensive than multiplication, then the rational function is the more efficient approximation. The derivation of rational approximations is, in general, a complicated process that we shall not discuss in this book. The above example is given only to show the importance of rational approximations. The interested reader should refer to the books on approximation theory at the end of Chapter 4.

3.3.2 Continued Fractions

The rational approximation given in (3.44b) can also be written in the forms

$$e^x \approx \cfrac{1}{1 + \cfrac{x}{-1 + \cfrac{x}{-2 + \cfrac{x}{3}}}} \qquad = \cfrac{2}{x - 7 + \cfrac{27}{x + 3.}} \qquad (3.44c)$$

With the first form, the approximation can be evaluated in $3A + 4D$ operations. The second form requires only $3A + 2D$ operations. Thus, depending on the relative speeds of the arithmetic operations, one of the forms given by (3.44c) may be more efficient than the rational function form (3.44b).

The expressions in (3.44c) are called *continued fractions*. Often it is possible to reduce the number of computations needed to evaluate a rational function by first converting it to a continued fraction. As an important example of this, many FORTRAN compilers compute e^{-x} by evaluating the rational function

$$R(x) = \frac{1680 - 840x + 180x^2 - 20x^3 + x^4}{1680 + 840x + 180x^2 + 20x^3 + x^4}. \qquad (3.45a)$$

It can be shown (Exercise 8) that (3.45a) is equivalent to

$$R(x) = 1 - \frac{2x}{.05x^2 + x + 6.9 - \dfrac{205.8}{x^2 + 42}}. \qquad (3.45b)$$

If $2x$ is computed as $x + x$, then this latter expression can be evaluated in only $6A + 2M + 2D$ operations, compared with $8A + 8M + 1D$ for (3.45a). The form (3.45b) is the one that is most frequently used.

More generally many rational functions of the form (3.42), in which $n = m$ or $n = m + 1$ can be written as a **continued fraction**:

$$c_0 x + d_0 + \cfrac{c_1}{x + d_1 + \cfrac{c_2}{x + d_2 + \cfrac{c_3}{\ddots \cfrac{}{\cfrac{c_m}{x + d_m}}}}} \qquad (3.46)$$

Here $c_0 = 0$ if $m = n$. To evaluate the function by using Algorithm (3.43) requires $(n + m)A + (n + m)M + 1D$ operations. On the other hand, the following algorithm evaluates (3.46) in only $(3m + 1)A + 1M + mD$ operations:

Algorithm to Evaluate Continued Fractions

[Algorithm to evaluate (3.46).]

1   Input $\{n, c_0, \ldots, c_m, d_0, \ldots, d_m\}$
2   Input $\{x\}$
3   $f \leftarrow d_m$
4   For $k = m - 1, m - 2, \ldots, 1, 0$ (3.47)
    4.1 $f \leftarrow d_k + c_{k+1}/(x + f)$
5   $f \leftarrow c_0 x + f$
6   Output $\{x, f\}$

This algorithm evaluates (3.46) "from the bottom up." That is, it starts with the quantities $d_{m-1} + c_m/(x + d_m)$, then computes $d_{m-2} + c_{m-1}/\{x + d_{m-1} + c_m/(x + d_m)\}$, etc. Generally, all continued fractions should be evaluated in this manner. The operation counts that were given for (3.44c) and (3.45b) assumed that this "bottom-up" evaluation was used.

A comparison of the operation counts for Algorithms (3.43) and (3.47) shows that whenever one division is cheaper than two multiplications, then (3.47) is the more efficient algorithm. Thus if a particular rational function is to be evaluated many times, it may be worthwhile first to transform it into a continued fraction and then apply an algorithm such as (3.47). Unfortunately, there is no method that can be applied to all rational functions to transform them into a continued fraction. For rational functions (3.42) with $n = m + 1$, however, the following long division process often works.

Algorithm to Convert Rational Functions into Continued Fractions

[q_m must have leading coefficient 1.0.]

1 Input $\{p_{m+1}(x), q_m(x)\}$
2 Divide $p_{m+1}(x)$ by $q_m(x)$ to get $c_0 x + d_0$ with remainder
   $s_0(x) = c_1 x^{m-1} + \cdots$
3 Let $q_{m-1} = (1/c_1)s_0(x) = x^{m-1} + \cdots$
4 For $k = 1, 2, \ldots, m - 1$
   4.1 Divide q_{m+1-k} by q_{m-k} to get $x + d_k$ with remainder
        $s_k(x) = c_{k+1} x^{m-k-1} + \cdots$
   4.2 Let $q_{m-k-1}(x) = [1/(c_{k+1})]s_k(x) = x^{m-k-1} + \cdots$
5 Let $q_1 = x + d_m$
6 Output $\{c_0, c_1, \ldots, c_m, d_0, d_1, \ldots, d_m\}$

This algorithm is based on the following series of reductions:

$$\frac{p_{m+1}(x)}{q_m(x)} = c_0 x + d_0 + \frac{s_0(x)}{q_m(x)}, \qquad s_0(x) = c_1 x^{m-1} + \cdots$$

$$= c_0 x + d_0 + c_1 \frac{(q_{m-1}(x))}{q_m(x)}, \qquad q_{m-1}(x) = x^{m-1} + \cdots$$

$$= c_0 x + d_0 + \frac{c_1}{\dfrac{q_m(x)}{q_{m-1}(x)}}$$

$$= c_0 x + d_0 + \frac{c_1}{x + d_1 + \dfrac{s_1(x)}{q_{m-1}(x)}}, \qquad s_1(x) = c_2 x^{m-2} + \cdots$$

$$= c_0 x + d_0 + \frac{c_1}{x + d_1 + \dfrac{c_2}{\dfrac{q_{m-1}(x)}{q_{m-2}(x)}}} = \cdots.$$

Clearly, this process will halt whenever $c_{k+1} = 0$ in step 4.1. However, even when this happens, the resulting *partial* reduction may be useful. For example, if the algorithm is applied to the rational function

$$\frac{2x^3 + 5x^2 - 3x - 7}{x^2 + x - 3},$$

the result is

$$\frac{2x^3 + 5x^2 - 3x - 7}{x^2 + x - 3} = 2x + 3 + \frac{2}{x^2 + x - 3}, \qquad (s_0(x) = 0.x + 2).$$

Thus the reduction stops at the first stage. However to evaluate this rational function with Algorithm (3.43) requires $5A + 5M + 1D$ operations. The partially reduced form given above can be evaluated in only $4A + 2M + 1D$ operations.

There are several modifications to this algorithm that can be used in special cases. For example, if only *even* powers of x occur in both polynomials in (3.42), then the continued fraction expansion can be carried out with only even powers of x (Exercise 11).

This entire discussion of continued fractions is motivated by a desire to evaluate rational functions as efficiently as possible. Unfortunately, sometimes this efficiency improvement is accompanied by a loss of accuracy. For example, the function $\cos x$ is approximated by the rational function

$$\frac{1 - .47059x^2 + .027388x^4 + .00037234x^6}{1 + .029494x^2 + .00042373x^4 + .0000032355x^6}. \qquad (3.48a)$$

If this is converted into a continued fraction, the result has the form

$$-115.078 \cdots + \cfrac{a}{x^2 + b + \cfrac{c}{x^2 + d + \cfrac{e}{x^2 + f}}} \qquad (3.48b)$$

for certain constants a, b, c, d, e, f. The last step in a bottom–up evaluation of (3.48b) will involve the addition of two numbers, one of which is $-115.078 \ldots$. *But* the result must be less than or equal to 1 since $|\cos x| \leq 1$ for all x. Hence, subtractive cancellation will cause the loss of two or more decimal digits of accuracy. Because of this, the less efficient form (3.48a) should be used in this case.

3.3.3 Transcendental Functions

Functions that cannot be evaluated exactly with only the four basic arithmetic operations are said to be *transcendental*. Examples of such functions are the trigonometric functions, the exponential and logarithm functions, square roots and cube roots, Bessel functions, etc. Such functions must be approximated by polynomials or rational functions before they can be evaluated with a computer. As shown in the preceding subsection, there may be several ways of approximating a particular function. Moreover, a particular approximation may have several forms in which it can be written. The comparison of accuracy and efficiency of these various approximations is a complicated problem that we shall not study in detail. Instead, we shall only mention a few general principals and refer the interested reader to specialized books on this subject.

We first note that it is often tempting to use a truncated power series to evaluate transcendental functions; however, the discussion in Section 3.2 indicated that this approach may lead to computational difficulties. Another method for dealing with such functions is to use a table of function values together with an *interpolation* process, as will be described in Chapter 4. This is a reasonable approach provided the function is understood well enough that proper decisions can be made concerning the form and degree of the interpolating process. Probably the main objection to the use of tables is the storage requirement and the rather complicated algorithm that is needed to search through the table in order to find the necessary values. One method for evaluating certain transcendental functions that often leads to disastrous consequences is the use of recursion formulas. For example, it is known that the Bessel functions of the first kind, $J_n(x)$, for fixed x and $n = 0, 1, \ldots$ satisfy the three-term recursion

$$y_{n+1} = (2n/x)y_n - y_{n-1}, \qquad n = 1, 2, \ldots . \tag{3.49}$$

That is, for a fixed value of x, $y_n = J_n(x)$ satisifies (3.49). But with $x = 1$ the computed value of $J_5(1)$ obtained by using (3.49) with $y_0 = J_0(1) = .7619769$, $y_1 = J_1 = .44005059$, is negative. This is nonsense because it is known that $J_n(1) > 0$ for all n. The recursion (3.49) is unstable.

Finally, it is worthwhile pointing out that transcendental functions occur frequently as "closed form" solutions to differential equations. It may be simpler and more accurate to solve the differential equation numerically, as discussed in Chapter 8. In particular, if the transcendental function is at all complicated (a power series whose coefficients are Bessel functions, for example), it may be *much* better to solve the differential equation than to worry about convergence, accuracy, cancellation, etc.

3.3.4 Error Estimates—Evaluating the Wrong Function

An error analysis of the algorithm for evaluating a rational function can be easily obtained from the analysis of Horner's method for polynomial evaluation. The computed value of $r(x)$, as given by (3.43) is just

$$\hat{r}(x) = \text{fl}\left\langle\frac{p_n(x)}{q_m(x)}\right\rangle = \frac{\hat{p}_n(x)}{\hat{q}_m(x)} \langle 1 \rangle,$$

where $\hat{p}_n$ (and $\hat{q}_m$) are defined by (3.24) and (3.26). By absorbing the $\langle 1 \rangle$ term into $\hat{p}_n$, we can write

$$\hat{r}(x) = \hat{p}_n(x)/\hat{q}_m(x). \tag{3.50a}$$

Here we have

$$\hat{p}_n(x) = \hat{a}_n x^n + \hat{a}_{n-1} x^{n-1} + \cdots + \hat{a}_0$$
$$\hat{q}_m(x) = \hat{b}_m x^m + \hat{b}_{m-1} x^{m-1} + \cdots + \hat{b}_0 \tag{3.50b}$$

with

$$|\hat{a}_k - a_k|/|a_k| \le (2k + 2)\mu', \qquad k = 0, 1, \ldots, n$$
$$|\hat{b}_k - b_k|/|b_k| \le (2k + 1)\mu', \qquad k = 0, 1, \ldots, m. \tag{3.50c}$$

(Estimate (3.50c) is obtained from (3.26) by using an obvious, but less accurate, bound.)

The analysis of the continued fraction evaluation is similar to the analysis of Horner's method. Algorithm (3.47) is very much like Algorithm (3.18), but with multiplication replaced by division. Since the errors are the same, the analysis will be the same.

For transcendental functions, the error analysis is complicated by the several sources of possible error: truncating an infinite series, subtractive cancellation, growth of errors, etc. In Chapter 4 the error in evaluating an interpolating polynomial will be analyzed. Other than this, a *general* analysis of the errors involved in approximating transcendental functions is not possible.

One important consequence of the error analyses of these various function evaluation algorithms is worth noting. If $f(x)$ denotes the function to be evaluated, then the *computed* value of $f(x)$ is the exact value of a different function $\hat{f}(x)$. The error analysis gives a relation between the function f and the function $\hat{f}$. For example, if f is a polynomial, so is $\hat{f}$, and the coefficients are related by (3.25) or (3.29). Similarly, if f is a rational function, so also is $\hat{f}$, and (3.50) gives the relationship between them. Thus generally we can write

$$\text{fl}[f(x)] = \hat{f}(x),$$

where f is related to f in some reasonable way. This observation is particularly important when solving nonlinear equations as in Chapter 7.

EXERCISES

1   Write out the expression for the efficiency measure of Algorithm (3.43).

2   For each of the following continued fractions give an equivalent rational function:

$$\cfrac{1+x}{\cfrac{2+x}{3+x}} \quad , \qquad \cfrac{3+x}{\cfrac{2+x}{1+x}} \quad , \qquad \cfrac{1+x}{\cfrac{2+x}{\cfrac{0+x}{3+x}}}.$$

3   For $m = n = 3$, show that (3.42) is equivalent to (3.46) with $c_0 = 0$. Which form is more efficient for use with *your* computer?

4   Write a program to evaluate $J_n(1)$ using the recursion (3.49) and the values of $y_0 = J_0(1)$, $y_1 = J_1(1)$, which are given following (3.49). For what values of n are the computed values positive?

5   Let $f(x) = (ax + b)/(cx + d)$, where a, b, c, d are floating point numbers. Determine a function $\hat{f}$ so that $\hat{f}(x) = \mathrm{fl}[f(x)]$, and show the relation between f and $\hat{f}$.

6   Repeat Exercise 5 for the function $f(x) = x^2 \sin x + \cos x$, assuming that $\sin x$ and $\cos x$ can be evaluated with three relative rounding errors; that is, $\mathrm{fl}(\sin x) = (\sin x)\langle 3 \rangle$ and $\mathrm{fl}(\cos x) = (\cos x)\langle 3 \rangle$.

7   Do an error analysis of (3.49) with $x = 1$ to show the unstability of this recursion. (Let ε_0 and ε_1 be the absolute errors in y_0 and y_1, respectively, and assume no other errors occur.) You may use the fact that $y_0 > y_1 > y_2 > y_3 > \cdots > 0$. Can you *prove* this fact?

8   Show that (3.45b) can be reduced to the form (3.45a).

9   Verify the operation counts given in the text for evaluating (3.44c), (3.45b), and for the general algorithm (3.47).

10   Do a backward error analysis of algorithm (3.47).
[*Hint*: Observe the similarities between this algorithm and Algorithm 3.18.]

11   Describe an algorithm for reducing a rational function of the form

$$\frac{a_n x^{2n} + a_{n-1} x^{2n-2} + \cdots + a_1 x^2 + a_0}{b_n x^{2n} + b_{n-1} x^{2n-2} + \cdots + b_1 x^2 + b_0}$$

to continued fraction form. When does the reduction process fail? Apply your algorithm to (3.48a).

3.4   Avoiding Subtractive Cancellation

Often it is possible to rewrite expressions so that subtractive cancellation can be avoided. In this section we present several methods for rewriting certain kinds of expressions so that they can be evaluated more accurately.

3.4.1 Isolating the Cancellation

In many cases where subtractive cancellation causes loss of accuracy, it is possible to pinpoint a specific subtraction operation that is at fault. For example, consider the solution

$$x = \frac{-b + \sqrt{b^2 - 4ac}}{2a} \tag{3.51}$$

to the quadratic equation $ax^2 + bx + c = 0$, as discussed in Section 1.3. It is clear that for the data (1.20) the operation of subtracting b from the square root caused the cancellation. Similarly, in evaluating the function $f(x) = \sin x - \cos x$ for x near $\pi/4$ (see Table 1.1), it is clear where the subtractive cancellation occurs. Generally, subtractive cancellation can be most easily examined if it is possible to write the computation in the form

$$G(x) - H(x). \tag{3.52}$$

Here x represents the data, and $G(x)$ and $H(x)$ are quantities that can be computed without danger of subtractive cancellation. Also, it should be realized that cancellation in (3.52) generally is possible only for values of x near some special value $\bar{x}$.

In many computations, however, it is not possible to attribute the cancellation to one particular subtraction. For example, when evaluating the series

$$\sin x = x - \frac{x^3}{3!} + \frac{x^5}{5!} - \cdots$$

for large x, individual terms may be very large. However, because of subtractive cancellation, the final result must be no larger than one (see Table 2.4). The only way that large numbers can be summed to obtain a small number is if cancellation occurs. Unfortunately, the cancellation occurs during several of the subtractions and hence cannot be analyzed by writing the expression in the form (3.52). In such a case it is not possible to control the cancellation by general techniques. Instead, the specific problem must be examined to see if some special result can be used, such as the fact that $\sin(x \pm 2k\pi) = \sin(x)$ in the above example.

3.4.2 Use of Taylor Series

There are two rather general techniques that can often be used to rewrite (3.52) in a more stable form. The first is to expand both G and H in a Taylor series *at* $\bar{x}$, and then cancel terms explicitly. To illustrate this, consider the evaluation of $F(x) = \sin x - \cos x$. For this problem, $G(x) = \sin x$, $H(x) = \cos x$, and $\bar{x} = \pi/4$. If we expand these functions in a **Taylor Series** at $\bar{x}$ (see the Appendix), we have

$$G(x) = \frac{\sqrt{2}}{2} + \frac{\sqrt{2}}{2}\left(x - \frac{\pi}{4}\right) - \frac{\sqrt{2}}{4}\left(x - \frac{\pi}{4}\right)^2 - \frac{\sqrt{2}}{12}\left(x - \frac{\pi}{4}\right)^3 + \cdots$$

$$H(x) = \frac{\sqrt{2}}{2} - \frac{\sqrt{2}}{2}\left(x - \frac{\pi}{4}\right) - \frac{\sqrt{2}}{4}\left(x - \frac{\pi}{4}\right)^2 + \frac{\sqrt{2}}{12}\left(x - \frac{\pi}{4}\right)^3 + \cdots$$

and hence

$$F(x) = \sqrt{2}\left(x - \frac{\pi}{4}\right) - \frac{\sqrt{2}}{6}\left(x\frac{\pi}{4}\right)^3 + \frac{\sqrt{2}}{120}\left(x - \frac{\pi}{4}\right)^5 - \frac{\sqrt{2}}{5040}\left(x - \frac{\pi}{4}\right)^7 + \cdots.$$

$$\tag{3.53}$$

For x near to $\pi/4$, $F(x)$ can be easily and accurately evaluated by taking, say, the first three or four terms in the series (3.53). As this example indicates, the use of this Taylor expansion method to eliminate subtractive cancellation leads to a formula involving an infinite summation whose terms contain $(x - \bar{x})^k$. If x is very near to $\bar{x}$, then these terms quickly become very small, and the infinite summation can be terminated after only a few terms. However, if x is *not* near $\bar{x}$, then the infinite expansion may converge very slowly or even not at all; *but* in this case subtractive cancellation will not be a problem and the expression (3.52) can be evaluated directly.

As another example, consider the evaluation of the function $F(x) = x \sin x/(1 - \cos x)$ for x near 0.0. In this case, the subtraction $1 - \cos x$ causes difficulty near $x = 0$. Expand $\cos x$ in a Taylor series at 0 and write

$$1 - \cos x = 1 - \left(1 - \frac{x^2}{2} + \frac{x^4}{4!} - \cdots\right) = \frac{x^2}{2} - \frac{x^4}{24} + \frac{x^6}{720} + \cdots,$$

so that

$$F(x) = \frac{1}{x^2}\left(\frac{x \sin x}{1/2 - x^2/24 + x^4/720 + \cdots}\right),$$

which can be evaluated without cancellation for very small x. However, small x may cause overflow problems when dividing by x^2. This in turn can be avoided by using $\sin x = x - x^3/3! + \cdots$, so that

$$F(x) = \frac{1 - x^2/6 + x^4/120 - \cdots}{1/2 + x^2/24 + x^4/720},$$

which will give an accurate value for small x.

3.4.3 Transforming the Expression

Another technique, which often eliminates the subtraction in (3.52), is to multiply and divide the expression by $G(x) + H(x)$ to obtain

$$G(x) - H(x) = [G(x)^2 - H(x)^2]/[G(x) + H(x)]. \qquad (3.54)$$

For example, suppose $G(x) = \sqrt{1 + x}$, $H(x) = \sqrt{1 - x}$. When x is near zero, cancellation will occur, but (3.54) expresses the difference as

$$\frac{\sqrt{1 + x}^2 - \sqrt{1 - x}^2}{\sqrt{1 + x} + \sqrt{1 - x}} = \frac{2x}{\sqrt{1 + x} + \sqrt{1 - x}},$$

which eliminates the danger of cancellation. This technique, in fact, can be applied to the quadratic formula (3.51) to give the *stable* formula

$$x = \frac{-2c}{b + \sqrt{b^2 - 4ac}},$$

which avoids cancellation when $b > 0$.

Another example where this kind of transformation is successful is

$$f(x) = 1 - \cos x, \qquad x \text{ near } 0.$$

From the fact that $\sin^2 x + \cos^2 x = 1$, we obtain

$$f(x) = (1 - \cos x)\frac{(1 + \cos x)}{1 + \cos x} = \frac{1 - \cos^2 x}{1 + \cos x} = \frac{\sin^2 x}{1 + \cos x},$$

which can be evaluated accurately for x near 0.

Generally it may take some ingenuity to find the proper transformation, if one exists, for avoiding cancellation. Unfortunately, some calculations cannot be rearranged in any manner. As an example, suppose one attempts to compute the derivative of a function $f(x)$ at $x = 1$, say, by using a very small value for h in the formula

$$f'(1) \approx [f(1 + h) - f(1)]/h. \qquad (3.55)$$

Clearly, if h is small, subtractive cancellation will occur. Since $f'(1)$ is unknown, the Taylor series method cannot be used to rewrite this formula. In fact, aside from very special functions f, which can be easily differentiated anyway, there is no way to eliminate the subtractive cancellation in (3.55)

EXERCISES

1 Use the Taylor series method to rewrite the following expressions so they can be evaluated without subtractive cancellation:

(a) $\tan x - \cot x$, x near $\pi/4$

(b) $1 - e^x$, x near 0

(c) $\sin^2 x - \cos^2(x + \pi/2)$, x near 0.

2 Rewrite the following expressions so that they can be evaluated without subtractive cancellation:

(a) $\sqrt{1 + x^2} - \sqrt{1 - x^2}$, x near 0

(b) $1/[\sqrt{1 + x^2} - \sqrt{1 - x^2}]$, x near 0

(c) $(1 + x)^2 - (1 - x)^2$, x near 0

(d) $[-b - \sqrt{b^2 - 4ac}]/2a$, $b < 0$, a and c very small.

3 For $f(x) = \sin(\pi x)$, use formula (3.55) with various values of h to approximate $f'(1) = -\pi$. How small must h be before subtractive cancellation becomes a problem?

4 Find a number M so that if $|x - \pi/4| \le M$, then the absolute error caused by using the first three terms of (3.53) to evaluate $F(x)$ will be less than 10^{-10}.

5 Use the first three terms in (3.53) to evaluate $F(x)$ for $x = (\pi/4) + 10^{-7}k$, $k = 1, 2, 3, 4, 5$. Compare your results with the exact values given in Table 1.1.

Suggestions for Further Reading

The book by Fike (1968) describes various methods that are used to approximate and evaluate functions. A thorough discussion of computer approximations is contained in Hart et al. (1968). A summary of functions that occur in numerical applications can be found in Abramowitz and Stegun (1964).

Chapter 4

INTERPOLATION AND
APPROXIMATION

4.1 Polynomial Interpolation

Interpolation is one of the most basic and most often used numerical techniques. Through the years many interpolation methods have been invented with the aim of simplifying the computations when performed by hand or with a desk calculator. A convenient interpolation technique for computers, however, is the original method given by Lagrange. This classic method and appropriate error estimates are derived in this section.

4.1.1 Lagrange Formula

Many numerical methods are based on the idea of interpolation; hence it is well to consider this process in detail. The process of interpolation is defined as follows: Given a table of values $f_1, f_2, \ldots, f_m$ of a function $f(x)$, for values of $x = x_1, x_2, \ldots, x_m$, find a "simple" function $p(x)$ that takes the same values as f at certain of the given values of x. That is, we want $p(x_k) = f_k$, for certain values of k. The function p is said to **interpolate** f at these values of x. The usual reason for wanting such a function $p(x)$ is to use the value of $p(\bar{x})$ to approximate $f(\bar{x})$, for some $\bar{x}$ different from the given values $x_1, x_2, \ldots, x_m$. Thus, it is reasonable to require the "simple" function p to be a function that can be evaluated easily. The use of rational functions for this purpose leads to difficult mathematical problems. Because of this, we shall restrict the discussion to *polynomial* interpolation.

A polynomial of degree n is defined by its $n + 1$ coefficients. Thus it is natural to expect that an interpolating polynomial of degree n would be completely determined by $(n + 1)$ function values f_k. This is indeed the case, as we shall now show. First of all, for simplicity, suppose that exactly $n + 1$ distinct points $x_1, x_2, \ldots, x_{n+1}$ are given. We are interested in a polynomial $p(x)$ of degree n that interpolates the function $f(x)$ at these points; that is, we want $p(x_k) = f_k$ for $k = 1, 2, \ldots, n, n + 1$. There are many formulas for finding this interpolating polynomial. A useful formula for computer applications is the **Lagrange formula**, which gives $p(x)$ explicitly in terms of f_k and x_k as

Lagrange interpolation formula	$p(x) = f_1 p_1(x) + f_2 p_2(x) + \cdots + f_{n+1} p_{n+1}(x)$ (4.1)

where

$$p_i(x) = \frac{(x - x_1) \cdots (x - x_{i-1})(x - x_{i+1}) \cdots (x - x_{n+1})}{(x_i - x_1) \cdots (x_i - x_{i-1})(x_i - x_{i+1}) \cdots (x_i - x_{n+1})},$$ (4.2)

$$i = 1, 2, \ldots, n + 1.$$

To verify that (4.1) solves the interpolation problem, note the following facts:

(i) $p_i(x)$ and hence $p(x)$ are polynomials of degree n

(ii) $p_i(x_j) = \begin{cases} 0 \text{ if } j \neq i \\ 1 \text{ if } j = i \end{cases}.$

From (ii) it follows immediately that $p(x_j) = f_j$ for $j = 1, 2, \ldots, n + 1$.

This proves the existence of the interpolating polynomial. It is sometimes useful to know that this polynomial is *unique*. Indeed, suppose $q(x)$ is another polynomial of degree n with $q(x_i) = f_i$, $i = 1, 2, \ldots, n + 1$. Then $r(x) = p(x) - q(x)$ is a polynomial of degree at most n and with $r(x_i) = 0$ for $i = 1, 2, \ldots, n + 1$. But a polynomial of degree n can have at most n zeros unless it is actually the *zero* polynomial. That is, $r(x) = 0$ for *all* x. This shows that, in fact, $q(x) = p(x)$. Thus $p(x)$, as given by (4.1), is the only polynomial of degree n that interpolates f at the $n + 1$ points $x_1, \ldots, x_{n+1}$. An important consequence of this uniqueness is that *any* method that produces an interpolating polynomial must produce the same polynomial as does (4.1), (4.2).

The polynomial, as given by (4.1), is not expressed in the " standard " form

$a_n x^n + a_{n-1} x^{n-1} + \cdots + a_0$. In general, it is not necessary to rearrange the interpolating polynomial into this standard form, and in fact it will be shown later that, in terms of rounding error and computational efficiency, nothing is gained from such a rearrangement. We shall assume that the interpolating polynomial for the points $x_1, \ldots, x_{n+1}$, with function values $f_1, \ldots, f_{n+1}$, is defined by formulas (4.1), (4.2). A suitable algorithm for evaluating this polynomial at $x = \bar{x}$, in order to obtain an approximation to $f(\bar{x})$ is the following.

Algorithm to Evaluate the Interpolating Polynomial

[Algorithm to evaluate the interpolating polynomial (4.1).]

$$
\begin{aligned}
&1 \quad \text{Input } \{x_1, \ldots, x_{n+1}, f_1, \ldots, f_{n+1}, \bar{x}\} \\
&2 \quad \bar{p} \leftarrow 0 \\
&3 \quad \text{For } i = 1, 2, \ldots, n+1 \\
&\quad\; 3.1 \quad p \leftarrow 1 \\
&\quad\; 3.2 \quad \text{For } j = 1, 2, \ldots, n+1 \\
&\quad\qquad 3.2.1 \quad \text{If } j \neq i,\, p \leftarrow p(\bar{x} - x_j)/(x_i - x_j) \\
&\quad\; 3.3 \quad \bar{p} \leftarrow \bar{p} + f_i p \\
&4 \quad \text{Output } \{\bar{x}, \bar{p}\}
\end{aligned}
\tag{4.3}
$$

Note that the subscripts i on the quantities p_i in (4.2) have been omitted in this algorithm. The reason for this is that it is not necessary to save the values of these variables once they have been used in step 3.3; hence, in an actual implementation of this algorithm, a simple variable name would be used for all of the p_i.

TABLE 4.1

x	-3.0	-1.0	1.0	2.0	2.5	3.0
$f(x)$	1.0	1.5	2.0	2.0	1.5	1.0

To illustrate this algorithm, consider the function values given in Table 4.1 and suppose a value for $f(.3)$ is desired. By letting $n = 3$, with $x_1 = -1.0$, $x_2 = 1.0$, $x_3 = 2.0$, $x_4 = 2.5$, we have as input to the algorithm the values $\{-1.0, 1.0, 2.0, 2.5, 1.5, 2.0, 2.0, 1.5, .3\}$. Step 3 then computes successively

$$p = .12466667, \quad p = 1.6206667, \quad p = -1.3346667, \quad p = .58933333$$

and finally gives $\bar{p} = 1.6430000$. On the other hand, if we had chosen $n = 2$, and let $x_1 = -1.0$, $x_2 = 1.0$, $x_3 = 2.0$, then the same algorithm gives the approximation $\bar{p} = 1.9008333$. Without further information about the function f it is, of course, impossible to determine which of these values is the more accurate approximation to $f(.3)$.

To determine the **efficiency of this algorithm**, we observe that step 3.2 involves n multiplications, n divisions, and $2n$ subtractions, and step 3.3 requires $n + 1$ multiplications and additions. These steps are executed $n + 1$ times because of statement 3, giving a total efficiency measure of $(3n^2 + 4n + 1)A + (2n^2 + 3n + 1)M + (n^2 + n)D$. Recall from Section 3.2 that the Horner method [Algorithm (3.18)] for evaluating a polynomial in standard form $a_n x^n + \cdots + a_1 x + a_0$ has efficiency $nA + nM$; however, to put the Lagrange polynomial (4.1), (4.2) into standard form would require approximately n^3 additions and multiplications (Exercise 4). Hence, an algorithm for evaluating the interpolating polynomial by first transforming it to standard form and then using Horner's method would be extremely wasteful.

4.1.2 Truncation Error

In order to approximate the value of the function $f(x)$ at some point $\bar{x}$, we can proceed as follows:

(1) Select n and the points $x_1, x_2, \ldots, x_{n+1}$.
(2) Use (4.1) and (4.2) to define $p(x)$.
(3) Evaluate $p(\bar{x})$ using Algorithm (4.3).

By replacing the value $f(\bar{x})$ with the value $p(\bar{x})$ we create a *truncation error*, which can sometimes be estimated with the help of the following result.

Theorem 4.1 Let $p(x)$ be the (unique) polynomial of degree n that satisfies the condition $p(x_i) = f_i, i = 1, 2, \ldots, n + 1$, where $f_i = f(x_i)$ and f is a function that is defined on an interval $[a, b]$ containing the $n + 1$ distinct points $x_1, \ldots, x_{n+1}$. If f is $(n + 1)$ times continuously differentiable on $[a, b]$, then for any $\bar{x} \in [a, b]$ there is a value $\xi \in [a, b]$ such that

$$f(\bar{x}) - p(\bar{x}) = [f^{(n+1)}(\xi)/(n+1)!](\bar{x} - x_1)(\bar{x} - x_2) \cdots (\bar{x} - x_{n+1}). \quad (4.4)$$

Proof Let $G(x) = (x - x_1)(x - x_2) \cdots (x - x_{n+1})$, let

$$E_T(x) = f(x) - p(x) \quad (4.5)$$

and let $H(t) = E_T(\bar{x})G(t) - E_T(t)G(\bar{x})$. Then clearly, $G(x_i) = E_T(x_i) = 0$, $i = 1, 2, \ldots, n + 1$ and hence $H(x_i) = 0$ for $i = 1, 2, \ldots, n + 1$. But also $H(\bar{x}) = 0$, so that $H(t)$ is a continuously differentiable function that has $n + 2$ zeros $x_1, x_2, \ldots, x_{n+1}, \bar{x}$, in $[a, b]$. By Rolle's theorem (see the Appendix) $H'(t)$ has $(n + 1)$ zeros, $H''(t)$ has n zeros, $\ldots$, and $H^{(n+1)}(t)$ has one zero, which will be called ξ. But

$$H^{(n+1)}(t) = E_T(\bar{x})G^{(n+1)}(t) - E_T^{(n+1)}(t)G(\bar{x}),$$

and it is easily verified (Exercise 9) that

$$G^{(n+1)}(t) = (n+1)!, \qquad E_T^{(n+1)}(t) = f^{(n+1)}(t),$$

so that

$$0 = H^{(n+1)}(\xi) = E_T(\bar{x})(n+1)! - f^{(n+1)}(\xi)G(\bar{x}).$$

By solving this equation for $E_T(\bar{x})$, formula (4.4) is obtained.

Thus, according to this theorem, the **truncation error** is

$$E_T(\bar{x}) = [f^{(n+1)}(\xi)/(n+1)!](\bar{x} - x_1)(\bar{x} - x_2) \cdots (\bar{x} - x_{n+1}). \qquad (4.6)$$

Provided that $f(x)$ has sufficiently many derivatives and these derivatives are uniformly bounded, the truncation error decreases as n increases. Furthermore, the factor

$$(\bar{x} - x_1)(\bar{x} - x_2) \cdots (\bar{x} - x_{n+1}), \qquad (4.6')$$

which occurs here, depends on the choice of $x_1, \ldots, x_{n+1}$ and is minimized in absolute value when $\bar{x}$ is as close as possible to all of the x_i (Exercise 8).

In this connection, it is important to note that even if $\bar{x}$ is *outside* an interval containing the x_i the truncation error formula (4.6) still holds. However, in this case, the factor (4.6') will be quite large. Evaluating the interpolating polynomial at such a point is called *extrapolation*, and the presence of (4.6') in the truncation error term indicates that, in general, extrapolation is less accurate than interpolation. The choice of n is somewhat more complicated because it affects not only the truncation error, but also the rounding error, as will be seen in the next section. Moreover, it is often the case that no information is available concerning the derivatives of f, and hence nothing is known about the factor $f^{(n+1)}(\xi)$. To illustrate these ideas, suppose it is known (e.g., from the fact that the function f describes a certain physical phenomenon) that the function given by Table 4.1 satisfies $|f^{(k)}(x)| \leq 1/k$, $k = 1, 2, \ldots$, all $x \in [-3, 3]$. Then the truncation error E_T, using $n = 3$, with $x_1 = -1.0$, $x_2 = 1.0$, $x_3 = 2.0$, $x_4 = 2.5$ satisfies $|E_T| \leq .0355$, whereas the truncation error E_T, using $n = 2$, with $x_1 = -1.0$, $x_2 = 1.0$, $x_3 = 2.0$ satisfies $|E_T| \leq .0860$.

To illustrate the importance of knowing something about the derivatives of f, suppose the function

$$f(x) = 1/(1 + 25x^2)$$

has been tabulated as

x	-1.0	$-.8$	$-.6$	$-.4$	$-.2$	0
f	.03846	.05882	.1000	.2000	.5000	1.000

Then Algorithm (4.3) gives the results shown in Table 4.2, where $f(-.5) = .1379310$. The *growth* of the truncation error as n increases is due to the fact

TABLE 4.2

n	$p(-.5)$	Absolute error
1	.15	.01207
2	.1250	.01293
3	.133824	.00411
4	.2008273	.062896

that the derivatives of this function are

$$f'(x) = \frac{-50x}{(1 + 25x^2)^2}, \qquad f''(x) = \frac{-50 + 3750x^2}{(1 + 25x^2)^3}, \qquad \text{etc.,}$$

which increase rapidly at $x = -.5$ as n increases.

4.1.3   Rounding Error

Algorithm (4.3) evaluates $p(\bar{x})$ by computing (4.2) for each i and then summing the terms in (4.1) from left to right. Let $\hat{p}_i$ denote the computed value of (4.2) for $x = \bar{x}$. Then by counting the number of floating point operations, we find that

$$\hat{p}_i = \frac{(\bar{x} - x_1) \cdots (\bar{x} - x_{i-1})(\bar{x} - x_{i+1}) \cdots (\bar{x} - x_{n+1})}{(x_i - x_1) \cdots (x_i - x_{i-1})(x_i - x_{i+1}) \cdots (x_i - x_{n+1})} \langle 4n \rangle.$$

That is,

$$\hat{p}_i = p_i \langle 4n \rangle. \tag{4.7}$$

If we assume that $f_1, \ldots, f_{n+1}$ are floating point numbers, then the computed value of (4.1) is given by

Application of the summation analysis of Chapter 3	$\hat{p}(\bar{x}) = \text{fl}[f_1 \hat{p}_1]\langle n + 1 \rangle + \text{fl}[f_2 \hat{p}_2]\langle n \rangle$ $+ \cdots + \text{fl}[f_{n+1} \hat{p}_{n+1}]\langle 1 \rangle.$

Here we have used (3.3c) but with $N = n + 1$. This leads to

$$\hat{p}(\bar{x}) = f_1 \hat{p}_1 \langle n + 2 \rangle + f_2 \hat{p}_2 \langle n + 1 \rangle + \cdots + f_{n+1} \hat{p}_{n+1} \langle 2 \rangle. \tag{4.8}$$

By combining (4.7) with (4.8) we have

$$\hat{p}(\bar{x}) = f_1 p_1 \langle 4n \rangle \langle n + 2 \rangle + f_2 p_2 \langle 4n \rangle \langle n + 1 \rangle + \cdots + f_{n+1} p_{n+1} \langle 4n \rangle \langle 2 \rangle$$
$$= f_1 p_1 \langle 5n + 2 \rangle + f_2 p_2 \langle 5n + 1 \rangle + \cdots + f_{n+1} p_{n+1} \langle 4n + 2 \rangle.$$

This can be written more compactly as

$$\hat{p}(\bar{x}) = \sum_{i=1}^{n+1} f_i p_i \langle 5n + 3 - i \rangle. \tag{4.9}$$

To derive a **forward error estimate**, we write this as

$$\hat{p}(\bar{x}) = \sum_{i=1}^{n+1} f_i p_i (1 + \langle 5n + 3 - i \rangle_0)$$

$$= p(\bar{x}) + \sum_{i=1}^{n+1} f_i p_i \langle 5n + 3 - i \rangle_0.$$

Hence,

$$|\hat{p}(\bar{x}) - p(\bar{x})| \leq \sum_{i=1}^{n+1} |f_i p_i| (5n + 3 - i)\mu'. \tag{4.10}$$

In order to simplify this bound, let $\bar{f} = \max\{|f_i|: i = 1, 2, \ldots, n + 1\}$, and let

$$M = \max\{|\bar{x} - x_i|: i = 1, 2, \ldots, n + 1\}$$
$$m = \min\{|x_i - x_j|: i, j = 1, 2, \ldots, n + 1, i \neq j\}.$$

Then $|p_i| \leq (M/m)^n$ for all i, and we have from (4.10) that

$$|(\hat{p}(\bar{x}) - p(\bar{x}))| \leq \bar{f} \left(\frac{M}{m}\right)^n \sum (5n + 3 - i)\mu'$$

$$= \bar{f} \left(\frac{M}{m}\right)^n (\tfrac{9}{2}n^2 + \tfrac{19}{2}n + 2)\mu'.$$

We can also write this as a kind of **relative error estimate**:

$$\left|\frac{\hat{p}(\bar{x}) - p(\bar{x})}{\bar{f}}\right| \leq \left(\frac{M}{m}\right)^n (\tfrac{9}{2}n^2 + \tfrac{19}{2}n + 2)\mu'. \tag{4.11}$$

There are two important parts to this estimate: the factor $(M/m)^n$ and the n^2 term. The former indicates that a uniform spacing of points, where $m \cong M$, is desirable. The latter term shows that the **relative rounding error increases** as n^2 when n grows. This increase is not serious for values of n that occur in practice.

4.1.4 Special Formulas and Hermite Interpolation

Next we shall consider some important special cases of the general formulas just derived, and also indicate some methods for deriving special-purpose interpolation formulas.

First, consider the Lagrange formula (4.1), (4.2) when $n = 1$:

$$p(x) = f_1 \frac{x - x_2}{x_1 - x_2} + f_2 \frac{x - x_1}{x_2 - x_1}$$

or, more simply,

> **First-**
> **degree**
> **interpolation**

$$p(x) = \frac{f_1(x - x_2) - f_2(x - x_1)}{x_1 - x_2}. \tag{4.12}$$

The **truncation error** is

$$E_T(x) = \tfrac{1}{2} f''(\xi)(x - x_1)(x - x_2) \tag{4.13}$$

and the **rounding error** estimates (4.10), (4.11) are

$$|\hat{p} - \bar{p}| \le (7|f_1 p_1| + 6|f_2 p_2|)\mu', \qquad |\hat{p} - p|/|\bar{f}| \le (M/m)16\mu'.$$

Thus in a decimal machine, for example, rounding error may cause a loss of one figure accuracy. Moreover, the truncation error may be quite large, especially if $|x_2 - x_1|$ is large. It is worth noting, geometrically, that the function $p(x)$ defined by (4.12) is simply the secant line connecting the two points $(x_1, f(x_1))$ and $(x_2, f(x_2))$ on the curve representing the function $f(x)$.

The Lagrange formula for $n = 2$ reduces to

> **Second-**
> **degree**
> **interpolation**

$$p(x) = f_1 \frac{(x - x_2)(x - x_3)}{(x_1 - x_2)(x_1 - x_3)} + f_2 \frac{(x - x_1)(x - x_3)}{(x_2 - x_1)(x_2 - x_3)}$$
$$+ f_3 \frac{(x - x_1)(x - x_2)}{(x_3 - x_1)(x_3 - x_2)} \tag{4.14a}$$

with the **truncation error** given by the formula

$$E_T(x) = \tfrac{1}{6} f'''(\xi)(x - x_1)(x - x_2)(x - x_3). \tag{4.14b}$$

Again, the round-off error is bounded by a multiple of μ, which indicates a loss of one or two decimal figures.

The Lagrange formula was motivated by a desire to obtain an approximation of the form $p(x) = a_n x^n + a_{n-1} x^{n-1} + \cdots + a_0$ to a function $f(x)$ whose values are known at certain points. In applications, it often happens that additional knowledge of the function f makes it reasonable to formulate the problem somewhat differently. For example, suppose it is known that f is an *even* function of x; that is, for any x, $f(x) = f(-x)$. Then a more reasonable type of approximation would be a function that involves only even powers of x:

| Approximation of even functions | $p(x) = a_0 + a_1 x^2 + a_2 x^4 + \cdots + a_n x^{2n}.$ | (4.15) |

Similarly, if f were an *odd* function $(f(-x) = -f(x))$, then an approximation using odd powers of x,

| Odd function | $p(x) = a_0 x + a_1 x^3 + a_2 x^5 + \cdots + a_n x^{2n+1},$ | (4.16) |

is indicated. General formulas for determining such special kinds of approximations can be derived; however, in simple cases it is often possible to derive the necessary formuas directly. Consider, for example, the approximation (4.15) with $n = 1$; that is, $p(x) = a_0 + a_1 x^2$. If this is to interpolate f at certain points, then we must require that $p(x_i) = f_i$ for certain values of i. Since there are two parameters, a_0, a_1, that define $p(x)$, it is reasonable to expect that $p(x)$ will interpolate at two points. That is, we can require

$$p(x_1) = f_1 \quad \text{and} \quad p(x_2) = f_2.$$

These conditions give, in fact, two equations:

$$a_0 + a_1 x_1^2 = f_1, \qquad a_0 + a_1 x_2^2 = f_2 \tag{4.17a}$$

for the unknowns a_0, a_1. If we solve them by elementary elimination, we find

$$a_0 = (f_1 x_2^2 - f_2 x_1^2)/(x_2^2 - x_1^2), \qquad a_1 = (f_2 - f_1)/(x_2^2 - x_1^2). \tag{4.17b}$$

This technique for deriving special types of approximations is called the **method of undetermined coefficients**. It is a useful device provided the system of equations, corresponding to (4.17a) for determining the coefficients, can be solved.

A somewhat different kind of interpolation is indicated when, in addition to values of f at certain points, the values of f' at certain points are also known. For example, suppose $f_1 = f(x_1)$, $f_2 = f(x_2)$, $f'_1 = f'(x_1)$, and $f'_2 = f'(x_2)$ are known. Then an interpolating function $p(x)$ should satisfy $p(x_1) = f_1$, $p(x_2) = f_2$, $p'(x_1) = f'_1$, and $p'(x_2) = f'_2$. If $p(x) = a_0 + a_1 x + a_1 x + a_2 x^2 + a_3 x^3$, then we have the four equations

$$a_0 + a_1 x_1 + \quad a_2 x_1^2 + \quad a_3 x_1^3 = f_1$$
$$a_0 + a_1 x_2 + \quad a_2 x_2^2 + \quad a_3 x_2^3 = f_2$$
$$a_1 \quad + 2a_2 x_1 + 3a_3 x_1^2 = f'_1$$
$$a_1 \quad + 2a_2 x_2 + 3a_3 x_2^2 = f'_2$$

to solve for the four coefficients a_0, a_1, a_2, and a_3. A function $p(x)$ that interpolates the function f, and whose derivative also interpolates the derivative of f is called a **Hermite interpolating function**. Elementary elimination can be used to find the coefficients a_0, a_1, a_2, a_3; however, it is easier simply to generalize the Lagrange formula. That is, for the $(n + 1)$ distinct points x_1, x_2, ..., x_{n+1}, we can define the polynomial

| Hermite formula |

$$p(x) = \sum_{k=1}^{n+1} \{1 - 2p'_k(x_k)(x - x_k)\}p_k(x)^2 f_k$$

$$+ \sum_{k=1}^{n+1} (x - x_k)p_k(x)^2 f'_k.$$

Here $p_k(x)$ is defined by (4.2). It can be verified that $p(x_i) = f_i$, $i = 1$, $2, \ldots, n + 1$ and also that $p'(x_i) = f'_i$, $i = 1, 2, \ldots, n + 1$. For the case $n = 1$, this gives

$$p(x) = [1 - 2p'_1(x_1)(x - x_1)]p_1(x)^2 f_1$$
$$+ [1 - 2p'_2(x_2)(x - x_2)]p_2(x)^2 f_2$$
$$+ (x - x_1)p_1(x)^2 f'_1 + (x - x_2)p_2(x)^2 f'_2$$

where

$$p_1(x) = (x - x_2)/(x_1 - x_2), \qquad p_2(x) = (x - x_1)/(x_2 - x_1).$$

Hence, after some simple arithmetic, we find

$$p(x) = (x_1 - x_2)^{-2}\{(1 - 2(x - x_1))(x - x_2)^2 f_1$$
$$+ (1 - 2(x - x_2))(x - x_1)^2 f_2$$
$$+ (x - x_1)(x - x_2)^2 f'_1 + (x - x_2)(x - x_1)^2 f'_2\}.$$

Note that the Hermite interpolating polynomial, which interpolates at $n + 1$ points, is of degree $2n + 1$. It can be shown (Exercise 20) that the **truncation error for Hermite interpolation** is given by

$$\frac{f^{(2n+2)}(\xi)}{(2n + 2)!}[(x - x_1)(x - x_2) \cdots (x - x_{n+1})]^2$$

In many applications (*most* applications, in fact), the data points x_1, $x_2, \ldots$ are **evenly spaced**; that is, there is some $h > 0$ so that

$$x_{i+1} - x_i = h \qquad \text{for all} \quad i.$$

When this is the case, the Lagrange (and Hermite) formula simplifies somewhat. Indeed, the factors that occur in the formula (4.2) for $p_i(x)$ can be written as:

$$x_i - x_k = \begin{cases} (x_i - x_{i+1}) + (x_{i+1} - x_{i+2}) + \cdots + (x_{k-1} - x_k), & k > i \\ (x_i - x_{i-1}) + (x_{i-1} - x_{i-2}) + \cdots + (x_{k+1} - x_k), & k < i \end{cases}$$

$$= \begin{cases} (-h) + (-h) + \cdots + (-h) = -(k-i)h, & k > i \\ (h) + (h) + \cdots + (h) = (i-k)h, & k < i \end{cases}$$

so that, for any i, k,

$$x_i - x_k = (i - k)h.$$

Furthermore, if we let $s = (x - x_1)/h$, then

$$x - x_k = (x - x_1) + (x_1 - x_k) = [s + (1 - k)]h = (s - k + 1)h;$$

hence

$$p_i(x) = \frac{s(s-1)(s-2) \cdots (s-i+2)(s-i) \cdots (s-n)}{(i-1)(i-2) \cdots 2 \cdot 1 \cdot (-1)(-2) \cdots (i-n+1)}, \qquad s = \frac{x - x_1}{h}.$$
(4.18)

The importance of this formulation is that the denominator in (4.18) is a product of *integers* and hence can be computed with no rounding error and with very little effort.

EXERCISES

1 Show, in detail, that $p_i(x_j) = \begin{cases} 0 & \text{if } j \neq i \\ 1 & \text{if } j = i \end{cases}$, where $p_i(x)$ is defined by (4.2).

2 Write a computer program to implement Algorithm (4.3). Test your program using the values given in Table 4.1.

3 For the values $f_1 = 1, f_2 = 3, f_3 = 5, x_1 = 1, x_2 = 3, x_3 = 5, n = 2$, write out the interpolating polynomial (4.1) in standard form $a_2 x^2 + a_1 x + a_0$. Evaluate the polynomial at $x = 2$ using this standard form, and also in the form (4.1). Which is easier?

4 Show that to rearrange the Lagrange polynomial (4.1), (4.2) into standard form $a_n x^n + a_{n-1} x^{n-1} + \cdots + a_0$ would require about n^3 additions and multiplications to determine the coefficients $a_n, a_{n-1}, \ldots, a_0$.

5 For the data of Table 4.1 and an interpolating polynomial with $n = 3$, what values should be used for x_1, x_2, x_3, x_4 when $\bar{x} = 0$? $\bar{x} = -2$? $\bar{x} = 2.75$?

6 Suppose each of the following functions has been tabulated at $x_1 = 0, x_2 = .1, x_3 = .2, x_4 = .3$. Estimate the truncation error at $\bar{x} = .15$ for the polynomial of degree 3 that interpolates at the given tabulated points

   (a) $f(x) = \sin x$ (b) $f(x) = 2x^3$ (c) $f(x) = 1/(x+1)$

7 Suppose f is a function defined for all $x > 0$ that has been tabulated at $x = 1, 2, 3, 4, \ldots,$

and suppose it is known that all derivatives of f are less than 1.0 in absolute value. What n should be used to guarantee a truncation error less than 10^{-5} for any $\bar{x} > 0$?

8 Show that the absolute value of the expression (4.6a) is minimized when $\bar{x}$ is the average of $x_1, x_2, \ldots, x_{n+1}$.

9 Show that, for $G(x)$ and $E_T(x)$ as defined in the proof of Theorem 4.1,

$$G^{(n+1)}(t) = (n+1)! \quad \text{and} \quad E_T^{(n+1)}(t) = f^{(n+1)}(t).$$

10 Discuss the possibility of subtractive cancellation during the evaluation of the Lagrange polynomial. For what kinds of functions f and what points $\bar{x}$ would subtractive cancellation occur?

11 In order to derive the simple estimate (4.11) from (4.10), we replaced the quantities $|f_i p_i|$ by the upper bound $\bar{f}(M/m)^n$. For the data in Table 4.1 how much is the estimate increased by making this substitution? That is, for this data, compute the bound in (4.10) and compare it to the bound in (4.11).

12 Write a computer program to implement Algorithm (4.3) and also compute the error estimate (4.10).

13 Suppose f is a function that has been tabulated at $x = 0, .1, .2, .3, .4, \ldots$ and suppose $|f^{(k)}(x)| \leq k$, $k = 0, 1, 2, \ldots$. What is the best value of n to use for interpolating f at $\bar{x} = .25$? That is, what value of n will result in the sum of the truncation error and rounding error [as given by (4.11)] to be as small as possible?

14 Write out the rounding error estimate (4.10), (4.11) for the special case (4.14a) when $n = 2$.

15 Find an interpolating function of the form $p(x) = a_1 x + a_2 x^2$.

16 Modify Algorithm (4.3) to determine p_i according to formula (4.18).

17 Modify the rounding error analysis of Section 4.3 in the case for which p_i is computed using (4.18), assuming no rounding error in computing the denominator.

18 Derive formulas for the coefficients a_1, a_2, a_3 so that the function $p(x) = a_1 \sin x + a_2 e^x + a_3/x$ has the same values as $f(x)$ for $x = x_1, x_2, x_3$.

19 Prove that the Hermite interpolation formula given in the text does, in fact, satisfy $p(x_i) = f_i$, $p'(x_i) = f_i'$, for $i = 1, 2, \ldots, n + 1$.

20 Derive the truncation error estimate for Hermite interpolation.
[*Hint*: Let $G(x)$ be as defined in the proof of Theorem 4.1, and let

$$E_T(x) = f(x) - p(x)$$

$$H(t) = E_T(\bar{x})G(t)^2 - E_T(t)G(\bar{x})^2.$$

Show that for some ξ, $H^{(2n+2)}(\xi) = 0$, so that

$$E_T(\bar{x}) = \frac{E_T(\xi)^{(2n+2)}G(\bar{x})^2}{[G(t)^2]^{(2n+2)}} = \frac{f^{(2n+2)}(\xi)G(\bar{x})^2}{(2n+2)!} \; . \;]$$

4.2 Least Squares Approximation

Least squares approximation occurs in a variety of applications, often under different names. Linear optimization, regression analysis, data smoothing, and curve fitting are a few of its disguises. The most commonly used technique for obtaining least squares approximations can, if used naively, lead to a serious computational pitfall. In this section we define the least squares approximation problem, show how to compute it, and discuss in some detail this pitfall and how to avoid it.

4.2.1 Least Squares Problems

The values of a tabulated function are usually given at a large number of points. Tables of trig functions, for example, give the values of the functions at several hundred angles. Similarly, physical experiments usually involve hundreds or thousands of measurements, each of which is the value of a certain unknown function. As was shown in Section 4.1, an attempt to use all of these values to determine an **interpolating function with very large** n generally leads to a small truncation error but a large rounding error. Even more serious, however, is the fact that an interpolating polynomial with $n = 1000$, for example, requires 1001 memory locations in the computer to store coefficients, plus many thousands of arithmetic operations to evaluate the polynomial at a single point. The technique of least squares approximation allows the use of many function values to obtain a simple approximation.

Suppose that $f_1, f_2, \ldots, f_m$ are values of some function f at distinct points $x_1, x_2, \ldots, x_m$, and an approximation $p(x) = a_0 + a_1 x + \cdots + a_n x^n$ is desired, where m is large and n is small. Clearly, it is not possible to make $p(x_i) = f_i$ for all i, but it may be possible to make $p(x_i) \sim f_i$ for all i. That is, suppose we try to find a polynomial p of degree n that agrees with f as nearly as possible at all the points $x_1, x_2, \ldots, x_m$. There are several ways of formulating this condition more precisely, but one that leads to a reasonable mathematical problem is to make the quantity

$$R = [f_1 - p(x_1)]^2 + [f_2 - p(x_2)]^2 + \cdots + [f_m - p(x_m)]^2 \qquad (4.19)$$

as small as possible. The reason for squaring the terms in (4.19) is to avoid possible cancellations of positive terms with negative terms. The use of absolute values for this purpose leads to a difficult mathematical problem.

Thus, we define the **least squares approximation problem** as follows:

Find the polynomial $p(x) = a_n x^n + a_{n-1} x^{n-1} + \cdots + a_0$ that minimizes the quantity (4.19).

Before discussing methods for solving this problem, we shall mention two useful modifications that lead to problems that can be solved by essentially the same techniques.

The first modification arises from the observation that in many applications there are one or more of the data points, x_1 for example, at which it is important for the approximating function $p(x)$ to be (almost) equal to the value f_1 of the function $f(x)$. That is, the term $[f_1 - p(x_1)]^2$ in (4.19) should be *very* small. In order to force this term to be smaller than the other terms in (4.19), it suffices to multiply it by a large positive *weighting factor* w_1 and then try to minimize the expression

$$R = w_1[f_1 - p(x_1)]^2 + [f_2 - p(x_2)]^2 + \cdots + [f_m - p(x_m)]^2.$$

w_1 is to be thought of as a fixed constant, say, 100 or 1000, depending on how close $p(x_1)$ should be to f_1. Notice that by forcing $p(x_1)$ to be very near f_1, we shall probably cause $p(x_i)$ to be farther from f_i for certain values of $i \neq 1$. More generally, we can associate with *each* data point $x_1, x_2, \ldots, x_m$ a positive weight $w_1, w_2, \ldots, w_m$, most of which will usually be 1, and then define the **weighted least squares approximation problem** by replacing (4.19) with

$$R = w_1[f_1 - p(x_1)]^2 + w_2[f_2 - p(x_2)]^2 + \cdots + w_m[f_m - p(x_m)]^2. \quad (4.20)$$

The second modification to the basic least squares problem is derived by noting that in fact R, as defined by either (4.19) or (4.20), is simply a function of the $(n + 1)$-variables, $a_0, a_1, \ldots, a_n$. Furthermore, these variables are just the coefficients of a linear combination of the simple functions 1, x, $x^2, \ldots, x^n$. This suggests that perhaps a better approximation could be obtained in certain cases by using other functions, such as e^x, $\sin x$, $1/x$, etc., in this linear combination. That is, instead of using *polynomial* approximations, it might be better to look for an approximation of the form, say,

$$p(x) = a_0 e^x + a_1(1/x) + a_2 \sin x.$$

In general, let $\varphi_0(x), \varphi_1(x), \ldots, \varphi_n(x)$ be some chosen set of functions, and $w_1, w_2, \ldots, w_m$ a set of weights. Then the **general weighted least squares approximation problem**, using basis functions $\varphi_0, \varphi_1, \ldots, \varphi_n$ is defined as follows:

Find the function $p(x) = a_0 \varphi_0(x) + a_1 \varphi_1(x) + \cdots + a_n \varphi_n(x)$ that minimizes the quantity (4.20).

Certain restrictions can be placed on the basis functions $\varphi_0, \varphi_1, \ldots, \varphi_n$ in order to ensure that this problem has a solution. We shall not discuss such restrictions, but rather shall always assume that a solution exists. In particular, it can be shown that if the basis functions are given by $\varphi_k(x) = x^k$, i.e., the approximating functions are just polynomials, then there will always exist a unique solution, provided only that the points $x_1, x_2, \ldots, x_m$ are not all identical.

The approximating functions $p(x) = a_0 \varphi_0(x) + \cdots + a_n \varphi_n(x)$ considered here have an important property: The unknown coefficients $a_0, a_1, \ldots, a_n$ occur *linearly*. Thus we cannot consider functions of the form $p(x) = e^{a_0 x} + \sin(a_1 x) + x^{a_2}$, where the unknowns occur nonlinearly. To be very precise, the problem defined above should be called a *linear* least squares approximation problem to distinguish it from this more general and much more difficult problem. Unfortunately, the term "linear least squares approximation" often means a least squares approximation by a linear function $p(x) = a_0 + a_1 x$.

4.2.2 Computing Least Squares Approximations

To solve the general least squares approximation problem, we shall derive a set of linear equations. The solution to these equations will be the set $a_0, a_1, \ldots, a_n$ of unknown coefficients, which defines the approximation $p(x)$ that minimizes (4.20). To illustrate this technique, consider the special polynomial case in which $n = 1$; i.e., we try to find a_0, a_1 so that $p(x) = a_0 + a_1 x$ minimizes (4.19). Observe that R in (4.19) is a function of the two unknowns a_0, a_1, so that, more precisely, we can write

$$
\boxed{\begin{array}{l} \text{Special} \\ \text{case} \\ n = 1 \end{array}} \quad
\begin{aligned}
R(a_0, a_1) = {}& [f_1 - (a_0 + a_1 x_1)]^2 + [f_2 - (a_0 + a_1 x_2)]^2 \\
& + \cdots + [f_m - (a_0 + a_1 x_m)]^2.
\end{aligned} \tag{4.21}
$$

A fundamental result from calculus says that such a function can have a minimum only where the partial derivatives are zero; that is, we must have

$$\partial R/\partial a_0 = 0 \quad \text{and} \quad \partial R/\partial a_1 = 0. \tag{4.22}$$

(The proof that (4.22) actually determines a *minimum* of (4.21) is left as an exercise.)
These derivatives are easily seen to be

$$
\begin{aligned}
(\partial R/\partial a_0) = {}& \{2[f_1 - (a_0 + a_1 x_1)](-1) + 2[f_2 - (a_0 + a_1 x_2)](-1) \\
& + \cdots + 2[f_m - (a_0 + a_1 x_m)](-1)\} \\
= {}& -2\{[f_1 - (a_0 + a_1 x_1)] + [f_2 - (a_0 + a_1 x_2)] \\
& + \cdots + [f_m - (a_0 + a_1 x_m)]\}
\end{aligned}
$$

and

$$
\begin{aligned}
(\partial R/\partial a_1) = {}& \{2[f_1 - (a_0 + a_1 x_1)](-x_1) + 2[f_2 - (a_0 + a_1 x_2)](-x_2) \\
& + \cdots + 2[f_m - (a_0 + a_1 x_m)](-x_m)\}.
\end{aligned}
$$

Thus (4.22) becomes

$$f_1 - (a_0 + a_1 x_1) + f_2 - (a_0 + a_1 x_2) + \cdots + f_m - (a_0 + a_1 x_m) = 0$$
$$x_1 f_1 - x_1(a_0 + a_1 x_1) + x_2 f_2 - x_2(a_0 + a_1 x_2)$$
$$+ \cdots + x_m f_m - x_m(a_0 + a_1 x_m) = 0$$

which, by collecting terms together, can be written as

$$(f_1 + f_2 + \cdots + f_m) - m a_0 - (x_1 + x_2 + \cdots + x_m)a_1 = 0$$
$$(x_1 f_1 + x_2 f_2 + \cdots + x_m f_m) - (x_1 + x_2 + \cdots + x_m)a_0$$
$$- (x_1^2 + x_2^2 + \cdots + x_m^2)a_1 = 0.$$

Finally, we write these in the standard form of two linear equations in two unknowns a_0 and a_1:

$$m a_0 + (x_1 + x_2 + \cdots + x_m)a_1 = (f_1 + f_2 + \cdots + f_m) \qquad (4.23)$$
$$(x_1 + x_2 + \cdots + x_m)a_0 + (x_1^2 + x_2^2 + \cdots + x_m^2)a_1$$
$$= (x_1 f_1 + x_2 f_2 + \cdots + x_m f_m).$$

The pair of equations (4.23) is called the *normal equations* for this least squares problem, and the solution is easily found to be

$$a_0 = \frac{(x_1^2 + \cdots + x_m^2)(f_1 + \cdots + f_m) - (x_1 + \cdots + x_m)(x_1 f_1 + \cdots + x_m f_m)}{m(x_1^2 + \cdots + x_m^2) - (x_1 + \cdots + x_m)^2}$$

$$a_1 = \frac{m(x_1 f_1 + \cdots + x_m f_m) - (x_1 + \cdots + x_m)(f_1 + \cdots + f_m)}{m(x_1^2 + \cdots + x_m^2) - (x_1 + \cdots + x_m)^2}. \qquad (4.24)$$

This solution can be written more easily using the standard summation notation $\sum$ as

$$a_0 = \frac{\sum_{i=1}^{m} x_i^2 \sum_{i=1}^{m} f_i - \sum_{i=1}^{m} x_i \sum_{i=1}^{m} x_i f_i}{m \sum_{i=1}^{m} x_i^2 - (\sum_{i=1}^{m} x_i)^2}$$

$$a_1 = \frac{m \sum_{i=1}^{m} x_i f_i - \sum_{i=1}^{m} x_i \sum_{i=1}^{m} f_i}{m \sum_{i=1}^{m} x_i^2 - (\sum_{i=1}^{m} x_i)^2}. \qquad (4.25)$$

The solution is always defined, provided only that the denominator is non-zero. But it can be shown (Exercise 8) that

$$m \sum_{i=1}^{m} x_i^2 - \left(\sum_{i=1}^{m} x_i \right)^2 = \frac{1}{2} \sum_{i=1}^{m} \sum_{j=1}^{m} (x_i - x_j)^2 \qquad (4.26)$$

so that the denominator is zero only if *all* of the points $x_1, \ldots, x_m$ are identical.

In the more **general case** when the approximation is to have the form

$$p(x) = a_0 \varphi_0(x) + a_1 \varphi_1(x) + \cdots + a_n \varphi_n(x)$$

the procedure is the same. To simplify the discussion, we shall use the $\sum$ notation, often without the summation index i, which will always go from 1 to m. If we write (4.20) as

$$R(a_0, a_1, \ldots, a_n) = \sum w_i[f_i - (a_0 \varphi_0(x_i) + a_1 \varphi_1(x_i) + \cdots + a_n \varphi_n(x_i)]^2,$$

$$(4.27)$$

then the derivatives $\partial R/\partial a_k$ are just given by

$$\partial R/\partial a_k = -2\{\sum w_i[f_i - (a_0 \varphi_0(x_i) + \cdots + a_n \varphi_n(x_i))]\varphi_k(x_i)\}.$$

By equating these to zero we obtain the equations

$$\sum[a_0 w_i \varphi_0(x_i)\varphi_k(x_i) + a_1 w_i \varphi_1(x_i)\varphi_k(x_i) + \cdots + a_n w_i \varphi_n(x_i)\varphi_k(x_i)]$$
$$= \sum w_i f_i \varphi_k(x_i)$$

or

$$\left(\sum_{i=1}^{m} w_i \varphi_k(x_i)\varphi_0(x_i)\right)a_0 + \left(\sum w_i \varphi_k(x_i)\varphi_1(x_i)\right)a_1 \qquad (4.28)$$

$$+ \cdots + \left(\sum w_i \varphi_k(x_i)\varphi_n(x_i)\right)a_n = \left(\sum w_i \varphi_k(x_i)f_i\right).$$

There will be one such equation for each a_k, $k = 0, 1, \ldots, n$, so that if we let

$$A_{kj} = \sum_{i=1}^{m} w_i \varphi_k(x_i)\varphi_j(x_i), \qquad \begin{cases} j = 0, 1, \ldots, n \\ k = 0, 1, \ldots, n, \end{cases} \qquad (4.29)$$

then the conditions

$$\partial R(a_0, a_1, \ldots, a_n)/\partial a_k = 0, \qquad k = 0, 1, \ldots, n \qquad (4.30)$$

imply the equations

$$
\boxed{\text{Normal equations}}
\quad
\begin{aligned}
A_{00}a_0 + A_{01}a_1 + \cdots + A_{0n}a_n &= \sum w_i \varphi_0(x_i)f_i \\
A_{10}a_0 + A_{11}a_1 + \cdots + A_{1n}a_n &= \sum w_i \varphi_1(x_i)f_i \\
&\vdots \\
A_{n0}a_0 + A_{n1}a_1 + \cdots + A_{nn}a_n &= \sum w_i \varphi_n(x_i)f_i.
\end{aligned}
\qquad (4.31)
$$

Equations (4.31) are a set of $(n + 1)$ linear equations for the $(n + 1)$ unknowns $a_0, a_1, \ldots, a_n$. This set of equations is called the **normal equations** for the least squares approximation problem. Numerical methods for solving these equations will be discussed in Chapter 6. Numerical difficulties with these equations will be discussed in subsection 4.2.4. Of interest here is the following algorithm, which can be used to determine the coefficients A_{ij} and the right-hand side b_i of these normal equations.

Computing the Coefficients of the Normal Equations

[Algorithm to determine the normal equations.]

1 Input $\{x_1, \ldots, x_m, f_1, \ldots, f_m, w_1, \ldots, w_m, \varphi_0, \varphi_1, \ldots, \varphi_n\}$
2 For $k = 0, 1, \ldots, n$
   2.1 For $j = k, k + 1, \ldots, n$
      2.1.1 $A_{kj} \leftarrow 0$
      2.1.2 For $i = 1, 2, \ldots, m$
         2.1.2.1 $A_{kj} \leftarrow A_{kj} + w_i \varphi_k(x_i)\varphi_j(x_i)$ (4.32)
      2.1.3 $A_{jk} \leftarrow A_{kj}$
   2.2 $b_k \leftarrow 0$
   2.3 For $i = 1, 2, \ldots, m$
      2.3.1 $b_k \leftarrow b_k + w_i \varphi_k(x_i) f_i$
3 Output $\{A_{kj}, b_k, k, j = 0, 1, \ldots, n\}$

In this algorithm, the "symmetry" property $A_{kj} = A_{jk}$ of the coefficients has been used to reduce the computation.

TABLE 4.3

x_i	1	2	4	6
f_i	10	5	2	1

As an example, suppose the data given by Table 4.3 is to be used to determine a function of the form

$$p(x) = a_0 + a_1 x + a_2 x^2$$

(we assume all weights $= 1$). Then the coefficients in the normal equations are determined, according to formula (4.29) with $\varphi_0(x) = 1$, $\varphi_1(x) = x$, $\varphi_2(x) = x^2$, by

$$A_{00} = \sum_{i=1}^{4} \varphi_0(x_i)\varphi_0(x_i) = \sum_{i=1}^{4} x_i^0 x_i^0 = \sum_{i=1}^{4} 1 = 4$$

$$A_{01} = A_{10} = \sum_{i=1}^{4} \varphi_1(x_i)\varphi_0(x_i) = \sum_{i=1}^{4} x_i = 1 + 2 + 4 + 6 = 13$$

$$A_{02} = A_{20} = \sum_{i=1}^{4} \varphi_2(x_i)\varphi_0(x_i) = \sum_{i=1}^{4} x_i^2 = 1 + 4 + 16 + 36 = 57$$

$$A_{11} = \sum_{i=1}^{5} \varphi_1(x_i)\varphi_1(x_i) = \sum_{i=1}^{5} x_i^2 = 57$$

$$A_{12} = A_{21} = \sum_{i=1}^{4} \varphi_2(x_i)\varphi_1(x_i) = \sum_{i=1}^{4} x_i^3 = 1 + 8 + 64 + 216 = 289$$

$$A_{22} = \sum_{i=1}^{4} \varphi_2(x_i)\varphi_2(x_i) = \sum_{i=1}^{4} x_i^4 = 1 + 16 + 256 + 1296 = 1569.$$

The right-hand sides of the equations are

$$\sum \varphi_0(x_i)f_i = \sum f_i = 10 + 5 + 2 + 1 = 18$$
$$\sum \varphi_1(x_i)f_i = \sum x_i f_i = 10 + 10 + 8 + 6 = 34$$
$$\sum \varphi_2(x_i)f_i = \sum x_i^2 f_i = 98.$$

Finally, the **normal equations** are

$$4a_0 + 13a_1 + 57a_2 = 18$$
$$13a_0 + 57a_1 + 289a_2 = 34$$
$$57a_0 + 289a_1 + 1569a_2 = 98.$$

The **solution** to these equations is

$$a_0 = 14.311527$$
$$a_1 = -5.2663096$$
$$a_2 = .51255987,$$

which gives the **approximating function**

$$p(x) = .51255987x^2 - 5.2663096x + 14.311527.$$

As a check on the accuracy of our approximation, we compute the **residual**

$$R(a_0, a_1, a_2) = \sum_{i=1}^{4} (f_i - p(x_i))^2$$
$$= (10 - 9.55777727)^2 + (5 - 5.82914728)^2$$
$$+ (2 - 1.4472465)^2 + (1 - 1.16582472)^2$$
$$= 1.21590087.$$

To illustrate the use of the more general kind of approximating functions, suppose we try to obtain an approximation of the form

$$p(x) = a_0 + (a_1/x).$$

Notice that a sketch of the data given in Table 4.3 suggests that this might be a more appropriate kind of function to use. The coefficients of the normal equations are now found by using $\varphi_0(x) = 1$, $\varphi_1(x) = 1/x$. This gives

$$A_{00} = \sum \varphi_0(x_i)\varphi_0(x_i) = 4$$

$$A_{01} = \sum \varphi_0(x_i)\varphi_1(x_i) = 1 + \tfrac{1}{2} + \tfrac{1}{4} + \tfrac{1}{6} = 1.91667$$

$$A_{11} = \sum \varphi_1(x_i)\varphi_1(x_i) = 1 + \tfrac{1}{4} + \tfrac{1}{16} + \tfrac{1}{36} = 1.340278$$

with right-hand sides

$$\sum \varphi_0(x_i)f_i = \sum f_i = 18$$

$$\sum \varphi_1(x_i)f_i = 10 + \tfrac{5}{2} + \tfrac{2}{4} + \tfrac{1}{6} = 13.16667.$$

The normal equations are

$$4a_0 + 1.91667a_1 = 18$$

$$1.91667a_0 + 1.34028a_1 = 13.16667,$$

whose solution gives the approximation

$$p(x) = -.658444 + 10.76543/x.$$

As a comparison with the previous approximation, we compute the **residual** for *this* function:

$$R(a_0, a_1) = \sum_{i=1}^{4} (f_i - p(x_i))^2$$

$$= (10 - 10.106986)^2 + (5 - 4.724271)^2$$

$$+ (2 - 2.0329135)^2 + (1 - 1.13579433)^2$$

$$= .10699288.$$

By comparing this with the previous residual of $1.2159\ldots$, we see that this latter function conforms to the given data much more closely and hence is a better approximation.

In a least squares approximation the **choice of basis functions** is often determined by the application that produced the data or, as in the above example, by a rough sketch of the data. Important numerical considerations that may influence this choice are discussed in subsection 4.2.4.

4.2.3   Rounding Errors in Least Squares Approximations

As the previous example illustrates, the form of the approximating function determines the accuracy of the approximation, as measured by the residual $R(a_0, a_1, \ldots, a_n)$. But once the basis functions $\varphi_0, \varphi_1, \ldots, \varphi_n$ have been selected, we have in effect defined our numerical problem to be the minimizing of the function $R(a_0, \ldots, a_n)$ given by (4.20). Now we are no longer concerned with an approximation problem but rather with the question of how nearly does the computed solution to the normal equations minimize R. There are, in fact, two distinct computational problems here. The first involves the computation of the coefficients of the normal equations; the second is concerned with the computation of a solution to these computed normal equations. This second problem will be discussed in detail in Chapter 6. To analyze the computation of the coefficients, we note that this consists of the summation of some floating point numbers, which have also been computed and hence contain rounding errors. That is, the **computed coefficients** have the form

$$\hat{A}_{kj} = \text{fl}\left[\sum_{i=1}^{m} \text{fl}[\varphi_k(x_i)\varphi_j(x_i)]\right]. \tag{4.33}$$

(For simplicity, we assume that all weights $w_i = 1$.)

Assume that the basis functions are evaluated with l rounding errors. That is,

$$\text{fl}[\varphi_k(x_i)] = \varphi_k(x_i)\langle l \rangle. \tag{4.34}$$

Then (4.33) becomes

$$\hat{A}_{kj} = \text{fl}\left[\sum_{i=1}^{m} \varphi_k(x_i)\varphi_j(x_i)\langle l + 1 \rangle\right]$$

$$= \sum_{i=1}^{m} \varphi_k(x_i)\varphi_j(x_i)\langle l + 1 \rangle\langle m - i + 1 \rangle$$

$$= \sum_{i=1}^{m} \varphi_k(x_i)\varphi_j(x_i)\langle l + m - i + 2 \rangle. \tag{4.35}$$

[We have used here the rounding error analysis for summation (3.3).] To derive a forward error estimate, we rewrite this as

$$\hat{A}_{kj} = \sum_{i=1}^{m} \varphi_k(x_i)\varphi_j(x_i)(1 + \langle l + m - i + 2 \rangle_0), \tag{4.36}$$

which gives

$$\hat{A}_{kj} - A_{kj} = \sum_{i=1}^{m} \varphi_k(x_i)\varphi_j(x_i)\langle l + m - i + 2 \rangle_0. \tag{4.37}$$

The resulting **forward error bound** is

$$|\hat{A}_{kj} - A_{kj}| \leq \sum_{i=1}^{m} |\varphi_k(x_i)|\,|\varphi_j(x_i)|(l + m - i + 2)\mu'. \tag{4.38}$$

There are several conclusions to be drawn from this estimate. First of all, the basis functions should not be extremely large; secondly, the error grows rather slowly as m increases; and thirdly, the accuracy cannot be improved much by evaluating the basis function with great accuracy. That is, there is not much to be gained by making l very small, especially if m is quite large.

A real difficulty that arises in connection with these normal equations is that, even when m is small, the equations may be very ill conditioned. That is, the rounding errors in the computed coefficients may have a tremendous influence on the solution to the equations. To illustrate this fact, consider the approximation $p(x) = a_0 + a_1 x$ to the data

x	10.0	10.1	10.2	10.3	10.4	10.5
$f(x)$	1	1.20	1.25	1.267	1.268	1.276

$$(4.39)$$

The normal equations are easily found to be

Ill conditioned normal equations

$$6a_0 + 61.5a_1 = 7.259$$
$$61.5a_0 + 630.55a_1 = 74.4843$$

$$(4.40)$$

and the solution is

$$a_0 = -3.44952372$$
$$a_1 = .45457142.$$

This gives the approximation $p(x) = -3.44952372 + .45457142x$.

To illustrate the extent of the ill-conditioning of the equations (4.40), we round all of the coefficients to four digits, i.e., change the second equation to

$$61.5a_0 + 630.6a_1 = 74.48.$$

$$(4.40')$$

This changes the solution to

Changed solution

$$a_0 = -2.21822217$$
$$a_1 = .3344444.$$

Thus a relative change of 10^{-4} in the coefficients of the normal equations has caused a relative change of 1 in the solution.

The changes in the normal equations (4.40), which produced the perturbed solution, could also result from small changes in the original data (4.39). Indeed, changes of relative size 10^{-3} in the data (4.39) will introduce relative changes of size 10^{-4} in the normal equation (4.40) that, in turn, cause changes in the solution of relative magnitude 1. Thus, the least squares approximation problem *itself* is ill conditioned. In many cases this ill-conditioning can be avoided by a careful choice of the basis functions. This will be discussed in more detail in the next subsection.

4.2.4   Choice of Basis Functions

The example just given shows that **the least squares problem may be ill conditioned**. That is, small changes in the data $\{x_i, f_i\}$ may cause large changes in the approximating function $p(x)$. In this case, solving the approximation problem by constructing and then solving the normal equations may introduce further complications. Indeed, the **normal equations may be even more ill conditioned** than the original approximation problem. General methods for computing least squares approximations *without* actually constructing the normal equations are beyond the scope of this book. [See, for example, Lawson and Hanson (1974)]. In many cases, however, the conditioning of the normal equations, and hence of the approximation problem, can be controlled by a careful choice of the basis functions. If the normal equations are well conditioned, then so is the approximation problem. Thus we consider here how the basis functions affect only the conditioning of the normal equations.

Roughly speaking, the ill-conditioning of the normal equations is caused by a kind of "dependence" among the basis functions. For example, the approximation $p(x) = a_0 + a_1 x$ to the data (4.39) used the basis functions $\varphi_0(x) = 1$ (i.e., the constant function), and the function $\varphi_1(x) = x$. But *for the values of x given in the table*, namely, $x = 10, 10.1, \ldots, 10.5$, $\varphi_1(x)$ is also nearly constant, i.e., $\varphi_1(x) \approx 10.0$. The resulting normal equations were ill conditioned. However, if the basis functions $\varphi_0(x) = 1$ and $\varphi_1(x) = x - 10$ are used, then the normal equations are

$$6a_0 + 1.5a_1 = 7.259 \tag{4.41}$$
$$1.5a_0 + .55a_1 = 1.8943.$$

These **equations are well conditioned** and can be easily solved for $a_0 = 1.09619047$ and $a_1 = .45457142$. This gives the approximation

$$p(x) = a_0 + a_1(x - 10) = 1.09619047 + .45457142(x - 10) \tag{4.42}$$
$$= -3.4495237 + .45457142x,$$

which is identical to the approximation obtained by *carefully* solving the ill-conditioned system (4.40).

As another example, suppose the function f were tabulated at small values of x, say $x = .001, .002, \ldots$, and the basis functions $\varphi_0(x) = 1$, $\varphi_1(x) = x$, $\varphi_2(x) = e^x$ were chosen. Then the normal equations will be ill conditioned because, for very small x, we have

$$\varphi_2(x) = e^x = 1 + x + \frac{x^2}{2} + \cdots \cong 1 + x = \varphi_0 + \varphi_1(x).$$

Thus, there is a near *dependence* among the chosen basis functions. A function $\varphi_n(x)$ is said to be *dependent* on the functions $\varphi_0, \varphi_1, \ldots, \varphi_{n-1}$ if there are coefficients $c_0, c_1, \ldots, c_{n-1}$ so that for all x

$$\varphi_n(x) = c_0 \varphi_0(x) + c_1 \varphi_1(x) + \cdots + c_{n-1} \varphi_{n-1}(x).$$

If this is the case, then the normal equations obtained by using these basis functions will be *singular*, i.e., may have *no* solution. In many applications it happens that, for the given values of x,

$$\varphi_n(x) \approx c_0 \varphi_0(x) + c_1 \varphi_1(x) + \cdots + c_{n-1} \varphi_{n-1}(x),$$

in which case the normal equations will be ill conditioned.

In some cases where a "near" dependence exists, the best thing to do is simply to discard one of the functions. In the above example where $\varphi_0 = 1$, $\varphi_1(x) = x$, $\varphi_2(x) = e^x$, it would be just as well to use φ_0 and either φ_1 or φ_2 but not both.

In other cases it may be possible to redefine the basis functions so as to **eliminate the near dependency** without changing the final approximating function. In the example for which we used (4.39), a linear approximation $p(x) = a + bx$ was desired. By letting a and b be the unknowns, we were led to the ill-conditioned system (4.40). However, we can also write $p(x) = (a + 10b) + b(x - 10) = c + d(x - 10)$ and can solve for c and d. This approach led to the well-conditioned system (4.41). More generally, any *polynomial* least squares approximation $p(x) = b_n x^n + b_{n-1} x^{n-1} + \cdots + b_0$ can be written as $p(x) = a_n \varphi_n(x) + a_{n-1} \varphi_{n-1}(x) + \cdots + a_0 \varphi_0(x)$ when $\varphi_k(x)$ is a polynomial of degree k. The set $\varphi_0, \ldots, \varphi_n$ of polynomials should be chosen so that the normal equations that define $a_n, a_{n-1}, \ldots, a_0$ are well-conditioned. One common approach is to choose the basis functions so that the normal equations are *diagonal*, i.e., in the notation of 4.2.2, $A_{ij} = 0$ if $i \neq j$. If this is done, then the equations (4.31) actually have the form

Diagonal normal equations	

$$\begin{aligned} A_{00} a_0 &= \sum w_i \varphi_0(x_i) f_i \\ A_{11} a_1 &= \sum w_i \varphi_1(x_i) f_i \\ \ddots \quad &\vdots \\ A_{nn} a_n &= \sum w_i \varphi_n(x_i) f_i \end{aligned}$$

(4.43)

and the solution can be written directly. By recalling the definition (4.29) of A_{kj}, we see that (4.43) holds if

$$\sum_{i=1}^{m} w_i \varphi_k(x_i) \varphi_j(x_i) = 0 \qquad \text{when} \quad k \neq j. \tag{4.44}$$

A set of polynomials $\varphi_0, \varphi_1, \ldots, \varphi_n$, where φ_k is a polynomial of degree k, which satisfies (4.44), is called a set of **orthogonal polynomials**, over the points $x_1, x_2, \ldots, x_m$, with respect to the weights $w_1, w_2, \ldots, w_m$.

Recursive formulas for finding sets of orthogonal polynomials exist and are used frequently. Unfortunately, these recursions are often unstable so that the computed normal equations are *not*, in fact, diagonal. Generally, however, they will be reasonably well conditioned. The most **commonly used recursion formulas** are given by

$$\varphi_0(x) = 1$$

$$\varphi_1(x) = x - \alpha_1 \tag{4.45a}$$

$$\varphi_{k+1}(x) = (x - \alpha_{k+1})\varphi_k(x) - \beta_{k+1}\varphi_{k-1}(x), \qquad k = 1, 2, \ldots,$$

where

$$\alpha_{k+1} = \frac{\sum_i w_i x_i [\varphi_k(x_i)]^2}{\sum_i w_i [\varphi_k(x_i)]^2}, \qquad k = 0, 1, \ldots$$

$$\tag{4.45b}$$

$$\beta_{k+1} = \frac{\sum_i w_i [\varphi_k(x_i)]^2}{\sum_i w_i [\varphi_{k-1}(x_i)]^2}, \qquad k = 1, 2, \ldots.$$

The proof of the fact that these polynomials actually satisfy (4.44) is left as an exercise (Exercise 9).

If the formulas (4.45) are applied to the points given in (4.39), we find $\varphi_0(x) = 1$ and

$$\varphi_1(x) = (x - \alpha_1)\varphi_0(x) = x - \alpha_1,$$

where

$$\alpha_1 = \sum x_i / \sum 1 = 6.5/6 = 10.25.$$

Hence, the new basis functions are $\varphi_0(x) = 1$ and $\varphi_1(x) = x - 10.25$. In this simple case, there has been **no rounding errors involved**, so that the normal equations are indeed diagonal:

$$6a_0 = 7.259$$

$$.175a_1 = .07955.$$

The solution is easily found to be $a_0 = 1.2098\ldots$, $a_1 = .45457\ldots$, which gives the approximation $p(x) = 1.2098\ldots + .45457\ldots(x - 10.25)$. This is the same as the approximation (4.42), written in a slightly different form.

In general, after computing the new basis functions from (4.45), the normal equations will not be diagonal. Large off-diagonal terms indicate that the recursion (4.45) was unstable. In this case more complicated techniques must be used. In many cases, however, especially when m and n

are not too large, the above process will lead to well-conditioned normal equations that can be solved by the methods of Chapter 6.

EXERCISES

1 How long (in seconds) will it take your computer to evaluate an interpolating polynomial with $n = 1000$? $n = 10,000$?

2 Plot the points $(0, 1), (10, 9), (20, 19), (30, 31), (40, 39)$, on a piece of graph paper and sketch the linear least squares approximation $p(x) = a_0 + a_1 x$ to these points. Next, sketch a least squares approximation, again with $n = 1$, but that passes through the point $(0, 1)$. Finally, sketch the linear approximation that comes as close as possible to $(10, 9)$.

3 Using basic functions from the collection $1, x, x^2, \sin x, 1/x, e^x$, describe *reasonable* approximating functions for each of the following functions:
  (a) A function that goes to ∞ as x goes to zero.
  (b) An oscillating function.
  (c) A function that goes to ∞ very rapidly as x goes to ∞.
  (d) A function that crosses the x axis twice.

4 Show that, if $n = 0$ so that the approximating function is just $p(x) = a_0$, then the solution to the least squares approximation problem is

$$a_0 = (1/m)(f_1 + f_2 + \cdots + f_m),$$

i.e., $p(x)$ is just the *average* of the function values.

5 Write out the normal equations for the data given by

x_i	1	2	3	4	5	6
f_i	-3	-2	0	1	3	4

with the approximating function $p(x) = a_0 + a_1 x + a_2 x^2 + a_3 x^3$, assuming all $w_i = 1$.

6 Repeat Exercise 5 with approximating function $p(x) = a_0 + a_1/x + a_2 x^2$.

7 Repeat Exercise 5 with the approximating function $p(x) = a_0 + a_1 x$ and with $w_1 = 100$, $w_i = 1$ for $i = 2, 3, \ldots, 6$.

8 Prove (4.26) by expanding $\sum_i \sum_j (x_i - x_j)^2$.

9 Prove that the polynomials $\varphi_0, \varphi_1, \ldots, \varphi_n$ defined by (4.45) satisfy (4.44).
  [*Hint*: Show that, for any $k < n$, $\sum w_i \varphi_{k+1}(x_i)\varphi_j(x_i) = 0$ for $j = 0, 1, \ldots, k$. Do this by induction on j.]

10 Repeat Exercise 5 using the orthogonal polynomial method.

11 Prove that a point (a_0, a_1) at which (4.22) holds is actually a minimum of (4.21).

4.3 Spline Interpolation

Spline interpolation has been used, in a mechanical formulation, for many years. Only recently, however, has a mathematical formulation been

developed and studied. The importance of the spline interpolation idea cannot be overemphasized. One of the most important developments in numerical analysis during the past 20 years is the use of splines (together with the so-called finite element method) to solve partial differential equations. In this section we define the basic spline approximation and discuss some of its important smoothness properties.

4.3.1 Cubic Splines

A major difficulty with Lagrange interpolation is that low degree approximations may have large truncation errors, whereas high degree polynomials tend to have a lot of "wiggles," i.e., are very erratic. In certain applications, such as when the approximating function is to be differentiated, it is important to obtain an approximation that is as smooth as possible, yet retains the interpolatory character of the Lagrange polynomial. One way to obtain such an approximation is to construct a mathematical model of a device used by draftsmen to draw smooth curves. This device, called a **mechanical spline**, is a thin flexible strip of wood or plastic that is anchored to the drawing table at certain points, called *knots*. The flexibility of the strip then allows it to assume a smooth shape. The mathematical model for this is derived as follows. (See Fig. 4.1.)

Let $x_1, x_2, \ldots, x_n$ be distinct points with

$$a = x_1 < x_2 < \cdots < x_n = b$$

and let $f_1, f_2, \ldots, f_n$ denote the function values at these points. The cubic

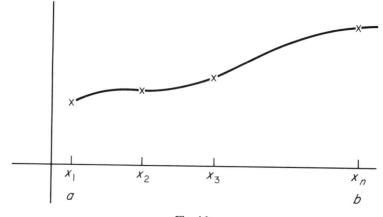

Fig. 4.1

spline that interpolates this data is a function $s(x)$ that is defined on $[a, b]$ and has the following properties:

(i) $s(x_k) = f_k$, $k = 1, 2, \ldots, n$.
(ii) $s(x)$, $s'(x)$, and $s''(x)$ are continuous on $[a, b]$.
(iii) In each subinterval $[x_k, x_{k+1}]$, $k = 1, 2, \ldots, n - 1$, $s(x)$ is a cubic polynomial. [But $s(x)$ is perhaps a *different* cubic polynomial in each subinterval.]
(iv) If $g(x)$ is any other function that satisfies (i), (ii), and (iii), then

$$\int_a^b s''(x)^2 \, dx \leq \int_a^b g''(x)^2 \, dx.$$

In terms of the mechanical spline, these properties can be described as follows:

(i)   The spline must pass through the knots.
(ii) The spline does not break, nor does it bend into sharp angles.
(iii) The theory of thin beams shows that, between the knots, the mechanical spline approximates a cubic polynomial.
(iv) The spline assumes the shape that minimizes its potential energy. (The potential energy is approximately proportional to the curvature, i.e., to the second derivative.)

Property (iv) is more difficult to formulate than are the other properties. It can be shown, however (subsection 4.3.5), that (iv) holds provided

(iv') $s''(a) = s''(b) = 0$.

We shall **define the cubic spline** to be the function $s(x)$ that satisfies (i), (ii), (iii), and (iv'). It can be shown that there is precisely one such function $s(x)$, and, in fact, the following construction of $s(x)$ will show this uniqueness property.

4.3.2 Computing Cubic Splines

Since $s(x)$ is a cubic polynomial for $x_k \leq x \leq x_{k+1}$, clearly $s''(x)$ must be linear. By using the formula for linear interpolation given in Section 4.1, we can write

$$s''(x) = s''(x_i) + \frac{x - x_i}{x_{i+1} - x_i} (s''(x_{i+1}) - s''(x_i)), \qquad x_i \leq x \leq x_{i+1}. \quad (4.46)$$

If this equation is integrated, the result is

$$s'(x) = s'(x_i) + \int_{x_i}^{x} s''(t)\, dt$$

$$= s'(x_i) + s''(x_i)(x - x_i) + \frac{s''(x_{i+1}) - s''(x_i)}{2(x_{i+1} - x_i)}(x - x_i)^2, \quad (4.47)$$

$$x_i \leq x \leq x_{i+1}.$$

By integrating again, and by using property (i) to replace $s(x_i)$ with f_i, we obtain

$$s(x) = f_i + s'(x_i)(x - x_i) + s''(x_i)\tfrac{1}{2}(x - x_i)^2$$

$$+ \frac{s''(x_{i+1}) - s''(x_i)}{6(x_{i+1} - x_i)}(x - x_i)^3, \quad x_i \leq x \leq x_{i+1}. \quad (4.48)$$

To simplify these equations write s_k'' for $s''(x_k)$ and let $h_k = x_{k+1} - x_k$, $k = 1, 2, \ldots, n - 1$. Then in (4.47) replace i by $i - 1$ and set $x = x_i$ to obtain

$$s'(x_i) = s'(x_{i-1}) + \tfrac{1}{2}(s_i'' + s_{i-1}'')h_{i-1}. \quad (4.49)$$

In (4.48) set $x = x_{i+1}$ and solve for $s'(x_i)$:

$$s'(x_i) = \frac{f_{i+1} - f_i}{h_i} - s_{i+1}''\frac{h_i}{6} - s_i''\frac{h_i}{3}. \quad (4.50)$$

From this equation and (4.48) we can see that if $s''(x_k)$ is known for $k = 1$, $2, \ldots, n$, then $s(x)$ can be evaluated for any value of $x \in [a, b]$. To derive formulas that will give the desired values of $s''(x_k)$ requires a fairly complicated manipulation of (4.49) and (4.50), taking into account property (ii). [Note that property (iii) was used to write (4.46).] First, we equate these two equations to obtain

$$s'(x_{i-1}) + \frac{s_i'' + s_{i-1}''}{2}h_{i-1} = \frac{f_{i+1} - f_i}{h_i} - s_{i+1}''\frac{h_i}{6} - s_i''\frac{h_i}{3}. \quad (4.51)$$

Next replace i by $i - 1$ in (4.50) and substitute the resulting expression for $s'(x_{i-1})$ into (4.51). After some simplifications, the resulting equation is

$$h_{i-1}s_{i-1}'' + 2(h_{i-1} + h_i)s_i'' + h_i s_{i+1}'' = 6\left(\frac{f_{i+1} - f_i}{h_i} - \frac{f_i - f_{i-1}}{h_{i-1}}\right), \quad (4.52)$$

which holds for $i = 2, 3, \ldots, n - 1$. Notice that this is a set of $n - 2$ linear equations for the unknowns $s_2'', s_3'', \ldots, s_{n-1}''$. The values for s_1'' and s_n'' are determined by property (iv'). Once these equations are solved for s_k'', $k = 2$, $3, \ldots, n - 1$, the spline function $s(x)$ can be evaluated at any $x \in [a, b]$ by using (4.48) and (4.51).

The system of equations (4.52) has several special properties that make it particularly easy to solve accurately and efficiently. Because of this, it is worthwhile giving a special algorithm that takes into account these properties even though the system can also be solved with the general methods described in Chapter 6. If we denote the right-hand side of (4.52) by d_i and use the conditions $s_1'' = s_n'' = 0$, then the equations, written out in full, are

$$2(h_1 + h_2)s_2'' + h_2 s_3'' \qquad\qquad\qquad\qquad\qquad = d_2$$

$$h_2 s_2'' + 2(h_2 + h_3)s_3'' + h_3 s_4'' \qquad\qquad\qquad = d_3$$

$$h_3 s_3'' + 2(h_3 + h_4)s_4'' + h_4 s_5'' \qquad\qquad = d_4$$

$$\ddots \qquad\qquad\qquad \vdots \quad (4.52')$$

$$h_{n-2}s_{n-2}'' + 2(h_{n-2} + h_{n-1})s_{n-1}'' = d_{n-1}.$$

The following recursion produces a system of equations that is equivalent to (4.52'):

$$a_2 = 2(h_1 + h_2)$$

$$a_k = 2(h_{k-1} + h_k) - (h_{k-1}^2/a_{k-1}), \qquad k = 3, 4, \ldots, n-1 \qquad (4.53)$$

$$b_2 = d_2$$

$$b_k = d_k - (h_{k-1} b_{k-1}/a_{k-1}), \qquad k = 3, 4, \ldots, n-1. \qquad (4.54)$$

(This is just the Gaussian elimination algorithm applied to a tridiagonal system. See Section 6.2 for the details.) The new system, however, has the form

$$a_2 s_2'' + h_2 s_3'' \qquad\qquad\qquad = b_2$$

$$a_3 s_3'' + h_3 s_4'' \qquad\qquad = b_3 \qquad (4.55)$$

$$\ddots \qquad\qquad \vdots$$

$$a_{n-1} s_{n-1}'' = b_{n-1},$$

which can be solved by the recursion

$$s_{n-1}'' = b_{n-1}/a_{n-1} \qquad\qquad\qquad\qquad\qquad (4.56)$$

$$s_k'' = (b_k - h_k s_{k+1}'')/a_k, \qquad k = n-2, n-3, \ldots, 2.$$

Notice that the recursion (4.53) involves only the knot spacing h_i, $i = 1$, $2, \ldots, n-1$, and is independent of the function values f_i. Hence, this part of the computation can be done independently of the particular function that is to be approximated.

Algorithms for Evaluating $s(\bar{x})$

$$\left[\begin{array}{c}\text{The following algorithm computes the various quantities needed}\\ \text{to evaluate the spline at a particular value of } x.\end{array}\right]$$

To compute $a_2, a_3, \ldots, a_{n-1}$:
  1   $a_2 \leftarrow 2(h_1 + h_2)$
  2   For $k = 3, 4, \ldots, n-1$
     2.1 $a_k \leftarrow 2(h_{k-1} + h_k) - h_{k-1}^2/a_{k-1}$

To compute $d_2, d_3, \ldots, d_{n-1}$:
  3   For $k = 2, 3, \ldots, n$
     3.1 $c_k \leftarrow (f_k - f_{k-1})/h_{k-1}$
  4   For $k = 2, 3, \ldots, n-1$
     4.1 $d_k \leftarrow 6(c_{k+1} - c_k)$

To compute $b_2, b_3, \ldots, b_{n-1}$:
  5   $b_2 \leftarrow d_2$
  6   For $k = 3, 4, \ldots, n-1$
     6.1 $b_k \leftarrow d_k - h_{k-1} b_{k-1}/a_{k-1}$

To solve for $s_2'', s_3'', \ldots, s_{n-1}''$:
  7   $s_{n-1}'' \leftarrow b_{n-1}/a_{n-1}$
  8   For $k = n-2, n-3, \ldots, 2$
     8.1 $s_k'' \leftarrow (b_k - h_k s_{k+1}'')/a_k$
  9   $s_1'' \leftarrow 0, \; s_n'' \leftarrow 0$

To evaluate s at a point $\bar{x} \in [a, b]$:
  10   $x_0 \leftarrow a + h_1$
  11   For $k = 1, 2, \ldots, n-1$
     11.1   If $\bar{x} \leq x_0$, go to step 13
     11.2 $x_0 \leftarrow x_0 + h_{k+1}$
  12   Error exit. $\bar{x}$ not in $[a, b]$
  13   $s' \leftarrow c_{k+1} - s_{k+1}'' h_k/6 - s_k'' h_k/3$
  14   $s \leftarrow f_k + s'(\bar{x} - x_k) + s_k''(\bar{x} - x_k)^2/2 + (s_{k+1}'' - s_k'')(\bar{x} - x_k)^3/(6h_k)$

Steps 10 and 11 of this algorithm determine the value of k for which $x_k \leq x \leq x_{k+1}$. A more efficient method is to use a binary search technique (Exercise 2). Notice that in addition to evaluating $s(\bar{x})$, the algorithm also determines $s'(\bar{x})$. Because of the smoothness property of the spline functions, $s'(\bar{x})$ will generally give a good approximation to $f'(\bar{x})$. This can be verified more rigorously using arguments similar to those of the next section (see Exercise 6).

The amount of work involved in this algorithm for $n \geq 3$ is

Efficiency of the spline evaluation	$(5n - 14)A + (5n - 13)M + (4n - 8)D$

to solve for s_k'', $k = 2, 3, \ldots, n-1$, and an additional $7A + 8M + 4D$ to evaluate $s(\bar{x})$ and $s'(\bar{x})$ once the proper interval $[x_k, x_{k+1}]$ has been located. The fact that these efficiency measures are *linear* in n implies that we need not be overly concerned about the size of n. That is, even for small n, the amount of work needed to evaluate $s(x)$ is fairly substantial. However, this amount grows only linearly with n, so that a small increase or decrease in n causes a similarly small change in the amount of work needed to evaluate the spline.

4.3.3   Truncation Error

In order to estimate the truncation error in cubic spline interpolation, we shall follow the idea used to prove Theorem 4.1. For simplicity, we shall consider only the case where the knots are *evenly spaced*; i.e., $h = h_i$, $i = 1, 2, \ldots, n-1$.

Let $\bar{x} \in [x_k, x_{k+1}]$, set

$$E(x) = f(x) - s(x)$$
$$G(x) = (x - x_k)(x - x_{k+1}),$$

and define

$$H(t) = E(\bar{x})G(t) - E(t)G(\bar{x}).$$

Then $H(t) = 0$ for $t = \bar{x}$, x_k, and x_{k+1}. Thus $H'(t) = 0$ for two values of t and $H''(t) = 0$ for at least one value, $\tau \in [x_k, x_{k+1}]$. Now

$$H''(t) = E(\bar{x})G''(t) - E''(t)G(\bar{x})$$
$$= 2E(\bar{x}) - (f''(t) - s''(t))G(\bar{x}),$$

so that $H''(\tau) = 0$ gives

$$f(\bar{x}) - s(\bar{x}) = E(\bar{x}) = \tfrac{1}{2}G(\bar{x})(f''(\tau) - s''(\tau)). \tag{4.57}$$

Furthermore, $|G(\bar{x})| = |(\bar{x} - x_k)(\bar{x} - x_{k+1})| \leq \tfrac{1}{4}(x_{k+1} - x_k)^2 = \tfrac{1}{4}h^2$, so that

$$|f(\bar{x}) - s(\bar{x})| \leq \tfrac{1}{8}h^2 |f''(\tau) - s''(\tau)| \leq \tfrac{1}{8}h^2(|f''(\tau)| + |s''(\tau)|). \tag{4.58}$$

To obtain a useful estimate, it is necessary to find a **bound for** $|s''(\tau)|$. But $s(x)$ is a cubic polynomial in $[x_k, x_{k+1}]$, so $s''(x)$ is linear on that interval. Hence,

$$|s''(\tau)| \leq \max\{|s''(x_k)|, |s''(x_{k+1})|\} \leq \max_{1 \leq k \leq n-1} |s_k''|. \tag{4.59}$$

Let $|s_i''|$ be the maximum of $|s_1''|$, $|s_2''|$, ..., $|s_{n-1}''|$. Then from (4.52) we have (still assuming $h = h_k$, $k = 1, 2, ..., n-1$),

$$hs_{i-1}'' + 4hs_i'' + hs_{i+1}'' = d_i.$$

But $|s_{i-1}''| \le |s_i''|$ and $|s_{i+1}''| \le |s_i''|$, so that

$$4h|s_i''| \le |d_i| + h|s_{i-1}''| + h|s_{i+1}''| \le \max_k |d_k| + 2h|s_i''|.$$

Hence,

$$2h|s_i''| \le \max_k |d_k|,$$

which gives

$$\max_k |s_k''| \le (1/2h) \max_k |d_k|. \tag{4.60}$$

Now,

$$d_k = 6\left[\frac{f_{k+1} - f_k}{h} - \frac{f_k - f_{k-1}}{h}\right],$$

where $f_k = f(x_k)$. By the mean value theorem,

$$(f_{k+1} - f_k)/h = f'(\eta_k) \qquad \text{some} \quad \eta_k \in [x_k, x_{k+1}]$$
$$(f_k - f_{k-1})/h = f'(\eta_{k-1}) \qquad \text{some} \quad \eta_{k-1} \in [x_{k-1}, x_k],$$

so that, for some $\xi_k \in [\eta_{k-1}, \eta_k]$,

$$d_k = 6[f'(\eta_k) - f'(\eta_{k-1})] = 6f''(\xi_k)(\eta_k - \eta_{k-1}).$$

We have assumed here that $f''(x)$ exists and is continuous. If we also assume $|f''(x)| \le M$ for all $x \in [a, b]$, then, from the fact that $|\eta_k - \eta_{k-1}| \le 2h$, we can write

$$\max_k |d_k| \le \max_k |6f''(\xi_k)(\eta_k - \eta_{k-1})| \le 12Mh. \tag{4.61}$$

This combines with (4.60) to give

$$\max_k |s_k''| \le 6M. \tag{4.61'}$$

Finally, by using this in (4.58) with (4.59), we obtain the **truncation error estimate**

$$|f(\bar{x}) - s(\bar{x})| \le \tfrac{1}{8}7h^2M, \tag{4.62}$$

where M is an upper bound on $|f''(x)|$ for $x \in [a, b]$. We summarize this result as follows:

Let f be a function that is twice continuously differentiable on the interval $[a, b]$, and let $M = \max_{a \leq x \leq b} |f''(x)|$. If $s(x)$ is the cubic spline approximation to f, with uniform knot spacing h, then for any $\bar{x} \in [a, b]$,

$$|f(\bar{x}) - s(\bar{x})| \leq \tfrac{1}{8}7h^2 M.$$

A similar argument shows that when the knots are *not* evenly spaced, but $2h_{min} > h_{max}$, where $0 < h_{min} \leq h_k \leq h_{max}$, $k = 1, 2, \ldots, n - 1$, then

$$|f(\bar{x}) - s(\bar{x})| \leq \tfrac{1}{8}h_{max}^2(2h_{min}/h_{max} - 1)^{-1}M. \tag{4.63}$$

Thus, in general, the truncation error can be made as small as desired by increasing the number of knots, provided that the knots are placed so that h_{max} becomes small, but always $h_{max}/h_{min} \leq \alpha < 2$. These estimates clearly show the nice accuracy provided by spline approximations.

4.3.4   Rounding Errors in Spline Computations

The major source of rounding errors in spline computations is the computation of the solution of the linear system (4.52). Once this system has been solved for the quantities s_k'', $k = 2, 3, \ldots, n - 1$, the only additional errors are those that occur during the evaluation of terms such as $(f_{i+1} - f_i)/h$ in (4.50), and during the evaluation of the expression for $s(x)$, as given by (4.48). The expression (4.48) is a cubic polynomial, and the rounding error in evaluating it has already been analyzed in Section 4.1. The remaining terms may contribute some error, due primarily to subtractive cancellation. However, if h is not too small, such cancellation should not affect the result very much.

To analyze the rounding error involved in computing the solution to the linear system (4.52) by using the algorithm given in subsection 4.3.2, we shall carry out a backward error analysis. That is, we shall show that the computed solution $\tilde{s}_k''$, $k = 2, 3, \ldots, n - 1$, is the *exact* solution to a system of equations that differs only slightly from the given system (4.52). As in the truncation error analysis, we shall assume, for simplicity, that the knots are evenly spaced; i.e., $h_k = h$, $k = 1, 2, \ldots, n - 1$. In this case equations (4.52) can be written, after dividing by h, as

$$s_{i-1}'' + 4s_i'' + s_{i+1}'' = (6/h^2)(f_{i+1} - 2f_i + f_{i-1}), \qquad i = 2, 3, \ldots, n - 1. \tag{4.64}$$

The recursions (4.53), (4.54), and (4.56) are

$$a_2 = 4, \qquad a_k = 4 - 1/a_{k-1}, \qquad k = 3, 4, \ldots, n-1 \quad (4.65a)$$

$$b_2 = \tilde{d}_2, \qquad b_k = \tilde{d}_k - b_{k-1}/a_{k-1}, \qquad k = 3, 4, \ldots, n-1 \quad (4.65b)$$

$$s''_{n-1} = b_{n-1}/a_{n-1}, \qquad s''_k = (b_k - s''_{k+1})/a_k, \qquad k = n-2, \ldots, 2, \quad (4.65c)$$

where $\tilde{d}_k$ denotes the computed right side of (4.64).

As in all **backward error analyses**, we begin with the computed solution, which we denote by $\tilde{s}''_k$, $k = 2, 3, \ldots, n-1$. Observe that, according to (4.65c),

$$\tilde{s}''_{n-1} = \mathrm{fl}[b_{n-1}/a_{n-1}] = (b_{n-1}/a_{n-1})\langle 1 \rangle$$

$$\tilde{s}''_k = \mathrm{fl}[(b_k - \tilde{s}''_{k+1})/a_k] = (b_k \langle 2 \rangle - s''_{k+1}\langle 2 \rangle)/a_k,$$

$$k = n-2, n-3, \ldots, 2.$$

Hence the computed solution satisfies the equations

$$\begin{aligned}
a_2 \tilde{s}''_2 + \tilde{s}''_3 \langle 2 \rangle &= b_2 \langle 2 \rangle \\
a_3 \tilde{s}''_3 + \tilde{s}''_4 \langle 2 \rangle &= b_3 \langle 2 \rangle \\
&\vdots \\
a_{n-1} \tilde{s}''_{n-1} &= b_{n-1} \langle 1 \rangle
\end{aligned} \qquad (4.66)$$

This system is very similar to the system (4.55). Next, we use (4.65) to express the computed values of a_k and b_k as

$$a_k = 4\langle 1 \rangle - \frac{1}{a_{k-1}} \langle 2 \rangle = 4\langle 1 \rangle - \frac{1}{a_{k-1}} \langle 2 \rangle_0 - \frac{1}{a_{k-1}},$$

$$k = 3, 4, \ldots, n-1$$

$$b_k = \tilde{d}_k \langle 1 \rangle - \frac{b_{k-1}}{a_{k-1}} \langle 2 \rangle = \tilde{d}_k (1 + \langle 1 \rangle_0) - \frac{b_{k-1}}{a_{k-1}} (1 + \langle 2 \rangle_0),$$

$$k = 3, 4, \ldots, n-1.$$

To convert the system (4.66) back into the form (4.64), multiply the $(k-1)$st equation in (4.66) by $1/a_{k-1}$ and add it to equation k. This gives the new equation

$$\tilde{s}''_{k-1} + \left(\frac{1}{a_{k-1}} \langle 2 \rangle + a_k \right) \tilde{s}''_k + \tilde{s}''_{k+1} \langle 2 \rangle = \frac{b_{k-1}}{a_{k-1}} \langle 2 \rangle + b_k \langle 2 \rangle.$$

Now, by using the above expressions for a_k and b_k, this can be rewritten as

$$\tilde{s}''_{k-1} + \left(4\langle 1 \rangle + \frac{1}{a_{k-1}} \langle 4 \rangle_0 \right) \tilde{s}''_k + \tilde{s}''_{k+1} \langle 2 \rangle = \tilde{d}_k + \tilde{d}_k \langle 3 \rangle_0 + \frac{b_{k-1}}{a_{k-1}} \langle 6 \rangle_0.$$

To simplify these equations, we write them in the form

> **Perturbed equations** $\tilde{s}''_{k-1} + (4 + \delta_k)\tilde{s}''_k + (1 + \varepsilon_k)\tilde{s}''_{k+1} = \tilde{d}_k + \gamma_k,$ (4.67)

where

$$\delta_k = 4\langle 1 \rangle_0 + \frac{\langle 4 \rangle_0}{a_{k-1}}, \qquad \varepsilon_k = \langle 2 \rangle_0, \qquad \gamma_k = \tilde{d}_k\langle 3 \rangle_0 + \frac{b_{k-1}}{a_{k-1}}\langle 6 \rangle_0. \quad (4.68)$$

Thus the computed solution $\tilde{s}''_2, \ldots, \tilde{s}''_{n-1}$ is the exact solution to the system (4.67). This latter system is the same as the original system except that the coefficients are perturbed by the quantities given by (4.68). To estimate the size of these perturbations, we shall ignore the rounding errors in a_k, b_k, and $\tilde{d}_k$. Then from (4.61) we have $|\tilde{d}_k| \leq 12M$ and (from Exercise 8) $1 \leq a_{k-1} \leq 4$. Furthermore, (4.61') implies that

$$|b_k| = |a_k s''_k + s_{k+1}| \leq |a_k|\,|s''_k| + |s''_{k+1}| \leq 30M.$$

By combining these estimates, we obtain the **bounds**

$$|\delta_k| \leq 8\mu', \qquad |\varepsilon_k| \leq 2\mu', \qquad |\gamma_k| \leq 216M\mu'. \quad (4.69)$$

These simple estimates for the perturbations in the system (4.67) point out the interesting fact that the rounding error in solving the system of equations for s''_k does *not* increase as h decreases. Therefore, as h decreases, the truncation error as estimated by (4.62) can be made arbitrarily small, while the bounds given by (4.69) do not increase. There is, of course, a point at which this argument breaks down, namely, when h becomes so small that subtractive cancellation occurs when computing the right-hand side of (4.64). More precisely, the *real* perturbation in the right side of (4.67) must include the rounding errors involved in the computation

$$\text{fl}[(f_{i+1} - 2f_i + f_{i-1})/h^2] \quad (4.70)$$

in (4.64). This is, in fact, just a discrete approximation to $f''(x_i)$. As will be shown in Chapter 5, the rounding error in computing this quantity *must* increase as h decreases. Nonetheless, this analysis shows that rounding error is *not* a serious problem when computing spline approximations in spite of the rather large amount of computing that is required.

4.3.5   Minimum Curvature Property of Splines

It was stated in subsection 4.3.1 that condition (iv'), which was actually used to define the cubic spline, also implies that the resulting approximation has minimal curvature as defined by the condition (iv). An intuitive interpretation of curvature is "smoothness"; hence equating (iv') with (iv) says, in effect, that the approximation derived in (4.3.2) is the smoothest such approximation. In many applications the smoothness of the spline approximation is the deciding factor in its favor; hence it is worthwhile verifying this property. More precisely, we shall prove the following fact.

Theorem 4.2   Let $s(x)$ be the cubic spline derived in (4.3.2), and let $g(x)$ be any other function which satisfies properties (i) and (ii). Then

$$\int_a^b [g''(t)]^2 \, dt = \int_a^b [s''(t)]^2 \, dt + \int_a^b [g''(t) - s''(t)]^2 \, dt.$$

Proof   A simple manipulation gives

$$\int_a^b [g''(t) - s''(t)]^2 \, dt = \int_a^b [g''(t)]^2 \, dt - \int_a^b [s''(t)]^2 \, dt$$
$$- 2 \int_a^b [g''(t) - s''(t)]s''(t) \, dt. \qquad (4.71)$$

Thus it suffices to show that the last integral in (4.71) is zero. But

$$\int_a^b [g''(t) - s''(t)]s''(t) \, dt = \sum_{i=1}^{n-1} I_i, \qquad (4.72)$$

where

$$I_i = \int_{x_i}^{x_{i+1}} [g''(t) - s''(t)]s''(t) \, dt,$$

and integration by parts gives

$$I_i = \{[g'(t) - s'(t)]s''(t)\}\Big|_{x_i}^{x_{i+1}} - \int_{x_i}^{x_{i+1}} [g'(t) - s'(t)]s'''(t) \, dt.$$

Since $s(x)$ is a cubic polynomial in $[x_i, x_{i+1}]$, $s'''(t)$ is constant in that interval, so

$$\int_{x_i}^{x_{i+1}} [g'(t) - s'(t)]s'''(t) \, dt = s'''(x_i) \int_{x_i}^{x_{i+1}} [g'(t) - s'(t)] \, dt$$
$$= s'''(x_i)[g(t) - s(t)]\Big|_{x_i}^{x_{i+1}} = 0.$$

This last equality is true because $g(t)$ and $s(t)$ are both equal to $f(t)$ when $t = x_i$ or $t = x_{i+1}$. Thus

$$I_i = [g'(x_{i+1}) - s'(x_{i+1})]s''(x_{i+1}) - [g'(x_i) - s'(x_i)]s''(x_i),$$

and $\sum I_i$ is a telescoping series with

$$\sum_{i=1}^{n-1} I_i = [g'(x_n) - s'(x_n)]s''(x_n) - [g'(x_1) - s'(x_1)]s''(x_1) = 0, \qquad (4.73)$$

where we have used property (iv'), i.e., $s''(x_n) = s''(x_1) = 0$. Thus the integral in (4.72) is zero and the theorem is true.

It is clear from this theorem that, unless $g(t) = s(t)$ at *every* t in $[a, b]$,

$$\int_a^b [g''(t)]^2 \, dt > \int_a^b [s''(t)]^2 \, dt, \tag{4.74}$$

so that the property (iv'), which was used to show (4.73), guarantees the smoothness conditions (iv).

EXERCISES

1 For the data given in the following table, write out the system of equations (4.52').

x	-1	1	2	2.5
f	1	0	.5	1

(You should have two equations.) Use the algorithm in Section 4.3.2 to solve this system.

2 Modify the algorithm in Section 4.3.2 to compute the spline approximation when the knots are evenly spaced; i.e., when $h_k = h$, $k = 1, 2, \ldots, n - 1$.

3 Write a computer program to compute the spline approximation to a function $f(x)$ on an interval $[a, b]$ with evenly spaced knots. The input should be a, b, and a table of values for f, i.e., values $f(a)$, $f(x_2)$, $f(x_3)$, $\ldots$, $f(x_n) = f(b)$, where $x_k = a + (k - 1)h$, $k = 1, 2, \ldots, n$, $h = (b - a)/(n - 1)$. The output should be the values s_2'', s_3'', $\ldots$, s_{n-1}''.

4 Rewrite the last part of the algorithm in Section 4.3.2 using a *binary search* method for finding the interval $[x_k, x_{k+1}]$ that contains $\bar{x}$.

5 Suppose you want to approximate the function

$$f(x) = x^2 \sin x + e^{x^2} \cos x$$

on the interval $[-\pi, \pi]$ using a spline function with evenly spaced knots. How many knots must be used so that the truncation error will be less than 10^{-4}?

6 Use an argument similar to that in (4.3.3) to estimate $|s'(x) - f'(x)|$.

7 Prove the estimate (4.63).
[*Hint:* Follow the derivation of (4.62).]

8 Prove that a_k as given by (4.65) satisfy $1 \le a_k \le 4$.
[*Hint:* Suppose $a_k < 1$ for some k. Let k' be the smallest such k and show that, in fact, $a_{k-1} \le \frac{1}{3} < 1$.]

Suggestions for Further Reading

Lawson and Hanson (1974) discuss the least squares approximation problem in great detail, including FORTRAN programs. The theory of spline-interpolation can be found in Ahlberg et al. (1967) and in Schultz (1975).

Chapter 5

DIFFERENTIATION AND
INTEGRATION

5.1 Numerical Differentiation

The mathematical process of differentiation is "easier" than integration in the sense that, of the functions that arise in applications, most can be differentiated analytically while few can be integrated analytically. In the numerical setting, the reverse is true. Numerical differentiation is very sensitive to rounding errors and hence is a more difficult problem than numerical integration. In this section techniques of numerical differentiation are discussed and the rounding error problems associated with them are analyzed.

5.1.1 Subtractive Cancellation

Suppose that the function $f(x)$ is defined and is differentiable for all x with $a < x < b$, and hence that

$$\lim_{h \to 0} \frac{f(x+h) - f(x)}{h} = f'(x)$$

exists for all x, $a < x < b$. An obvious way to approximate $f'(x)$ is to choose a small value for h and use the formula

| Simple approximation to $f'(x)$ | $$f'(x) \cong \frac{f(x+h)-f(x)}{h}.$$ | (5.1) |

As was pointed out in Section 2.4, if h is very small, such an attempt will lead to subtractive cancellation and large rounding error. In fact, Table 5.1, which was computed using the Univac 1108 with the function $f(x) = e^x$, illustrates this point. [Recall that for this function $f'(0) = 1$.]

TABLE 5.1

h	$(f(h)-f(0))/h$	h	$(f(h)-f(0))/h$
.1	$.10517092 \times 10^1$	.00001	$.99986792 \times 10^0$
.01	$.10050163 \times 10^1$	.000001	$.99837780 \times 10^0$
.001	$.10004937 \times 10^1$	.0000001	$.10430813 \times 10^1$
.0001	$.10000169 \times 10^1$	.00000001	0

5.1.2 More Accurate Formulas

The basic difficulty with using (5.1) is that in order for this to be a good approximation, the value of h must be so small that cancellation occurs. The precise meaning of the statement that (5.1) is a " good approximation" is that the **truncation error** is small. Thus, one attempt to avoid this difficulty is to find approximations that have small truncation error for values of h that do not cause cancellation. To this end, consider a polynomial p that approximates f in some sense. In Chapter 4 many such polynomials were given, such as the linear interpolating polynomial:

$$p(x) = f(x_1)\frac{x-x_2}{x_1-x_2} + f(x_2)\frac{x-x_1}{x_2-x_1}. \tag{5.2}$$

Here x_1 and x_2 are certain values in the interval where f is defined. A reasonable approximation to $f'(x)$ is then given by differentiating (5.2):

$$p'(x) = f(x_1)\frac{1}{x_1-x_2} + f(x_2)\frac{1}{x_2-x_1} = \frac{f(x_2)-f(x_1)}{x_2-x_1}. \tag{5.3}$$

Furthermore, the truncation error for (5.2) was also derived and can be expressed by the equation

$$f(x) = p(x) + \tfrac{1}{2}f''(\xi(x))(x-x_1)(x-x_2).$$

We note explicitly here that the point ξ at which f'' must be evaluated depends on x, as well as on x_1 and x_2. Differentiating this formula gives

$$f'(x) = p'(x) + \tfrac{1}{2}f''(\xi(x))[(x - x_1) + (x - x_2)]$$
$$+ \tfrac{1}{2}(x - x_1)(x - x_2) \, df''(\xi(x))/dx.$$

Even if $f''(\xi(x))$ is differentiable with respect to x so that this last term is defined, the term cannot be evaluated because the dependence of ξ on x is not known. If, however, $x = x_1$ *or* $x = x_2$, then this last term is zero no matter what $df''(\xi(x))/dx$ is. Hence, we have

$$f'(x_1) = p'(x_1) + \tfrac{1}{2}f''(\xi(x_1))(x_1 - x_2)$$
$$= [f(x_2) - f(x_1)]/(x_2 - x_1) + \tfrac{1}{2}f''(\xi_1)(x_1 - x_2).$$

Thus the **truncation error in (5.3)** as an approximation to $f'(x_1)$ is

$$\tfrac{1}{2}f''(\xi_1)(x_1 - x_2), \tag{5.4}$$

where ξ_1 is some point between x_1 and x_2.

We summarize the formulas just derived as follows:

If $f(x)$ is a function that is defined and can be differentiated three times for all x with $a < x < b$, then for any x_1 and x_2 in (a, b), there is a ξ between them such that

$$f'(x_1) = [f(x_2) - f(x_1)]/(x_2 - x_1) + \tfrac{1}{2}f''(\xi)(x_1 - x_2).$$

Note that (5.3) with $h = x_2 - x_1$ and $x = x_1$ is just (5.1) and hence we have succeeded in deriving an explicit expression for the truncation error in (5.1).

If $f(x)$ is **approximated by a second degree interpolating polynomial** and then differentiated, we find from (4.14a,b) that

$$f'(x) = p'(x) + \tfrac{1}{6}f'''(\xi(x))[(x - x_2)(x - x_3) + (x - x_1)(x - x_3)$$
$$+ (x - x_1)(x - x_2)]$$
$$+ [df'''(\xi(x))/dx]\tfrac{1}{6}(x - x_1)(x - x_2)(x - x_3). \tag{5.5}$$

Again we ignore the possibility that $f'''(\xi(x))$ may not be differentiable with respect to x; that is, this last term may not be defined for certain functions f. In any case by differentiating (4.14a) we obtain

$$p'(x) = f(x_1)\frac{[(x - x_2) + (x - x_3)]}{(x_1 - x_2)(x_1 - x_3)} + f(x_2)\frac{[(x - x_1) + (x - x_3)]}{(x_2 - x_1)(x_2 - x_3)}$$
$$+ f(x_3)\frac{[(x - x_1) + (x - x_2)]}{(x_3 - x_1)(x_3 - x_2)}. \tag{5.6}$$

Once again, in order to obtain a reasonable expression for the truncation error in (5.5) we set x equal to one of the data points x_1, x_2, or x_3. An examination of the factor $[(x - x_2)(x - x_3) + (x - x_1)(x - x_3) + (x - x_1)(x - x_2)]$ that occurs in (5.5) shows that it is minimized when $x = x_2$, and thus we write

$$f'(x_2) = p'(x_2) + \tfrac{1}{6}f'''(\xi_2)[(x_2 - x_1)(x_2 - x_3)]$$

$$p'(x_2) = \frac{f(x_1)(x_2 - x_3)}{(x_1 - x_2)(x_1 - x_3)}$$

| Approximation to $f'(x_2)$ |

$$+ \frac{f(x_2)[(x_2 - x_1) + (x_2 - x_3)]}{(x_2 - x_1)(x_2 - x_3)} \qquad (5.7)$$

$$+ \frac{f(x_3)(x_2 - x_1)}{(x_3 - x_1)(x_3 - x_2)}.$$

In most applications, the points x_1, x_2, x_3 can be chosen with some freedom, or else these points are fixed but **evenly spaced**. Thus it is not too restrictive to assume that

$$x_2 = x_1 + h, \qquad x_3 = x_2 + h, \qquad h > 0.$$

in which case (5.7) simplifies to

$$f'(x_2) = p'(x_2) + \tfrac{1}{6}h^2 f'''(\xi_2)$$

$$p'(x_2) = \frac{f(x_1)(-h)}{(-h)(-2h)} + \frac{f(x_2)[h + (-h)]}{h(-h)} + \frac{f(x_3)(h)}{(2h)(h)}$$

$$= [f(x_3) - f(x_1)]/2h.$$

That is, we have the following result:

If $f(x)$ is defined and four times differentiable for $a < x < b$, then for any x and any $h > 0$ such that $a < x - h < x + h < b$, the approximation

$$f'(x) \cong [f(x + h) - f(x - h)]/2h \qquad (5.8)$$

has truncation error

$$\tfrac{1}{6}h^2 f'''(\xi) \qquad (5.9)$$

for some ξ between $x - h$ and $x + h$.

If f is a "smooth" function—so that neither $f''(\xi)$ nor $f'''(\xi)$ is large, then for $h < 1$ the truncation error (5.9) will be considerably smaller than the corresponding truncation error (5.4) for the formula (5.1). Hence a larger value of h can be used in (5.8) to obtain a reasonably good approximation.

In exactly the same manner other numerical differentiation formulas and their corresponding truncation errors can be derived. For example, if $x = x_1$ is used in (5.5), we obtain the result

---

If f is four times differentiable in (a, b), then for any x and h, with $a < x < x + 2h < b$, there is a value ξ with $x - h < \xi < x + h$ such that

$$f'(x) = (1/2h)[-3f(x) + 4f(x + h) - f(x + 2h)] + \tfrac{1}{3}h^2 f'''(\xi).$$
$$(5.10)$$

The use of a third degree interpolating polynomial gives

If f is six times differentiable in (a, b), then for any x and h, with $a < x - 2h < x + 2h < b$, there is a ξ with $x - 2h < \xi < x + 2h$ such that

$$f'(x) = (1/12h)[f(x - 2h) - 8f(x - h) + 8f(x + h)$$
$$- f(x + 2h)] + \tfrac{1}{30}h^4 f^{(5)}(\xi).$$
$$(5.11)$$

5.1.3 Rounding Error

The instability problem that is illustrated by the results in Table 5.1 for the numerical differentiation formula (5.1) is common to all formulas derived by the method presented here. In fact, any such formula [for example, (5.8), (5.10), (5.11)] has the form

---

General form of differentiation formulas

$$f'(x_k) \cong p'(x_k) = (1/h)(a_1 f(x_1)$$
$$+ a_2 f(x_2) + \cdots + a_n f(x_n)),\qquad (5.12)$$

---

where $a_1, a_2, \ldots, a_n$ are fixed numbers that define the formula, and $x_1, x_2, \ldots, x_n$ are points with $x_{i+1} - x_i = h, i = 1, 2, \ldots, n - 1$. The truncation error has the form

---

General form of the truncation error

$$f^{(j)}(\xi)h^p/\alpha, \qquad j \geq 1 \qquad (5.13)$$

---

for some integer p. Now, for the special function $f(x) \equiv 1$, the truncation error (5.13) is zero since $f'(x) = f''(x) = \cdots = 0$. Hence (5.12) must be an *equation* for this particular function f. That is, if $f(x) = 1$ is substituted into (5.12), we have

$$0 = f'(x_k) = p'(x_k) = (1/h)(a_1 + a_2 + \cdots + a_n).$$

Hence, in any numerical differentiation formula, such as (5.12), the coefficients must sum to zero. If f is continuous, then this is exactly one of the **unstable calculations** that was discussed in Section (2.5.5).

To illustrate this point further, consider formula (5.11) applied to the function $f(x) = \sin x$ with $x = \pi/4$ and various values of h. Table 5.2 gives computed values of the approximations so obtained.

TABLE 5.2   *Values of $f'(\pi/4)$, for $f(x) = \sin x$, Computed from (5.11) (Correct Digits Are Underlined)*

h	Computed values[a]	h	Computed values[a]
.1	.70710435	.0001	.70707251
.01	.70710647	.00001	.70699800
.001	.70709921	.000001	.70470075

[a] Exact value .707107.

As before, we see that the rounding error begins to increase as h decreases.

Formula (2.48) gives an estimate that shows the magnitude of the rounding error in (5.12) when h is very small. In fact, since we are assuming that f is differentiable, the analysis that gave (2.48) can be made more precise. Let $\tilde{p}'$ be the computed value of the formula in (5.12), where we assume rounding errors occur only during the multiplications $a_k f(x_n)$. That is

$$\tilde{p}' = (1/h)\{a_1 f(x_1)\langle 1\rangle + a_2 f(x_2)\langle 1\rangle + \cdots + a_n f(x_n)\langle 1\rangle\}$$
$$= (1/h)\{(a_1 f(x_1) + a_2 f(x_2) + \cdots + a_n f(x_n)) + (a_1 f(x_1)\langle 1\rangle_0$$
$$+ \cdots + a_n f(x_n)\langle 1\rangle_0)\}.$$

Thus,

$$\tilde{p}' - p' = (1/h)(a_1 f(x_1)\langle 1\rangle_0 + a_2 f(x_2)\langle 1\rangle_0 + \cdots + a_n f(x_n)\langle 1\rangle_0).$$

Now, if $f'(x)$ is assumed to be continuous, then by the mean value theorem (see the Appendix), we have

$$f(x_k) = f(x_1) + (x_k - x_1)f'(\xi_k)$$

for some ξ_k between x_1 and x_k. Thus, because $x_k - x_1 = (k-1)h$, we have

$$\tilde{p}' - p' = (1/h)[a_1 f(x_1)\langle 1\rangle_0 + a_2(f(x_1) + hf'(\xi_2))\langle 1\rangle_0$$
$$+ a_3(f(x_1) + 2hf'(\xi_3))\langle 1\rangle_0$$
$$+ \cdots + a_n(f(x_1) + (n-1)hf'(\xi_n))\langle 1\rangle_0]$$
$$= (1/h)(a_1\langle 1\rangle_0 + a_2\langle 1\rangle_0 + \cdots + a_n\langle 1\rangle_0)f(x_1)$$
$$+ a_2 f'(\xi_2)\langle 1\rangle_0 + 2a_3 f'(\xi_3)\langle 1\rangle_0$$
$$+ \cdots + (n-1)a_n f'(\xi_n)\langle 1\rangle_0.$$

Since $p'(x) \cong f'(x)$, we can write this as a **relative error estimate** in the form

$$\frac{\tilde{p}' - p'}{f'(x_1)} = \frac{(a_1\langle 1\rangle_0 + a_2\langle 1\rangle_0 + \cdots + a_n\langle 1\rangle_0)}{h}\frac{f(x_1)}{f'(x_1)}$$
$$+ \frac{a_2 f'(\xi_2)\langle 1\rangle_0 + 2a_3 f'(\xi_3)\langle 1\rangle_0 + \cdots + (n-1)a_n f'(\xi_n)\langle 1\rangle_0}{f'(x_1)}.$$

This gives the **forward error bound**

$$\left|\frac{\tilde{p}' - p'}{f'(x_1)}\right| \le (|a_1| + |a_2| + \cdots + |a_n|)\left|\frac{f(x_1)}{f'(x_1)}\right|\left|\frac{\mu'}{h}\right|$$
$$+ \frac{|a_2|\,|f'(\xi_2)| + \cdots + (n-1)|a_n|\,|f'(\xi_n)|}{|f'(x_1)|}\mu'.$$

Since for small h we have $f'(\xi_k) \cong f'(x_1)$, this bound can be written less precisely but more simply as:

$$\left|\frac{\tilde{p}' - p'}{f'(x_1)}\right| \le (|a_1| + |a_2| + \cdots + |a_n|)\left[\left|\frac{f(x_1)}{f'(x_1)}\right|\frac{1}{h} + (n-1)\right]\mu'. \quad (5.14)$$

This estimate clearly shows the growth of error that accompanies a decrease in h. Notice also that if $|f'(x_1)|$ is very small compared to $|f(x_1)|$, then the rounding error may be large. This fact can also be deduced directly from the formula (5.12). That is, if $f(x)$ is large, and $f'(x)$ is small, then (5.12) determines a small value $(p'(x_k))$ by summing together large values. Thus subtractive cancellation must occur.

The truncation error and the rounding error, as functions of h, can be described as in Fig. 5.1. The optimum value to use for h is when the sum of these errors is smallest. Unfortunately in practice it is impossible even to estimate this optimum value. Sometimes it is necessary to compute several approximations with different values of h or even with different formulas. A comparison of these results may provide an indication of their accuracy.

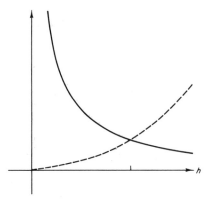

Fig. 5.1   Rounding error ——, truncation error – – –.

5.1.4   Extrapolation to the Limit

As the previous discussion shows, it is not possible in general to reduce the total error in a numerical differentiation formula by simply decreasing the size of h. It is possible, however, to use a more subtle technique, called **extrapolation to the limit**, to improve the accuracy without having to resort to a more complicated formula. Suppose we denote by $D(h)$ an approximate value for the derivative of some function $f(x)$ at $x = \bar{x}$. We think of $D(h)$ as having been obtained from a particular formula, such as (5.8), with exact arithmetic and spacing $h > 0$. If we could find $D(h)$ for any value of h, then we could draw a graph such as in Fig. 5.2. The point at which this curve intersects the vertical axis is the exact value of $f'(\bar{x})$. As has been shown, however, it is not possible to *compute* values of $D(h)$ for small h because of instability caused by rounding error. Suppose instead that values for $D(h_1)$ and $D(h_2)$ are computed, where h_1 and h_2 are two values that are not so small as to cause serious cancellation error. Then we may *extrapolate* these

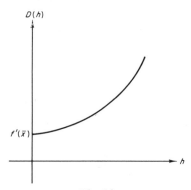

Fig. 5.2

values to obtain an approximation to $D(0)$. Generally, the truncation error analysis gives some information about the curve $D(h)$ that can be used in this extrapolation. For example, the formulas (5.8), (5.9) can be written as

$$[f(\bar{x} + h) - f(\bar{x} - h)]/2h = f'(\bar{x}) - \tfrac{1}{6}h^2 f'''(\xi(h)).$$

Thus if the left side of this equation is denoted by $D(h)$, then $D(h) = a + bh^2$ with $a = f'(\bar{x})$ and $b = -\tfrac{1}{6}f'''(\xi(h))$. If b were independent of h, the interpolation formula (4.17b) would give

Extrapolated value	$f'(\bar{x}) = D(0) = a = \dfrac{h_2^2 D(h_1) - h_1^2 D(h_2)}{h_2^2 - h_1^2}.$	(5.15)

But b *does* depend on h so that (5.15) is not an exact formula. However, it is reasonable to expect that (5.15) will give a better approximation to $f'(\bar{x})$ than do either $D(h_1)$ or $D(h_2)$. To simplify these formulas, choose a value for h and let $h_1 = h$, $h_2 = \tfrac{1}{2}h$. Then (5.15) becomes

$$f'(\bar{x}) \cong \frac{(\tfrac{1}{2}h)^2 D(h) - h^2 D(\tfrac{1}{2}h)}{(\tfrac{1}{2}h)^2 - h^2} = \frac{4D(\tfrac{1}{2}h) - D(h)}{3}.$$

If we define

Better approximation to $f'(\bar{x})$	$D_1(h) = \tfrac{1}{3}[4D(\tfrac{1}{2}h) - D(h)],$	(5.16)

then $D_1(h)$ may be expected to be a better approximation to $f'(\bar{x})$ than is $D(h)$ or $D(\tfrac{1}{2}h)$. It is important to realize that if h is *too* small, then the computed values of $D(h)$ and $D(\tfrac{1}{2}h)$ will be inaccurate. In this case $D_1(h)$ will, of course, also be inaccurate. Table 5.3 however, shows the effectiveness of this technique. Table 5.3 shows that if h is not too small, then $D_1(h)$ is a better approximation than $D(h)$. Nevertheless, it is still necessary to examine the rounding error in order to avoid subtractive cancellation.

TABLE 5.3 *Approximations to d/dx sin x at x = π/4 (Accurate Digits Underlined)*

h	$D(h)^a$	$D_1(h)^a$
.4	.68840087	.70706921
.2	.70240212	.70710437
.1	.70592880	.70710659
.01	.70709474	.70710617
.001	.70710107	.70709611
.0001	.70709734	.70704769

[a] Exact value .70710678.

The procedure used here to derive the extrapolated value $D_1(h)$ given by (5.16) can also be applied to other approximation formulas provided an appropriate interpolation formula is used. For example, if $G(h)$ denotes the approximation to $f'(\bar{x})$ as given by formula (5.11), then $G(h) \cong a + bh^4$ for some constants a, b, and hence (see Exercise 11) we obtain the extrapolated value

$$G_1(h) = \tfrac{1}{15}[16G(\tfrac{1}{2}h) - G(h)]. \tag{5.17}$$

Extrapolation to the limit is especially important in connection with numerical integration. More will be said about this in Section 5.4.

EXERCISES

1 For the function $f(x) = e^x$ and values of h as given in Table 5.1, compute the approximations to $f'(0)$ using formula (5.8) instead of (5.1).

2 Repeat Exercise 1 using formula (5.11).

3 Derive formulas (5.10) and (5.11) by differentiating appropriate interpolation formulas as given in Chapter 4.

4 Let $f(x)$ be a function whose derivative $f'(x)$ is to be approximated for values of x between 0 and 1, with truncation error less than 10^{-8}. Suppose it is known that $|f'''(x)| \le 1$ for all x in $[0, 1]$. What value of h is needed in (5.8) for this approximation?

5 Show that the instability in formula (5.12) cannot be eliminated by first dividing each a_i by h and then computing the sum of the products $(a_i/h)f(x_i)$.

6 Verify that all of the numerical differentiation formulas given in Section 5.1 have the property that the coefficients sum to zero.

7 Use the relative error estimate (5.14) to estimate the rounding error in evaluating formulas (5.10) and (5.11).

8 Derive an estimate for the rounding error in formula (5.12) that *includes* the errors caused by the arithmetic operations involved in the formula.

9 Use the first two values given in Table 5.1 to determine an extrapolated value for $f'(0)$. Repeat, using the second two and then the third two.

10 According to (5.14) the relative rounding error in the computed value of $D(h)$ is approximately

$$\left(\frac{1}{h}\frac{f(x_1)}{f'(x_1)} + (n - 1)\right)\mu'.$$

Using this fact, estimate the total rounding error in the computed value of $D_1(h)$.

11 Derive formula (5.17) by determining the value for the coefficient a such that $G(h) = a + bh^4$ and $G(\tfrac{1}{2}h) = a + b(\tfrac{1}{2}h)^4$, where $G(h)$ is the approximation to $f'(\bar{x})$ given by formula (5.11).

12 From formulas (5.8) and (5.9) we can write

$$D(h) = f'(\bar{x}) - \tfrac{1}{6}h^2 f'''(\xi(h))$$

$$D(h/2) = f'(\bar{x}) - \tfrac{1}{6}(\tfrac{1}{2}h)^2 f'''(\xi(\tfrac{1}{2}h)).$$

If we assume that $f'''(\xi(h)) = f'''(\xi(\tfrac{1}{2}h))$, then we have, in fact, two equations in the two unknowns $f'(\bar{x})$ and $f'''(\xi(h))$, which can be solved for $f'(\bar{x})$. Show that the solution is just the right-hand side of (5.16).

5.2 Numerical Integration—Newton–Cotes Formulas

Because of the difficulties inherent in the *mathematical* process of integration, approximate methods for integration are frequently required. Fortunately, numerical integration is not plagued with the instability problems associated with numerical differentiation. In this section, we derive some of the most commonly used integration formulas, and examine their stability.

5.2.1 Trapezoidal Rule

In this section we shall study the problem of determining an approximation to the value of the **definite integral**

$$\int_a^b f(x)\,dx, \tag{5.18}$$

where f is some continuous function whose integral is not known. We assume that f can be evaluated at any x in the interval $[a, b]$.

The classical method for solving this problem is to replace the function f with a simpler function p whose integral can be easily found. This replacement introduces a truncation error that must be estimated. The use of polynomials as these simpler functions leads to a variety of numerical integration formulas, called the **Newton–Cotes formulas**. The most obvious such formula results from letting p be the linear polynomial that interpolates f at a and b. That is, we use formula (4.12) to write

$$f(x) \approx p(x) = f(a)(x - a)/(b - a) + f(b)(x - b)/(a - b).$$

Then (5.18) can be approximated by

$$\int_a^b p(x)\,dx = \frac{f(a)}{b - a}\int_a^b (x - a)\,dx + \frac{f(b)}{a - b}\int_a^b (x - b)\,dx$$

$$= \frac{f(a)}{b - a}\frac{(x - a)^2}{2}\Big|_a^b + \frac{f(b)}{a - b}\frac{(x - b)^2}{2}\Big|_a^b$$

$$= f(a)\tfrac{1}{2}(b - a) + f(b)\tfrac{1}{2}(b - a).$$

Thus we have

$$\int_a^b f(x)\, dx \approx \frac{b-a}{2}(f(a)+f(b)).$$

This formula is just a simple form of the **trapezoidal rule** that was derived in Section 1.1.

5.2.2 Truncation Error for Trapezoidal Rule

As the geometric interpretation of the trapezoidal rule clearly shows, the truncation error may be quite large, especially if $b-a$ is large. To show this same fact analytically, let f be twice continuously differentiable on $[a, b]$. Then by (4.13) the truncation error in the interpolating polynomial $p(x)$ is given by $\frac{1}{2}f''(\xi)(x-a)(x-b)$. That is,

$$f(x) = p(x) + \tfrac{1}{2}f''(\xi(x))(x-a)(x-b) \qquad \text{for all} \quad x,$$

hence

$$\int_a^b f(x)\, dx = \int_a^b p(x)\, dx + \int_a^b \frac{f''(\xi(x))}{2}(x-a)(x-b)\, dx.$$

Thus the trapezoidal approximation can be expressed exactly by

$$\int_a^b f(x)\, dx = \frac{b-a}{2}(f(a)+f(b)) + E_{\mathrm{T}}, \tag{5.19}$$

where the **truncation error in the integration formula** is

$$E_{\mathrm{T}} = \frac{1}{2}\int_a^b f''(\xi(x))(x-a)(x-b)\, dx. \tag{5.20}$$

This truncation error is complicated by the fact that the point ξ at which f'' is to be evaluated is a function of x, and hence the integral cannot be evaluated directly. However, the mean value theorem for integrals (see the Appendix) can be used to simplify (5.20). We apply this theorem by observing that the function $g(x)=(x-a)(x-b)$ is always negative for x in $[a, b]$; hence if $h(x)=\frac{1}{2}f''(\xi(x))$, then the theorem states that for some η in $[a, b]$,

$$E_{\mathrm{T}} = \int_a^b g(x)h(x)\, dx = h(\eta)\int_a^b g(x)\, dx = \frac{f''(\xi(\eta))}{2}\int_a^b (x-a)(x-b)\, dx.$$

This simplifies to

$$E_{\mathrm{T}} = -\tfrac{1}{12}f''(\xi)(b-a)^3 \qquad \text{for some} \quad \xi \quad \text{in} \quad [a, b]. \tag{5.21}$$

These results can be summarized as follows:

> If $f(x)$ is twice continuously differentiable in the interval (a, b), then the **simple trapezoidal rule**
>
> $$\tfrac{1}{2}(b - a)[f(a) + f(b)]$$
>
> approximates the integral (5.18) with **truncation error**
>
> $$-\tfrac{1}{12}f''(\xi)(b - a)^3,$$
>
> where ξ is some value between a and b.

5.2.3 Composite Trapezoidal Rule

The formula (5.21) shows analytically that the truncation error may be quite large if the interval $[a, b]$ is large. To avoid this difficulty, we use the **additivity property** of the integral:

$$\int_a^b f(x)\,dx = \int_{x_0}^{x_1} f(x)\,dx + \int_{x_1}^{x_2} f(x)\,dx + \cdots + \int_{x_{n-1}}^{x_n} f(x)\,dx, \quad (5.22)$$

where $a = x_0 < x_1 < x_2 < \cdots < x_{n-1} < x_n = b$. In fact, for simplicity, let us assume that the points $x_0, x_1, \ldots, x_n$ are evenly spaced:

$$x_k = x_0 + kh, \qquad k = 0, 1, 2, \ldots, n, \quad h = (b - a)/n.$$

If the trapezoid rule (5.19) is applied to *each* of the integrals on the right-hand side of (5.22), then we have

$$\int_a^b f(x)\,dx = \frac{x_1 - x_0}{2}\left(f(x_0) + f(x_1)\right) + E_{T_1}$$

$$+ \frac{x_2 - x_1}{2}\left(f(x_1) + f(x_2)\right) + E_{T_2}$$

$$+ \cdots + \frac{x_n - x_{n-1}}{2}\left(f(x_{n-1}) + f(x_n)\right) + E_{T_n}$$

$$= \tfrac{1}{2}h[f(x_0) + f(x_1) + f(x_1) + f(x_2) + \cdots + f(x_{n-1}) + f(x_n)]$$

$$+ E_{T_1} + E_{T_2} + \cdots + E_{T_n}, \qquad (5.23)$$

where $x_k - x_{k-1} = h$ was used to obtain the last expression. The truncation errors E_{T_k} are given by (5.21) as

$$E_{T_k} = -\tfrac{1}{12}f''(\xi_k)(x_k - x_{k-1})^3 = -\tfrac{1}{12}f''(\xi_k)h^3.$$

Combining terms in (5.23) we find that

$$\int_a^b f(x)\,dx = h[\tfrac{1}{2}f(x_0) + f(x_1) + f(x_2) + \cdots + f(x_{n-1}) + \tfrac{1}{2}f(x_n)] + E_T^{(c)},$$

where the **total (composite) truncation error** $E_T^{(c)}$ is

$$E_T^{(c)} = E_{T_1} + E_{T_2} + \cdots + E_{T_n}$$
$$= [-\tfrac{1}{12}f''(\xi_1)h^3] + [-\tfrac{1}{12}f''(\xi_2)h^3] + \cdots + [-\tfrac{1}{12}f''(\xi_{n-1})h^3]$$
$$= -\tfrac{1}{12}h^3(f''(\xi_1) + f''(\xi_2) + \cdots + f''(\xi_n)). \tag{5.24}$$

This expression for the composite truncation error can be further simplified by noting that

$$(1/n)[f''(\xi_1) + f''(\xi_2) + \cdots + f''(\xi_n)]$$

is just the *average* of the n function values $f''(\xi_1), f''(\xi_2), \ldots, f''(\xi_n)$. As such, it lies between the smallest and the largest of these values. But since f'' was assumed to be a continuous function, it must take this value at some η between a and b; that is, $f''(\eta) = (1/n)[f''(\xi_1) + f''(\xi_2) + \cdots + f''(\xi_n)]$. This expression substituted into (5.24) gives

$$E_T^{(c)} = -\tfrac{1}{12}h^3 n f''(\eta) = -\tfrac{1}{12}h^2(b-a)f''(\eta),$$

where $b - a = hn$, and η lies somewhere in the interval $[a, b]$. Thus, we have derived the **composite trapezoidal rule**:

If f is twice continuously differentiable in the interval (a, b), then

$$\int_a^b f(x)\,dx = h\left[\frac{1}{2}f(x_0) + f(x_1) + \cdots + f(x_{n-1}) + \frac{1}{2}f(x_n)\right] + E_T,$$
$$\tag{5.25a}$$

where the **truncation error** E_T is

$$E_T = -\tfrac{1}{12}h^2(b-a)f''(\eta) \tag{5.25b}$$

for some value η between a and b.

Note here that a *bound* on the truncation error can be determined provided a bound on $|f''(\eta)|$ for all η in $[a, b]$ is known.

5.2.4   More Accurate Formulas

In a similar manner, other numerical integration formulas and their truncation errors can be derived. As another example, we approximate f by the second degree polynomial (4.14) that interpolates f at a, b, and $c = \tfrac{1}{2}(a + b)$.

This produces the well-known **Simpson's rule**:

$$\int_a^b f(x)\, dx$$

$$= \int_a^b \left\{ f(a)\frac{(x-b)(x-c)}{(a-b)(a-c)} + f(b)\frac{(x-a)(x-c)}{(b-a)(b-c)} + f(c)\frac{(x-a)(x-b)}{(c-a)(c-b)} \right\} dx$$

$$+ E_S.$$

After some manipulation, it becomes

Simpson's rule (simple form)	$$\int_a^b f(x)\, dx = \frac{b-a}{6}[f(a) + 4f(c) + f(b)] + E_S$$	(5.26)

where

Truncation error for (5.26)	$$E_S = \int_a^b \frac{f'''(\xi(x))}{6}(x-a)(x-b)(x-c)\, dx.$$	(5.27)

Unfortunately, the mean-value theorem, which was used to simplify the truncation error formula for the trapezoidal rule, cannot be used here because the function $(x-a)(x-b)(x-c)$ *does* change sign in the interval $[a, b]$. A more complicated argument must be used [see, for example, Ralston (1965)] to show that (5.27) can be written as

Simplified (5.27)	$$E_S = -[f^{(4)}(\xi)/2880](b-a)^5.$$	(5.28)

As with the trapezoidal rule, a **composite formula** is obtaining by subdividing the interval $[a, b]$ into subintervals $[x_k, x_{k+1}]$, where $x_k = a + kh$, $h = (b-a)/n$, and then applying (5.26) to each *double* interval $[x_k, x_{k+2}]$:

$$\int_{x_k}^{x_{k+2}} f(x)\, dx = \frac{x_{k+2} - x_k}{6}[f(x_k) + 4f(x_{k+1}) + f(x_{k+2})] + E_S^{(k)}$$

$$= \tfrac{1}{3}h(f(x_k) + 4f(x_{k+1}) + f(x_{k+2})) + E_S^{(k)}.$$

By summing these, we obtain

Composite Simpson's rule	$$\int_a^b f(x)\, dx = \frac{h}{3}[f(x_0) + 4f(x_1) + 2f(x_2) \\ + 4f(x_3) + \cdots + 4f(x_{n-1}) \\ + f(x_n)] + E_S^{(c)},$$	(5.29a)

where

Truncation error for composite Simpson's rule

$$E_S^{(c)} = (h^4/180)f^{(4)}(\xi)(b - a). \tag{5.29b}$$

Notice that since (5.29a) is obtained by integrating over a double subinterval, the interval $[a, b]$ must be divided into an even number of subintervals $[x_k, x_{k+1}]$; that is, n in (5.29) *must be even*. We summarize this as follows:

> Let f have four continuous derivatives in the interval (a, b), and let $h = (b - a)/n$, where n is even. Then (5.29a) holds, with $E_S^{(c)}$ given by (5.29b) for some value ξ between a and b.

Examples of other numerical integration formulas, with their truncation errors, are given in Table 5.4. In each of these truncation error expressions, the derivative that occurs is assumed to exist and be continuous, and the value ξ is some (unknown) point in the interval of integration. Composite formulas can be derived from these exactly as was done above for the simple trapezoidal and Simpson formulas. Notice that the more complicated the formula, the smaller, in general, is the truncation error.

TABLE 5.4 *Some Newton–Cotes Formulas*

$$\int_{x_0}^{x_3} f(x)\, dx \approx \frac{3h}{8} (f(x_0) + 3f(x_1) + 3f(x_2) + f(x_3)),$$
$$E_T = (-3h^5/80)f^{(4)}(\xi)$$

$$\int_{x_0}^{x_4} f(x)\, dx \approx \frac{2h}{45} (7f(x_0) + 32f(x_1) + 12f(x_2) + 32f(x_3) + 7f(x_4)),$$
$$E_T = (-8h^7/945)f^{(6)}(\xi)$$

$$\int_{x_0}^{x_5} f(x)\, dx \approx \frac{5h}{288} (19f(x_0) + 75f(x_1) + 50f(x_2) + 50f(x_3) + 75f(x_4) + 19f(x_5)),$$
$$E_T = (-275h^7/12096)f^{(6)}(\xi)$$

To illustrate the use of these formulas, consider the integral

$$\int_0^1 e^x \, dx$$

whose exact value is $e - 1 = 1.7182818\dots$. The **trapezoidal rule** (5.25) with

$h = .5$ gives the approximation

$$.5(\tfrac{1}{2}e^0 + e^{.5} + \tfrac{1}{2}e^1) = 1.753925\ldots.$$

Simpson's rule, also with $h = .5$, produces the value

$$(.5/3)(e^0 + 4e^{.5} + e^1) = 1.71885\ldots.$$

Table 5.5 gives other approximations to the value of this integral, obtained from the trapezoidal rule, Simpson's rule, and formula 2 of Table 5.4. For the function $f(x) = e^x$, all of the derivative $f'(x)$, $f''(x)$, etc., are bounded by $e = 2.718\ldots$ for $0 \leq x \leq 1$. Hence, the results in Table 5.5 reflect the fact that the truncation error decreases as h decreases and as the power of h in the truncation formula increases.

TABLE 5.5   *Approximations to $\int_0^1 e^x \, dx =$* 1.7182818...

h	Trapezoid	Simpson	Formula 2 Table 5.4
.25	1.7272219	1.7183188	1.7408548
.125	1.7205186	1.7182841	1.7182818
.0625	1.7188411	1.7182820	1.7182818
.03125	1.7184216	1.7182818	1.7182818

It is important to realize, however, that sometimes a very innocuous-looking function can have small first and second derivatives but *very* large higher derivatives. For such functions the trapezoidal and Simpson's rules will give better results than the higher order formulas.

5.2.5   Rounding Error in Integration

To illustrate the effect of rounding error on the computational evaluation of numerical integration formulas, as derived above, we shall consider the composite trapezoidal rule (5.25):

$$\int_a^b f(x) \, dx \approx S \equiv h(\tfrac{1}{2}f_0 + f_1 + f_2 + \cdots + f_{n-1} + \tfrac{1}{2}f_n),$$

where for simplicity we have used f_k to denote $f(x_k)$. The algorithm for evaluating S is a simple modification of the summation algorithm (3.1) analyzed in Chapter 3:

Composite Trapezoidal Algorithm

$$\left[\begin{array}{l}\text{The input to the algorithm consists of the values } a, b, \text{ and } n, \text{ as}\\ \text{well as a subroutine for evaluating } f(x) \text{ for any } x \text{ in } (a, b).\end{array}\right]$$

1   $h \leftarrow (b - a)/n$
2   $x \leftarrow a$
3   Evaluate f at x
4   $S \leftarrow f(x)/2$
5   For $k = 1, 2, \ldots, n - 1$
   5.1 $x \leftarrow x + h$
   5.2   Evaluate f at x (5.30)
   5.3 $S \leftarrow S + f(x)$
6   $x \leftarrow x + h$
7   Evaluate f at x
8   $x \leftarrow h(S + f(x)/2)$
9   Output $\{S\}$

Let $\hat{f}_k$ be the computed value of $f(x_k)$, with relative error ε_k. In a binary computer the divisions by 2 in Steps 4 and 8 entail no rounding error, only an adjustment of the exponent; hence we will ignore the errors, if any, caused by this division. Then, the **computed approximation** is given by

$$\begin{aligned}\hat{S} &= \text{fl}[hfl[(\tfrac{1}{2}\hat{f}_0) + \hat{f}_1 + \cdots + \hat{f}_{n-1} + (\tfrac{1}{2}\hat{f}_n)]]\\ &= h[\tfrac{1}{2}\hat{f}_0\langle n + 1\rangle + \hat{f}_1\langle n\rangle + \cdots + \hat{f}_{n-1}\langle 2\rangle + \tfrac{1}{2}\hat{f}_n\langle 1\rangle]\langle 1\rangle\\ &= h(\tfrac{1}{2}\hat{f}_0\langle n + 2\rangle + \hat{f}_1\langle n + 1\rangle + \cdots + \hat{f}_{n-1}\langle 3\rangle + \tfrac{1}{2}\hat{f}_n\langle 2\rangle).\end{aligned}$$

We have used here the general formula (3.3c), with one additional error counter to take into account the multiplication by h. If $\hat{f}_k$ is replaced by $f_k(1 + \varepsilon_k)$, then

$$\hat{S} = h(\tfrac{1}{2}f_0(1 + \varepsilon_0)\langle n + 2\rangle + f_1(1 + \varepsilon_1)\langle n + 1\rangle + \cdots + \tfrac{1}{2}f_n(1 + \varepsilon_n)\langle 2\rangle).$$

In order to simplify this formula, assume that $(1 + \varepsilon_k) = \langle l\rangle, k = 0, 1, \ldots, n,$ where l is a fixed integer. That is, we assume that $f(x)$ can be evaluated, at any x, with l rounding errors. Then

$$\begin{aligned}\hat{S} &= h(\tfrac{1}{2}f_0\langle n + 2 + l\rangle + f_1\langle n + 1 + l\rangle + \cdots + \tfrac{1}{2}f_n\langle 2 + l\rangle)\\ &= h(\tfrac{1}{2}f_0 + f_1 + \cdots + \tfrac{1}{2}f_n) + h(\tfrac{1}{2}f_0\langle n + 2 + l\rangle_0\\ &\quad + f_1\langle n + 1 + l\rangle_0 + \cdots + \tfrac{1}{2}f_n\langle 2 + l\rangle_0).\end{aligned}$$

Finally, let $\bar{f} = \max\{|f(x)|, a \leq x \leq b\}$, and use (2.37), under the assumption that $(n + 2 + l)r\mu \leq .1$, to write the **absolute error estimate** as

$$|\hat{S} - S|$$
$$\leq h|\bar{f}|\{\tfrac{1}{2}(n + 2 + l) + (n + 1 + l) + (n + l) + \cdots + (3 + l) + \tfrac{1}{2}(2 + l)\}\mu'.$$

After some simple manipulations, this reduces to

$$|\hat{S} - S| \le h\,|\,\bar{f}\,|\,\{\tfrac{1}{2}n^2 + 2n + n\}\mu'. \tag{5.31}$$

Since $nh = b - a$, this may be written as

$$|\hat{S} - S| \le |\,\bar{f}\,|\,[(b - a)^2/2h + 2(b - a) + (b - a)l]\mu'. \tag{5.32}$$

The dominant part of this bound is the term $(b - a)^2/2h$, which goes to ∞ as h goes to zero. Thus we expect the rounding error eventually to increase as h decreases. As with most discrete approximates, one cannot expect the *total* error (truncation plus rounding) to continue to decrease as the discretization parameter h decreases. The estimate (5.32) also shows the effect of the rounding errors during the evaluation of $f(x)$. In particular, we see that errors in these evaluations can be ignored whenever h is so small that $(b - a)/2h > l$.

The most important consequence of (5.32), however, is the fact that numerical integration is a stable process. That is, this (pessimistic) bound on the accumulated rounding error increases as h decreases only because the *number* of computations increases. Since this increase in the error bound is linear, we can conclude that rounding error is not a serious problem in numerical integration.

In view of the statement just made, it is interesting to compare the error bounds for the differentiation methods with the rounding error estimate (5.32). Consider Table 5.6. Notice that in both cases the truncation error decreases as h^2, while the rounding error increases as $1/h$, when h goes to zero. A naive interpretation of these results suggests that the total errors in these processes should act the same when h decreases. Experience, however, shows that this is not true. The explanation for the discrepancy between this theoretical conclusion and actual experience is as follows. The rounding error estimate for the differentiation formula is mostly due to the subtraction cancellation that occurs during the single subtraction operation. The error bound is a fairly realistic estimate of the error in this operation, and as h decreases this error will increase just as predicted by the error bound. The rounding error estimate for the integration formula, however, has a quite different meaning. As already observed, this estimate is obtained by assuming that the worst possible errors are committed during a sequence of operations. As h decreases, the number of operations increases. That is, there are more and more floating point operations, each of which is assumed to contribute an error of magnitude μ'. Thus as h decreases this error estimate becomes less and less realistic. In fact, for *very* small h, the estimate may be in error by many orders of magnitude (Exercise 11).

Generally, error estimates of computations that involve only a few

TABLE 5.6   *Error Estimate Comparison*

	Differentiation	Integration
Formula	(5.8) $f'(x) \approx [f(x + h) - f(x - h)]/2h$	(5.25) $\int_a^b f(x)\, dx$ $\approx h(\frac{1}{2}f_0 + f_1 + \cdots + f_{n-1} + \frac{1}{2}f_n)$
Truncation error	(5.9) $\frac{1}{6}f'''(\xi)h^2$	(5.25) $-f''(\xi)\frac{1}{12}h^2(b - a)$
Absolute rounding error	(5.14) $(|f(x)|/h + |f'(x)|)\mu'$	(5.32) $|\bar{f}|((b - a)^2/2h + 2(b - a)$ $+ (b - a)l)\mu'$

floating point operations are fairly realistic. As the *number* of operations increases, these estimates become less and less reliable.

EXERCISES

1   Compute the (composite) trapezoidal rule approximations to the following integrals using $n = 3$.

  (a) $\displaystyle\int_{-1}^{1} \frac{1}{(x^2 + 1)^2}\, dx$ (b) $\displaystyle\int_{0}^{1} \left(x^2 + \frac{1}{x + 1}\right) dx$ (c) $\displaystyle\int_{-\pi}^{2\pi} \sin^2 x\, dx$.

2   Determine upper bounds on the truncation errors for the approximations in Exercise 1.

3   Compute the Simpson rule approximations to the integrals in Exercise 1, using $n = 4$.

4   Show, in detail, how the composite truncation error (5.29b) is derived from (5.28).

5   Derive the first formula in Table 5.4 by integrating the third degree polynomial that interpolates f at x_0, x_1, x_2, x_3. Then determine the appropriate composite formula.

6   Derive the simple rectangle rule:

$$\int_a^b f(x)\, dx = (b - a)f(a) + \frac{(b - a)^2}{2} f'(\xi).$$

7   Derive a composite formula based on the rectangle rule in Exercise 6.

8   Determine a rounding error estimate of the type (5.31) for the composite Simpson's rule. (Before starting the error analysis, an appropriate algorithm must be given.)

9   One way of approximating the "optimal" value for h that will minimize the *sum* of the truncation and rounding errors is to use that value that makes these two errors approximately equal. Use formula (5.25) (for the truncation error) and formula (5.32) (for the rounding error) to determine such a value for h, under the assumption that $l = 5$.

10   Supply the details for the derivation of (5.32) and the two preceding inequalities.

11   Write a program to compute the composite trapezoidal approximation to $\int_0^1 e^x\, dx$ in both single and double precision. Also compute the rounding error estimate (5.32). Run your program for values of $h = .1, .01, .001, \dots$. Compare the estimate of rounding error with

the "true" rounding error obtained by subtracting the single-precision value from the double-precision value.

5.3 Numerical Integration—Gaussian Methods

The Newton–Cotes formulas derived in the previous section are especially recommended for "uncomplicated" integrals. These formulas may not be suitable when the interval of integration is infinite or when the integrand is very difficult to evaluate or has a singularity. In these cases, it may be better to use the more complicated Gauss formulas that will be introduced in this section.

5.3.1 General Formulation

Numerical methods generally are expected to give only an approximate solution to the problem. However, most methods provide the *exact* answer when applied to certain special cases (ignoring, for the moment, the existence of rounding errors). For example, the trapezoidal rule [formula (5.25)] gives the exact value whenever the integrand $f(x)$ is a linear function. Similarly, Simpson's rule gives the exact value when applied to any polynomial of degree *three* or less. Often the accuracy of a method can be judged on the basis of how large is the class of special cases for which the method produces the exact answer. Thus Simpson's rule may be judged more accurate than the trapezoidal rule.

Consider the general form of the Newton–Cotes formulas derived in Section 5.2:

$$\int_a^b f(x)\, dx \approx a_1 f(x_1) + a_2 f(x_2) + \cdots + a_n f(x_n). \qquad (5.33)$$

There, the points $x_1, x_2, \ldots, x_n$ were assumed to be fixed and evenly spaced over the interval $[a, b]$. As just observed, these formulas are *exact* whenever $f(x)$ is a polynomial of degree less than a certain value. The idea behind the Gauss integration methods is to derive formulas of the form (5.33) without assuming the points $x_1, x_2, \ldots, x_n$ to be fixed in advance. Instead, these points, as well as the coefficients $a_1, a_2, \ldots, a_n$, are chosen so that the formula gives the exact value when applied to polynomials of as high a degree as possible. Note that a polynomial of degree m is determined by its $m + 1$ coefficients, and there are now $2n$ parameters in formula (5.33). Thus it seems reasonable to expect that, by proper choice of all of these parameters, we can obtain a formula of the form (5.33) that is exact whenever f is a polynomial of degree less than or equal to $m = 2n - 1$. This is indeed the case, as we shall indicate in the next subsection.

Before proceeding with the derivation of particular Gauss formulas, we observe two facts that influence these derivations. First of all, if $p(x) = b_m x^m + b_{m-1} x^{m-1} + \cdots + b_1 x + b_0$ is any polynomial of degree m, then

$$\int_a^b p(x)\, dx = b_m \int_a^b x^m\, dx + b_{m-1} \int_a^b x^{m-1}\, dx + \cdots + b_0 \int_a^b 1\, dx. \quad (5.34)$$

Thus, a particular numerical integration formula can be used to approximate $\int_a^b p(x)\, dx$ in two ways: by applying the formula to $p(x)$ directly or by applying it to each of the integrals on the right side of (5.34), and then combining these results as indicated. Because all integration formulas of the form (5.33) are linear, it follows that these two approximations will be identical. Hence, if an integration formula gives the exact answer when applied to the *special* polynomials x^m, x^{m-1}, ..., x, 1, then it will give the exact result when applied to any polynomial of degree less than or equal to m.

Next, when deriving the Gauss formulas, it becomes clear immediately that the points $x_1, x_2, \ldots, x_n$ are rather complicated functions of the limits of integration a and b (Exercise 7). However, a simple change of variable can be used to convert any finite definite integral into an integral over some fixed interval, say $[-1, 1]$. That is, if x is an integration variable that goes from a to b, and if y is another variable related to x by

$$\begin{aligned} y &= [2x - (a+b)]/(b-a) \\ x &= \tfrac{1}{2}(b-a)y + \tfrac{1}{2}(a+b), \end{aligned} \quad (5.35)$$

then y goes from -1 to $+1$. Thus it follows that

$$\int_a^b f(x)\, dx = \frac{b-a}{2} \int_{-1}^1 F(y)\, dy,$$

where $F(y)$ is obtained by substituting the second expression in (5.35) for x in the function $f(x)$. For example, with $a = 0$, $b = 1$, (5.35) gives $x = \tfrac{1}{2}y + \tfrac{1}{2}$, so that

$$\int_0^1 e^x\, dx = \frac{1}{2} \int_{-1}^1 \exp\left(\frac{1}{2}y + \frac{1}{2}\right) dy \quad (5.36)$$

Thus, the integral on the left in (5.36) can be approximated by applying a numerical integration formula to the integral on the right of (5.36). All of the Gauss formulas mentioned in the next two subsections will apply to integrals over the interval $[-1, 1]$.

5.3.2   Two-Point Formula

Consider the formula (5.33) for $n = 2$ and with $b = -a = 1$. As noted above, by a proper choice of a_1, a_2 and x_1, x_2, the formula should give the

exact value of the integral whenever $f(x)$ is a polynomial of degree $m = 2n - 1 = 3$, or less. Moreover, it suffices to consider only the special polynomials $1, x, x^2, x^3$. Thus we have the following relations:

$$\int_{-1}^{1} 1 \, dx = a_1 + a_2 \qquad [f(x) = 1 \text{ in } (5.33)] \qquad (5.37\text{a})$$

$$\int_{-1}^{1} x \, dx = a_1 x_1 + a_2 x_2 \qquad [f(x) = x \text{ in } (5.33)] \qquad (5.37\text{b})$$

$$\int_{-1}^{1} x^2 \, dx = a_1 x_1^2 + a_2 x_2^2 \qquad [f(x) = x^2 \text{ in } (5.33)] \qquad (5.37\text{c})$$

$$\int_{-1}^{1} x^3 \, dx = a_1 x_1^3 + a_2 x_2^3 \qquad [f(x) = x^3 \text{ in } (5.33)]. \qquad (5.37\text{d})$$

The left sides of these equations are the constants $2, 0, \frac{2}{3}, 0$, respectively. Hence, we have four (nonlinear) equations

$$a_1 + a_2 = 2$$

$$a_1 x_1 + a_2 x_2 = 0$$

$$a_1 x_1^2 + a_2 x_2^2 = \tfrac{2}{3} \qquad\qquad (5.37\text{e})$$

$$a_1 x_1^3 + a_2 x_2^3 = 0$$

in four unknowns a_1, a_2, x_1, x_2. These equations are not easily solved; however, it is easily *verified* that a solution is

$$a_1 = a_2 = 1, \qquad x_2 = -x_1 = 1/\sqrt{3}.$$

Hence, the integration formula has the form

$$\int_{-1}^{1} f(x) \, dx \approx f(-1/\sqrt{3}) + f(1/\sqrt{3}). \qquad (5.38)$$

This is called the *two-point Gauss integration formula.*

Conditions (5.37) guarantee that (5.38) is an *equation* whenever $f(x)$ is a polynomial of degree three or less. For other functions, (5.38) provides only an approximation. The (truncation) error in (5.38) will be examined in the next subsection. To illustrate the use of the formula, consider the integral in (5.36) that was approximated by various Newton–Cotes formulas in Table 5.5. In order to apply (5.38), the interval $[0, 1]$ must be transformed into $[-1, 1]$, as shown in (5.36). Formula (5.38) then gives the approximation

1.717905, as follows:

$$\int_0^1 e^x \, dx = \frac{1}{2} \int_{-1}^1 \exp\left(\frac{1}{2} y + \frac{1}{2}\right) dy$$

$$\approx \frac{1}{2} \left\{ \exp\left[\frac{1}{2}\left(\frac{-1}{\sqrt{3}}\right) + \frac{1}{2}\right] + \exp\left[\frac{1}{2}\left(\frac{1}{\sqrt{3}}\right) + \frac{1}{2}\right] \right\} = 1.717905.$$

By examining Table 5.5, we see that this is more accurate than the trapezoidal rule with $h = .125$.

5.3.3   More Accurate Formulas

More complicated (but more accurate) Gauss integration formulas can be derived by exactly the same technique as was used to derive (5.38). The solutions to the nonlinear systems corresponding to (5.37) are, however, rather difficult to obtain. An alternative derivation of these formulas that avoids the necessity of solving systems of nonlinear equations will now be described. For this, recall the **Hermite interpolation formula** as given in paragraph 4.1.4. That is, if f is any function that is defined and has $2n$ derivatives on an interval $[a, b]$, then for any points $x_1, x_2, \ldots, x_n$ in $[a, b]$

$$f(x) = \sum_{k=1}^n [1 - 2p_k'(x_k)(x - x_k)]p_k(x)^2 f(x_k)$$

$$+ \sum_{k=1}^n (x - x_k)p_k(x)^2 f'(x_k)$$

$$+ [f^{2n}(\xi)/(2n)!][(x - x_1)(x - x_2) \cdots (x - x_n)]^2, \quad (5.39a)$$

where

$$p_k(x) = \frac{(x - x_1) \cdots (x - x_{k-1})(x - x_{k+1}) \cdots (x - x_n)}{(x_k - x_1) \cdots (x_k - x_{k-1})(x_k - x_{k+1}) \cdots (x_k - x_n)}. \quad (5.39b)$$

The first two terms in (5.39a) define a polynomial of degree $2n - 1$ that interpolates f at $x_1, x_2, \ldots, x_n$, and whose *derivative* equals f' at these same points. By integrating (5.39a) we find

$$\int_{-1}^1 f(x) \, dx = \sum_{k=1}^n a_k f(x_k) + \sum_{k=1}^n b_k f'(x_k) + E_T, \quad (5.40a)$$

where

$$a_k = \int_{-1}^{1} [1 - 2p_k'(x_k)(x - x_k)]p_k(x)^2 \, dx, \qquad (5.40b)$$

$$b_k = \int_{-1}^{1} (x - x_k)p_k(x)^2 \, dx, \qquad (5.40c)$$

and

$$E_T = \frac{1}{(2n)!} \int_{-1}^{1} f^{(2n)}(\xi(x))[(x - x_1)(x - x_2) \cdots (x - x_n)]^2 \, dx. \quad (5.40d)$$

Now (5.40a) will be of the desired form (5.33) provided that the b_k in (5.40c) are all zero. Also, the truncation error, as given by (5.40d), will be zero whenever f is a polynomial of degree $2n - 1$ or less. In order to see what the points $x_1, x_2, \ldots, x_n$ must be so that b_k is zero, we write (5.40c) as

$$b_k = \frac{1}{c_k} \int_{-1}^{1} q_k(x)l_n(x) \, dx, \qquad (5.41a)$$

where

$$q_k(x) = (x - x_1) \cdots (x - x_{k-1})(x - x_{k+1}) \cdots (x - x_n)$$
$$l_n(x) = (x - x_1)(x - x_2) \cdots (x - x_n) \qquad (5.41b)$$
$$c_k = [q_k(x_k)]^2.$$

Thus l_n is a polynomial of degree n whose roots are the values $x_1, \ldots, x_n$, and q_k is a polynomial of degree *less* than n. From (5.41a) it follows that $b_k = 0$ for $k = 1, 2, \ldots, n$ if l_n has the property

Orthogonality condition	$\int_{-1}^{1} r(x)l_n(x) = 0$	for any polynomial $r(x)$ of degree less than n.	(5.41c)

A polynomial l_n that satisfies (5.41c) is said to be *orthogonal to all polynomials of degree less than* n on the interval $[-1, 1]$. Orthogonal polynomials arise in many applications and have been the subject of much study. In particular, it is well known that the so-called *Legendre polynomials* $l_0, l_1, \ldots,$ as defined by the recursion

Legendre polynomials	$l_0(x) = 1, \qquad l_1(x) = x$ $l_{n+1}(x) = [(2n + 1)/(n + 1)]xl_n(x)$ $\quad - [n/(n + 1)]l_{n-1}(x), \qquad n = 1, 2, \ldots,$	(5.41d)

satisfy (5.41c) for each n. We summarize this as follows:

> The numerical integration formula
>
> $$\int_{-1}^{1} f(x)\, dx = \sum_{k=1}^{n} a_k f(x_k) + E_T$$
>
> will have $E_T = 0$ whenever f is a polynomial of degree less than $2n$, provided $x_1, x_2, \ldots, x_n$ are the roots of the Legendre polynomials (5.41d) and $a_1, a_2, \ldots, a_n$ are defined in (5.40b).

Table 5.7 gives values of a_k and x_k for several values of n.

TABLE 5.7   *Gauss Integration Formulas—Coefficients and Arguments*

n	a_k	x_k
2	$a_1 = a_2 = 1.00000000$	$x_2 = -x_1 = .57735027$
3	$a_2 = .88888889$ $a_1 = a_3 = .55555556$	$x_2 = 0.00000000$ $x_3 = -x_1 = .77459667$
4	$a_2 = a_3 = .65214515$ $a_1 = a_4 = .34785485$	$x_3 = -x_2 = .33998104$ $x_4 = -x_1 = .86113631$
5	$a_2 = a_4 = .47862867$ $a_1 = a_5 = .23692689$ $a_3 = .56888889$	$x_4 = -x_2 = .53846931$ $x_5 = -x_1 = .90617985$ $x_3 = 0.00000000$

By using the mean value theorem for integrals, the truncation error (5.40d) can be simplified to

> Truncation error for formulas in Table 5.7
>
> $$E_T = \frac{2^{2n+1}(n!)^4}{(2n+1)[(2n)!]^3} f^{(2n)}(\xi)$$
> (5.41e)

when ξ is some value in $[-1, 1]$. Thus if the derivatives of f are all of the same magnitude, then an n point Gauss formula is more accurate than a comparable Newton–Cotes formula. By "comparable" we mean here a formula that uses the value of f at n points. If f is very difficult (i.e., expensive) to evaluate, then a Gauss formula may be much more efficient than a Newton–Cotes method. It must be kept in mind, however, that a preliminary change of variables, as given by (5.35), may be necessary before the Gauss formula can be applied.

5.3.4 Special-Purpose Formulas

The method that was used to derive the n-point Gauss integration formula can also be used to derive some special-purpose formulas. Consider, for example, an integral of the form

$$\int_{-1}^{1} \frac{f(x)}{\sqrt{1 - x^2}} \, dx,$$

where $f(x)$ is a well-behaved function. The Newton–Cotes formulas of Section 5.1 cannot be used here because the integrand cannot be evaluated at $x = -1$ or $x = 1$. The Gauss formulas derived above may not give good results because the integrand $f(x)/\sqrt{1 - x^2}$ may have large derivatives. Instead, suppose we consider a formula of the type

Gauss–Chebyshev formulas	$\displaystyle\int_{-1}^{1} \frac{f(x)}{\sqrt{1 - x^2}} \, dx = a_1 f(x_1) + a_2 f(x_2)$ $\qquad\qquad + \cdots + a_n f(x_n) + E_T.$ (5.42a)

Note that on the right of (5.42a) only the "nice" part of the integrand occurs. The idea here is to choose $a_1, a_2, \ldots, a_n$ and $x_1, x_2, \ldots, x_n$ so that the truncation error E_T is zero whenever f is a polynomial of degree $2n - 1$ or less. That is, the formula will be accurate whenever $f(x)$ is a smooth function, in spite of the fact that the complete integrand goes to ∞ at the endpoints. If we multiply (5.40a) by $1/\sqrt{1 - x^2}$ and then integrate, the result will be a formula of the type (5.42a), provided

$$\int_{-1}^{1} \frac{(x - x_k)p_k(x)^2}{\sqrt{1 - x^2}} \, dx = 0, \qquad k = 1, 2, \ldots, n.$$

Thus the orthogonality condition (5.41c) now becomes

Orthogonality condition for (5.42a)	$\displaystyle\int_{-1}^{1} \frac{r(x)}{\sqrt{1 - x^2}} \, l_n(x) \, dx = 0$ (5.42b) for any polynomial r of degree less than n.

The **Chebyshev polynomials**, $T_0, T_1, \ldots$, defined by

$$T_0 = 1, \qquad T_1(x) = x$$
$$T_{n+1} = 2x T_n(x) - T_{n-1}(x), \qquad n = 1, 2, \ldots, \tag{5.42c}$$

have the orthogonality property (5.42b). Thus, formula (5.42a) will have zero truncation errors for f a polynomial of degree less than $2n$, provided $x_1, x_2,$

$\ldots, x_n$ are the roots of $T_n(x)$, and the coefficients $a_1, \ldots, a_n$ are given by

$$a_k = \int_{-1}^{1} \frac{[1 - 2p_k'(x_k)(x - x_k)]p_k(x)^2}{\sqrt{1 - x^2}} \, dx.$$

It can be shown that these **coefficients and points** are given by

$$x_k = \cos\left(\frac{(2k - 1)\pi}{2n}\right), \qquad k = 1, 2, \ldots, n$$

$$a_1 = a_2 = \cdots = a_n = \pi/n.$$

(5.42d)

If f is $(2n)$ times continuously differentiable, the **truncation error** may be shown to be

$$E_T = \frac{2\pi}{2^{2n}(2n)!} f^{(2n)}(\xi) \qquad \text{for some} \quad \xi \in [-1, 1].$$

These integration formulas are called the *Gauss–Chebyshev* formulas. For $n = 2$, (5.42a) is

$$\int_{-1}^{1} \frac{f(x)}{\sqrt{1 - x^2}} \, dx = \frac{\pi}{2} \left\{ f\left(\frac{\sqrt{2}}{2}\right) + f\left(\frac{-\sqrt{2}}{2}\right) \right\} + E_T,$$

where

$$E_T = (\pi/192)f^{(4)}(\xi).$$

To illustrate the usefulness of this formula, consider the integral

$$\int_{-1}^{1} \frac{x \sin x}{\sqrt{1 - x^2}} \, dx.$$

The Gauss–Chebyshev formula, with $n = 2$, gives the approximate value

$$\frac{\pi}{2} \left[\frac{\sqrt{2}}{2} \sin\left(\frac{\sqrt{2}}{2}\right) + \left(\frac{-\sqrt{2}}{2}\right) \sin\left(\frac{-\sqrt{2}}{2}\right) \right] = \pi \frac{\sqrt{2}}{2} \sin\left(\frac{\sqrt{2}}{2}\right) = 1.4480.$$

The truncation error estimate

$$|E_T| = (\pi/192)|\xi \sin \xi - 4 \cos \xi| \leq (5\pi/192) \approx .082$$

indicates that this approximation has two-digit accuracy. For comparison, the two-point Gauss formula gives the value

$$\frac{(-1/\sqrt{3}) \sin(-1/\sqrt{3})}{\sqrt{1 - (1/\sqrt{3})^2}} + \frac{(1/\sqrt{3}) \sin(1/\sqrt{3})}{\sqrt{1 - (1/\sqrt{3})^2}} = \sqrt{2} \sin(1/\sqrt{3}) = .775017.$$

This approximation has no accuracy at all. We again stress the fact that the Newton–Cotes formulas, such as the trapezoidal rule, cannot be used here because of the singularities at the points $x = \pm 1$.

This same technique can also be used to derive formulas for numerical integration over infinite intervals.

5.3.5 Rounding Errors in Gauss Formulas

The effect of rounding errors in Gaussian integration is somewhat different than in the Newton–Cotes formulas analyzed in Section 5.2. Generally, only low-order Gaussian formulas are used; that is, n is usually small. Thus the rounding error due to the arithmetic operations needed to evaluate the formula is quite small. On the other hand, the points $x_1, \ldots, x_n$ are usually irrational numbers that must be approximated by floating point numbers $\tilde{x}_i$. Thus in place of the terms $f(x_i)$ in these formulas, quantities of the form $\mathrm{fl}[f(\tilde{x}_i)]$ will actually be used. Clearly these are related by

$$\mathrm{fl}[f(\tilde{x}_i)] = f(x_i)(1 + \varepsilon_i).$$

Now, however, ε_i represents both the error caused by the floating point evaluation of f and the error in $\tilde{x}_i$. In many cases ε_i will be quite large. Since the *form* of the Gaussian formulas is the same as the Newton–Cotes formulas, the error estimate (5.32) will still hold, but generally the term corresponding to $(b - a)^2/2h$ will be small, whereas l may be quite large. In particular, if the evaluation of the function f, which appears in the integrand, is very sensitive to rounding errors, the Gauss formulas should be avoided if possible.

EXERCISES

1 Use the two-point Gaussian integration formula to approximate the values of the integrals

(a) $\displaystyle \int_{-1}^{1} \frac{1}{(x^2 + 1)^2}\, dx$ (b) $\displaystyle \int_{-1}^{1} e^x\, dx$ (c) $\displaystyle \int_{0}^{1} e^{x^2}\, dx.$

2 Estimate the truncation errors for each of the approximations in Exercise 1.

3 Write out the Legendre polynomials l_2, l_3, l_4, and l_5 as defined by (5.41d).

4 Show that the roots $x_1, x_2, \ldots, x_n$ of the Legendre polynomials (5.41d) all lie in the interval $[-1, 1]$.
[*Hint:* Suppose not, and let $y_1, y_2, \ldots, y_n$ be those roots that *do* lie in $[-1, 1]$. Then $s(x) = (x - y_1)(x - y_2) \cdots (x - y_m)l_n(x)$ does not change sign in $[-1, 1]$, hence $\int_{-1}^{1} s(x)\, dx \neq 0$. Show that this contradicts the orthogonality property (5.41c).]

5 Use a Gauss–Chebyshev formula to evaluate the integrals

(a) $\displaystyle \int_{-1}^{1} \frac{x^3}{\sqrt{1 - x^2}}\, dx$ (b) $\displaystyle \int_{-1}^{1} \frac{\cos \pi x}{\sqrt{1 - x^2}}\, dx$ (c) $\displaystyle \int_{0}^{1} \frac{x^2}{\sqrt{x(1 - x)}}\, dx.$

6 Estimate the truncation errors for the values obtained in Exercise 5.

7 Derive a two-point Gauss formula of the form

$$\int_a^b f(x) \approx a_0 f(x_0) + a_1 f(x_1),$$

where a and b are not fixed.
[*Hint*: Apply the two-point formula derived in the text to the integral

$$\frac{b-a}{2} \int_{-1}^1 F(y) \, dy$$

when $F(y) = f[\frac{1}{2}(b-a)y + \frac{1}{2}(b+a)]$. Rewrite the resulting expression in the form

$$a_0 f(x_0) + a_1 f(x_1)$$

when a_0, a_1, x_0, x_1 are functions of a and b.]

8 Use the mean value theorem for integrals (Appendix) to show that (5.40d) can be written in the form (5.41e).
[*Hint*: To evaluate $\int_{-1}^1 [(x - x_1) \cdots (x - x_n)]^2 \, dx$, use the recursion (5.41d) to evaluate $\int_{-1}^1 l_n(x)^2 \, dx$.]

9 Prove by induction that the polynomials defined by (5.41d) have the orthogonality property (5.41c).

10 Use the trigonometric identities

$$\cos(n+1)\theta = \cos n\theta \cos \theta - \sin n\theta \sin \theta$$

$$\cos(n-1)\theta = \cos n\theta \cos \theta + \sin n\theta \sin \theta$$

to show that

$$T_n(x) = \cos n\theta, \qquad \theta = \arccos x$$

defines a set of polynomials that satisfy the recursion (5.42c). Use this to show that the roots of T_n are given by (5.42d).

5.4   Romberg Integration

The discussion of the Newton–Cotes formulas showed that there are two ways to reduce truncation error: The value of h may be decreased, or a higher order formula can be used. For computer applications, changing the formula is difficult since it requires a new program to evaluate the new formula. In this section we discuss the Romberg method that automatically produces approximations based on progressively higher order formulas.

5.4.1   Extrapolation to $h = 0$

In order to increase the accuracy of a computed approximation to a definite integral, the same technique can be applied to numerical integration formulas as was used for numerical differentiation—namely, extrapolation to the limit, as discussed in Section 5.1.4.

Let $T(h)$ be the (composite) trapezoidal rule approximation to the value of a definite integral (5.18) with spacing h. Then formula (5.25) says that

$$T(h) = I + (\tfrac{1}{12}(b-a)f''(\eta(h)))h^2,$$

where we use I to denote the integral (5.18). As before, we approximate $T(h)$ by the polynomial $A + Bh^2$ and then evaluate this approximation at $h = 0$. By using (5.15), we find that

$$T(0) \approx \frac{h_2^2 T(h_1) - h_1^2 T(h_2)}{h_2^2 - h_1^2},$$

where h_1 and h_2 are two values of h for which T can be evaluated with reasonable accuracy. If $h_2 = \tfrac{1}{2}h_1$, then this gives the new approximation

$$T_1(h) = \tfrac{1}{3}[4T(\tfrac{1}{2}h) - T(h)]. \tag{5.43}$$

The formula (5.25) on which this entire process is based was obtained quite easily by examining the truncation error in the linear interpolating polynomial. A much more complicated analysis [Ralston (1965)] shows that, in fact, (5.25) is just the first part of an expansion of $T(h)$ in even powers of h. That is,

$$T(h) = I + a_2 h^2 + a_4 h^4 + a_6 h^6 + \cdots,$$

where $a_2, a_4, a_6, \ldots$ are certain constants. Thus, by assuming that $T(h) = I + a_2 h^2$, which led to (5.42), we are left with an error involving only $h^4, h^6, \ldots$, that is,

$$T_1(h) = I + b_4 h^4 + b_6 h^6 + \cdots. \tag{5.44}$$

5.4.2 Repeated Extrapolations

We repeat the same process with (5.44). In other words, assume that $T_1(h) = I + b_4 h^4$. Then it follows easily that

$$T_2(h) = I + c_6 h^6 + c_8 h^8 + \cdots, \tag{5.45}$$

where

$$T_2(h) = \tfrac{1}{15}[16T_1(\tfrac{1}{2}h) - T_1(h)]. \tag{5.46}$$

[Compare this with formula (5.17).] Similarly, we have

$$T_3(h) = I + d_8 h^8 + d_{10} h^{10} + \cdots, \tag{5.47}$$

where

$$T_3(h) = \tfrac{1}{63}[64T_2(\tfrac{1}{2}h) - T_2(h)] \tag{5.48}$$

and so on. A simple way to derive these formulas is to write (5.44) as

$$T_1(h) = I + b_4 h^4 + b_6 h^6 + \cdots$$

$$T_1(\tfrac{1}{2}h) = I + b_4 \tfrac{1}{16} h^4 + b_6 \tfrac{1}{64} h^6 + \cdots .$$

Now, multiply the second equation by 16, and subtract the first to get

$$16 T_1(\tfrac{1}{2}h) - T_1(h) = 15I + (\tfrac{1}{4} b_6 - b_6)h^6 + \cdots .$$

A simple rearrangement gives

$$T_2(h) \equiv \tfrac{1}{15}[16 T_1(\tfrac{1}{2}h) - T_1(h)] = I + c_6 h^6 + \cdots .$$

Thus, by repeating this process, we obtain approximations to the integral I with smaller and smaller truncation errors.

5.4.3   Romberg Table

Notice that to find $T_2(h)$, we must find $T_1(\tfrac{1}{2}h)$. But from (5.43) we see that

$$T_1(\tfrac{1}{2}h) = \tfrac{1}{3}[4 T(\tfrac{1}{4}h) - T(\tfrac{1}{2}h)],$$

which means that we must first find $T(\tfrac{1}{4}h)$. Similarly, $T_3(h)$ requires $T_2(\tfrac{1}{2}h)$, which involves $T_1(\tfrac{1}{4}h)$, which is determined from $T(\tfrac{1}{8}h)$. To show this in a more systematic way, we define the **Romberg table**:

$$T(h)$$
$$T(\tfrac{1}{2}h) \quad T_1(h)$$
$$T(\tfrac{1}{4}h) \quad T_1(\tfrac{1}{2}h) \quad T_2(h)$$
$$T(\tfrac{1}{8}h) \quad T_1(\tfrac{1}{4}h) \quad T_2(\tfrac{1}{2}h) \quad T_3(h)$$
$$\vdots$$

Here, each entry in columns 2–4 are obtained by combining the two entries to the left according to formulas (5.43), (5.46), or (5.48), etc.

As an **example**, consider again the integral

$$\int_0^1 e^x \, dx,$$

which is approximated in various ways in Table 5.5. With $h = .25$, we have

$$T(h) = 1.7272219$$
$$T(\tfrac{1}{2}h) = T(.125) = 1.7205186$$
$$T(\tfrac{1}{4}h) = T(.0625) = 1.7188411$$
$$T(\tfrac{1}{8}h) = T(.03125) = 1.7184216$$

and hence

$$T_1(h) = \tfrac{1}{3}[4T(\tfrac{1}{2}h) - T(h)] = 1.7182841$$

$$T_1(h/2) = \tfrac{1}{3}[4T(\tfrac{1}{4}h) - T(\tfrac{1}{2}h)] = 1.7182819$$

$$T_1(h/4) = \tfrac{1}{3}[4T(\tfrac{1}{8}h) - T(\tfrac{1}{4}h)] = 1.7182818$$

$$T_2(h) = \tfrac{1}{15}[16T_1(\tfrac{1}{2}h) - T_1(h)] = 1.7182818$$

$$T_2(h/2) = \tfrac{1}{15}[16T_1(\tfrac{1}{4}h) - T_1(\tfrac{1}{2}h)] = 1.7182818$$

$$T_3(h) = \tfrac{1}{63}[64T_2(\tfrac{1}{2}h) - T_2(h)] = 1.7182818.$$

TABLE 5.8   *Romberg Table for $\int_0^1 e^x\, dx$*

1.7272219			
1.7205186	1.7182841		
1.7188411	1.7182819	1.7182818	
1.7184216	1.7182818	1.7182818	1.7182818

This gives the Romberg table (Table 5.8). Observe the interesting coin-cidence that the diagonal entries in Table 5.8 are identical to various entries of Table 5.5. That is, $T_1(.25)$ is the same as Simpson's rule with $h = .125$, and $T_2(.25)$ is the same as formula 2 with $h = .125$. In fact, this is not just a coincidence, but rather is a result of the way in which the Romberg table is built up. For example, if $f_k = f[x_0 + k(h/2)]$, then

$$T_1(h) = \tfrac{1}{3}[4T(\tfrac{1}{2}h) - T(h)]$$

$$= \tfrac{1}{3}[4(\tfrac{1}{2}h)(\tfrac{1}{2}f_0 + f_1 + f_2 + \cdots + f_{2n-1} + \tfrac{1}{2}f_{2n})$$

$$\quad - h(\tfrac{1}{2}f_0 + f_2 + f_4 + \cdots + f_{2n-2} + \tfrac{1}{2}f_{2n})]$$

$$= \tfrac{1}{3}(\tfrac{1}{2}h)(f_0 + 4f_1 + 2f_2 + 4f_3 + 2f_4 + \cdots + f_{2n}).$$

This is just **Simpson's rule with spacing** $\tfrac{1}{2}h$; that is, $T_1(.25)$ is the same as Simpson's rule with $h = .25/2$. In a similar manner, $T_2(h)$ can be identified with formula 2 in Table 5.5. Thus the Romberg table provides a systematic way of obtaining increasingly accurate approximations.

5.4.4   Romberg Algorithm

In order to compute $T(h/2)$ after $T(h)$ has been determined, it is only necessary to evaluate f at certain "new" points (see Fig. 5.3). That is, let

$$x_k = x_0 + kh, k = 0, 1, \ldots, n, \text{ and } x_k^{(1/2)} = x_0 + k(\tfrac{1}{2}h), k = 0, 1, \ldots, 2n.$$

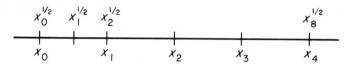

Fig. 5.3   New points introduced when h is replaced by $\frac{1}{2}h$.

Then

$$T(\tfrac{1}{2}h) = (\tfrac{1}{2}h)(\tfrac{1}{2}f(x_0) + f(x_1^{(1/2)}) + f(x_2^{(1/2)})$$
$$+ \cdots + f(x_{2n-1}^{(1/2)}) + \tfrac{1}{2}f(x_{2n}^{(1/2)}))$$
$$= (\tfrac{1}{2}h)(\tfrac{1}{2}f(x_0) + f(x_2^{(1/2)}) + \cdots + f(x_{2n-2}^{(1/2)}) + \tfrac{1}{2}f(x_{2n}^{(1/2)}))$$
$$+ f(x_1^{(1/2)}) + f(x_3^{(1/2)}) + \cdots + f(x_{2n-1}^{(1/2)})),$$

from which it follows that

$$T(\tfrac{1}{2}h) = \tfrac{1}{2}T(h) + (\tfrac{1}{2}h)(f(x_1^{(1/2)}) + f(x_3^{(1/2)}) + \cdots + f(x_{2n-1}^{(1/2)})) \quad (5.49)$$

since $x_{2k}^{(1/2)} = x_k$. With this fact, the algorithm for determining the Romberg table is very efficient. The following algorithm uses this technique to compute successive values of $T(h/2^k)$, $k = 0, 1, \ldots$. The general form of (5.43), (5.46), (5.48),

$$T_k(h) = [4^k T_{k-1}(\tfrac{1}{2}h) - T_{k-1}(h)]/(4^k - 1) \quad (5.50)$$

is used to determine the remaining entries in the *Romberg table*.

Romberg Integration Algorithm

> This algorithm computes the first J column of the Romberg table for the integral of $f(x)$ over the interval $[a, b]$. The input consists of a, b, and J, together with a value of n.

1   $h \leftarrow (b - a)/n$
2   Compute $T_{0,1} = T(h)$ using (5.30)
3   For $j = 2, 3, \ldots J$
    3.1 $h \leftarrow \tfrac{1}{2}h$
    3.2 $S \leftarrow 0$
    3.3   For $i = 1, 3, 5, \ldots, 2n - 1$
          3.3.1 $S \leftarrow S + f(a + ih)$ (5.51)
    3.4 $T_{0,j+1} \leftarrow .5T_{0,j} + hS$
    3.5   FOUR $= 4$
    3.6   For $k = 1, 2, \ldots, j$
          3.6.1 $T_{k,j} \leftarrow (\text{FOUR } T_{k-1,j} - T_{k-1,j-1})/(\text{FOUR} - 1)$
          3.6.2   FOUR $\leftarrow$ FOUR 4
4   Output $\{(T_{k,j}, k = 0, 1, \ldots, j), j = 1, 2, \ldots, J\}$

This algorithm computes the first J column of the Romberg table. It should be understood, however, that successive entries will be more and more accurate only if the higher derivatives of the function f are well behaved. It may happen that such is not the case, and that the most accurate values are found in the first few columns. Usually, by scanning the values in the table, we can quickly discover such a behavior.

EXERCISES

1 Write a computer program to compute the Romberg table for the integral $\int_{-1}^{1} (\sin x)^2 \, dx$, starting with $h = .25$.

2 Write out a description of an algorithm to find $T_2(h)$ and determine the efficiency of it. Compare this with the efficiency of Simpson's rule, using $\frac{1}{2}h$. (Your efficiency estimate will depend on $n = (b - a)/h$.)

3 Show that the second formula in Table 5.4 is the same as the approximation $T_2(4h)$ to the integral $\int_{x_0}^{x_0 + 4h} f(x) \, dx$.

4 It was shown in Section 5.2 (equation (5.32)) that the rounding error in the computed value of $T(h/2^k)$ is (approximately) bounded by

$$| \bar{f} | \{ [(b - a)^2/h]2^{k-1} + (2 + l)(b - a) \} \mu',$$

which for large k is approximately equal to

$$| \bar{f} | [(b - a)^2/h]2^{k-1} \mu'.$$

Using this estimate, analyze the rounding error in the computed value of $T_3(h)$.

5 To compute the trapezoidal approximation with, say $h = \frac{1}{64} = 1/2^6$, two methods could be used: Compute $T(\frac{1}{64})$ directly, or compute the first column of the Romberg table: $T(1)$, $T(\frac{1}{2})$, $T(\frac{1}{4})$, ..., $T(\frac{1}{64})$ from formula (5.49). Compare the efficiency of these two methods. Based on this comparison, is it worthwhile to worry very much about the choice of n in Algorithm (5.51)?

Suggestions for Further Reading

Formulas for numerical integration can be found in Krylov (1962) and Stroud and Secrest (1966). For a more complete discussion of Romberg integration, together with programs and further references, see Davis and Rabinowitz (1975).

Chapter 6

SYSTEMS OF LINEAR EQUATIONS

6.1 Basic Concepts

In this section we introduce the notation that will be used throughout the chapter. Also some elementary facts about linear equations are reviewed.

6.1.1 Notation

A *system of n linear equations in n unknowns* is usually expressed in the general form

$$
\begin{aligned}
a_{11}x_1 + a_{12}x_2 + \cdots + a_{1n}x_n &= b_1 \\
a_{21}x_1 + a_{22}x_2 + \cdots + a_{2n}x_n &= b_2 \\
&\ \ \vdots \\
a_{n1}x_1 + a_{n2}x_2 + \cdots + a_{nn}x_n &= b_n.
\end{aligned}
\tag{6.1}
$$

The *coefficients* $a_{11}, a_{12}, \ldots, a_{nn}$ are assumed to be floating point numbers, as are the *right-hand sides* $b_1, b_2, \ldots, b_n$. The problem is to find numbers x_1, $x_2, \ldots, x_n$ so that each of the equations in (6.1) is satisfied.

Often it is convenient to express (6.1) by using **matrix–vector notation**. That is, we write

$$
AX = b
\tag{6.2}
$$

where A is the $n \times n$ *matrix*

| Coefficient matrix | $$A = \begin{pmatrix} a_{11} & a_{12} & \cdots & a_{1n} \\ a_{21} & a_{22} & \cdots & a_{2n} \\ \vdots & & & \\ a_{n1} & a_{n2} & \cdots & a_{nn} \end{pmatrix}$$ | (6.3) |

and b the *column vector*

| Right-hand side | $$b = \begin{pmatrix} b_1 \\ b_2 \\ \vdots \\ b_n \end{pmatrix}.$$ |

The solution to equation (6.2) is again a column vector

| Solution vector | $$X = \begin{pmatrix} x_1 \\ x_2 \\ \vdots \\ x_n \end{pmatrix}.$$ |

Here AX denotes the usual matrix–vector multiplication

$$AX = \begin{pmatrix} \sum_{k=1}^{n} a_{1k} x_k \\ \sum_{k=1}^{n} a_{2k} x_k \\ \vdots \\ \sum_{k=1}^{n} a_{nk} x_k \end{pmatrix}.$$

6.1.2 Nonsingular Systems

Recall from linear equation theory that there are three possibilities concerning the solution of (6.1). Either the system has *no* solution, as, for example, the system

$$2x_1 + 3x_2 = 5$$
$$4x_1 + 6x_2 = 1,$$

in which case the equations are said to be **inconsistent**; or the system may have infinitely many solutions, such as

$$2x_1 + 3x_2 = 5$$
$$4x_1 + 6x_2 = 10.$$

Here the equations are called **dependent**. The third alternative is that the system has exactly one solution, in which case the equations are said to be **nonsingular**. Actually the nonsingularity of the system is completely determined by the coefficients a_{ij}. That is, the system is nonsingular if and only if it has a unique solution for *every* right-hand side $b_1, \ldots, b_n$. Thus we usually speak of the nonsingularity of the coefficient matrix (6.3). If A is not nonsingular, that is, if A is *singular*, then for certain right-hand sides the system will be inconsistent and for others the equations will be dependent.

There are several methods for determining theoretically whether a given system is singular. Computationally, however, the problem of detecting singularity is unstable. That is, small changes in the elements of a singular system will cause it to become nonsingular. More will be said about this later; for now we shall simply assume that the systems we are solving are nonsingular.

6.1.3   Triangular Systems

There are several special forms of (6.1) for which the solution can be determined very easily. The simplest such case is that of a **diagonal matrix**

$$A = \begin{pmatrix} a_{11} & 0 & 0 & \cdots & 0 \\ 0 & a_{22} & 0 & \cdots & 0 \\ \vdots & & & & \vdots \\ & & & & 0 \\ 0 & & & 0 & a_{nn} \end{pmatrix}.$$

Clearly the solution to (6.1) in this case is simply

$$x_1 = b_1/a_{11}, \qquad x_2 = b_2/a_{22}, \ldots, x_n = b_n/a_{nn},$$

provided all the a_{kk} are nonzero. If for some i, $a_{ii} = 0$, then that equation has no solution if $b_i \neq 0$ and has infinitely many solutions if $b_i = 0$.

Slightly more general and also more useful is the **upper triangular form** of the system (6.1) in which $a_{ij} = 0$ if $i > j$. That is, the coefficient matrix (6.3) has the form

$$A = \begin{pmatrix} a_{11} & a_{12} & \cdots & a_{1n} \\ 0 & a_{22} & \cdots & a_{2n} \\ \vdots & & & \vdots \\ 0 & & 0 & a_{nn} \end{pmatrix}.$$

In this case the equations are as follows:

$$a_{11}x_1 + a_{12}x_2 + a_{13}x_3 + \quad \cdots \quad + a_{1n}x_n = b_1$$
$$a_{22}x_2 + a_{23}x_3 + \quad \cdots \quad + a_{2n}x_n = b_2$$
$$a_{33}x_3 + \quad \cdots \quad + a_{3n}x_n = b_3$$
$$\ddots \qquad\qquad\qquad \vdots \qquad\qquad (6.4)$$
$$a_{n-1,\,n-1}x_{n-1} + a_{n-1,\,n}x_n = b_{n-1}$$
$$a_{nn}x_n = b_n.$$

The solution to such a system is easily determined by *backsolving*. That is, the last equation is solved for x_n,

$$x_n = b_n/a_{nn};$$

then this value is used in the next to last equation to determine x_{n-1},

$$x_{n-1} = \left(b_{n-1} - a_{n-1,\,n}x_n\right)/a_{n-1,\,n-1};$$

which in turn is used in the $(n-2)$nd equation to determine x_{n-2}, etc. The complete backsolving process is described by the following INFL program:

Backsolving Algorithm

$$\left[\begin{array}{c} a_{ij} \text{ and } b_i \text{ are coefficients of an upper triangular system} \\ \text{of linear equations.} \end{array} \right]$$

1 Input $\{n, a_{ij}, b_i\}$
2 $x_n \leftarrow b_n/a_{nn}$
3 For $k = n - 1, n - 2, \ldots, 1$
  3.1 $x_k \leftarrow b_k$
  3.2 For $i = k + 1, \ldots, n$ (6.5a)
    3.2.1 $x_k \leftarrow x_k - a_{ki}x_i$
  3.3 $x_k \leftarrow x_k/a_{kk}$
4 Output $\{x_1, \ldots, x_n\}$

Steps 3.1, 3.2, 3.3 of this program determine x_k from the formula

$$x_k = \left(b_k - a_{k,\,k+1}x_{k+1} - a_{k,\,k+2}x_{k+2} - \cdots - a_{kn}x_n\right)/a_{kk}. \qquad (6.5b)$$

If any one of the a_{ii} is zero, then it can be shown (Exercise 4) that the system is singular.

If the coefficients in (6.1) satisfy $a_{ij} = 0$ when $i < j$, then the system is called *lower triangular* and can be solved directly by *forward-solving*, in which equation 1 is first used to determine x_1, then equation 2 is solved for x_2, etc. Examples of both these special triangular forms are given in Table 6.1 together with the backward or forward solutions.

TABLE 6.1

System	Solution
$x_1 - 2x_2 + x_3 = 1$ $2x_2 + 4x_3 = 10$ $-2x_3 = 8$ (upper triangular)	$x_3 = (-\frac{1}{2})8 = -4$ $x_2 = \frac{1}{2}(10 - 4x_3) = 13$ $x_1 = 1 + 2x_2 - x_3 = 31$ (backsolving)
$2x_1 \qquad\qquad = -4$ $-x_1 + x_2 \qquad = 1$ $5x_1 + 3x_2 + 5x_3 = 2$ (lower triangular)	$x_1 = \frac{1}{2}(-4) = -2$ $x_2 = 1 + x_1 = -1$ $x_3 = \frac{1}{5}(2 - 5x_1 - 3x_2) = 3$ (forward-solving)

EXERCISES

1 Write an INFL description of an algorithm for forward-solving a lower triangular system of equations.

2 Determine which of the following systems are nonsingular. For those that are not, tell whether there are *no* solutions or infinitely many solutions.

(a) $2x_1 + x_2 = 5$
    $4x_1 + 2x_2 = 10$

(b) $x_1 - x_2 = 1$
    $x_1 + x_2 = 0$

(c) $x_1 - x_2 = 1$
    $-x_1 + x_2 = 1$

(d) $x_1 + 2x_2 - x_3 = 1$
    $2x_1 + x_2 + 4x_3 = 2$
    $x_1 + x_2 + x_3 = 1$

(e) $x_1 + 2x_2 - x_3 = 1$
    $2x_1 + x_2 + 4x_3 = 2$
    $x_1 + x_2 + x_3 = 0$

(f) $x_1 + 2x_2 - x_3 = 1$
    $x_3 = 2$
    $x_2 + 4x_3 = 1.$

3 Solve the following upper triangular systems using Algorithm (6.5a):

(a) $x_1 + 2x_2 + x_3 = 5$
    $2x_2 + 4x_3 = -4$
    $2x_3 = 4$

(b) $-x_1 + x_2 + x_3 = 0$
    $2x_2 - x_3 = 0$
    $x_3 = -1$

(c) $2x_1 + x_2 - x_3 + x_4 = 6$
    $x_2 + 2x_3 + x_4 = 1$
    $-2x_3 + 4x_4 = -2$
    $3x_4 = -6.$

4 Show that if, for some i, $a_{ii} = 0$ in the upper triangular system (6.4), then the system cannot be nonsingular.

6.2   Gaussian Elimination

The most frequently used method for solving moderately sized linear systems is also one of the oldest such methods, namely, Gaussian elimination. In this section we analyze this important algorithm in detail and discuss some of its properties.

6.2.1   Elementary Operations

The idea of Gaussian elimination is to transform a linear system of the general form (6.1) into a system of the special upper triangular form (6.4). The solution is then found directly by backsolving. The transformation can be done in such a way that, if exact arithmetic is used, the solution of the triangular system will be the same as the solution to the original system. Of course, computationally the transformation will involve rounding errors. Hence the transformed system will have a solution that may differ somewhat from the solution to the given system. Note, however, that the only error is due to rounding; that is, there is no truncation error in this method.

The transformation referred to above is actually a series of transformations in which the coefficients in the lower part of the system are systematically replaced by zeros. First of all, note that the following **elementary operations** on the system (6.1) have no effect on the *solution* of the system:

(i)   An equation can be multiplied by a nonzero constant. (That is, each coefficient and the right-hand side can be multiplied by the same nonzero number.)

(ii)   Two equations can be added together and either of the equations replaced by the sum.

(iii)   Two equations can be interchanged. (That is, the equations can be written down in any order.)

To verify that (i) has no effect on the solution is trivial since such an operation really does not change the equation. Also it is clear that (iii) has no effect on the solution. Operation (ii), however, is not quite so simple. Suppose that equations 1 and 2 in (6.1) are added together and that equation 2 is replaced by the sum. The resulting system is

$$
\begin{aligned}
a_{11}x_1 + \quad\quad a_{12}x_2 + \cdots + \quad\quad a_{1n}x_n &= b_1 \\
(a_{11} + a_{21})x_1 + (a_{12} + a_{22})x_2 + \cdots + (a_{1n} + a_{2n})x_n &= b_1 + b_2 \\
a_{31}x_1 + \quad\quad a_{32}x_2 + \cdots + \quad\quad a_{3n}x_n &= b_3 .
\end{aligned}
$$
$$\vdots \qquad\qquad\qquad \vdots$$

Now if $x_1^*, x_2^*, \ldots, x_n^*$ is a solution to these equations, then equations 1, 3, 4, ..., n of the original system (6.1) are also satisfied. But from the second equation above we have

$$a_{21}x_1^* + a_{22}x_2^* + \cdots + a_{2n}x_n^* + a_{11}x_1^* + a_{12}x_2^* + \cdots + a_{1n}x_n^* = b_1 + b_2$$

and since equation 1 ensures that $a_{11}x_1^* + a_{12}x_2^* + \cdots + a_{1n}x_n^* = b_1$, it follows that

$$a_{21}x_1^* + a_{22}x_2^* + \cdots + a_{2n}x_n^* = b_2.$$

Thus equation 2 of (6.1) is also satisfied. Hence any solution to the above equations is a solution of (6.1).

Finally note that operations (i) and (ii) can be combined into a single operation:

(iv)   Multiply one equation by a nonzero constant, add it to another equation, then replace either equation by this new equation.

Operations (i), (ii), (iii), and (iv) are called *elementary* (row) *operations* on the system (6.1). They will be used to transform the system to the upper triangular form (6.4).

6.2.2   Reduction to Triangular Form

The first step in transforming (6.1) into upper triangular form is to eliminate all coefficients of x_1 except for that in equation 1. This is accomplished by using operation (iv) as follows:

First Reduction Step

$$
\begin{array}{ll}
1 & \text{For } i = 2, 3, \ldots, n \\
& 1.1 \quad m_{i1} \leftarrow -a_{i1}/a_{11} \\
& 1.2 \quad \text{For } j = 1, 2, \ldots, n \\
& \qquad 1.2.1 \quad a_{ij} \leftarrow a_{ij} + m_{i1}a_{1j} \\
& 1.3 \quad b_i \leftarrow b_i + m_{i1}b_1
\end{array}
\qquad (6.6a)
$$

In this program step 1.2 adds to each coefficient in equation i the constant m_{i1} multiplied by the corresponding coefficient of equation 1. Step 1.3 does the same to the right-hand side, and hence steps 1.2 and 1.3 together multiply equation 1 by m_{i1} and add it to equation i. But the *multiplier* m_{i1} has been so chosen that the new value for a_{i1} is *zero*; that is, in step 1.2.1 with $j = 1$:

$$a_{i1} \leftarrow a_{i1} + m_{i1}a_{11} = a_{i1} + (-a_{i1}/a_{11})a_{11} = 0.$$

Thus after doing this for $i = 2, 3, \ldots, n$, the new set of equations has the form:

Equations after the first reduction

$$a_{11}x_1 + a_{12}x_2 + \cdots + a_{1n}x_n = b_1$$
$$a_{22}x_2 + \cdots + a_{2n}x_n = b_2$$
$$\vdots$$
$$a_{n2}x_2 + \cdots + a_{nn}x_n = b_n.$$

Now the only nonzero coefficient of x_1 is a_{11}. Next the process is repeated on the subset that consists of equations 2–n. That is, all coefficients of x_2 in equations 3, 4, ..., n are eliminated by:

Second Reduction Step

1 For $i = 3, 4, \ldots, n$
  1.1 $m_{i2} \leftarrow -a_{i2}/a_{22}$
  1.2 For $j = 2, 3, \ldots, n$ (6.6b)
    1.2.1 $a_{ij} \leftarrow a_{ij} + m_{i2}a_{2j}$
  1.3 $b_i \leftarrow b_i + m_{i2}b_2$

Steps 1.2 and 1.3 multiply equation 2 by m_{i2} and add the result to equation i. By continuing this process with equations 3, 4, ..., $n - 1$, we finally obtain the reduced form (6.4).

The complete algorithm for this reduction is

Gaussian Elimination Algorithm

[Reduction to triangular form by Gaussian elimination.]

1 Input $\{n, a_{ij}, b_i\}$
2 For $k = 1, 2, \ldots, n - 1$
  2.1 For $i = k + 1, k + 2, \ldots, n$
    2.1.1 $m_{ik} \leftarrow -a_{ik}/a_{kk}$
    2.1.2 For $j = k + 1, k + 2, \ldots, n$ (6.7)
      2.1.2.1 $a_{ij} \leftarrow a_{ij} + m_{ik}a_{kj}$
    2.1.3 $b_i \leftarrow b_i + m_{ik}b_k$
3 Output $\{n, a_{ij}, b_i\}$

The main parts of this program (steps 2.1) are essentially the Algorithms (6.6a) and (6.6b) with the subscript k replacing subscripts 1 and 2, respectively. One new feature of (6.7) is that since $a_{i,k} = 0$ in step 2.1.2.1, there is no need to *compute* this zero, and hence the second subscript begins with $k + 1$.

6.2.3 Failure of the Reduction Method

The first thing to observe about this algorithm is that it can only work when $a_{kk} \neq 0$ in step 2.1.1. As an example, consider the set of equations

$$5x_2 + x_3 = 2$$
$$2x_1 + x_2 - x_3 = 1$$
$$4x_1 - x_2 + x_3 = 0.$$

Clearly the algorithm fails at step 2.1.1 when $k = 1$, $i = 2$. However, if the first and second **equations are interchanged** so that the system is written as

$$2x_1 + x_2 - x_3 = 1$$
$$5x_2 + x_3 = 2 \qquad\qquad (6.8)$$
$$4x_1 - x_2 + x_3 = 0,$$

then the algorithm works fine (Exercise 2). Since (6.8) was obtained by an application of one of the elementary operations [operation (iii)], the solution is unchanged. It can be seen that if the equations are nonsingular, then it is *always* possible to interchange equations in this fashion so as to make $a_{kk} \neq 0$ in step 2.1.1. Indeed, suppose that the algorithm has completed several of the reduction steps so that the transformed equations have the form

Equations at the kth reduction step

$$a_{11}x_1 + a_{12}x_2 + \cdots + a_{1k}x_k + \cdots + a_{1n}x_n = b_1$$
$$a_{22}x_2 + \cdots + a_{2k}x_k + \cdots + a_{2n}x_n = b_2$$
$$\vdots$$
$$a_{kk}x_k + \cdots + a_{kn}x_n = b_k \qquad\qquad (6.9)$$
$$\vdots$$
$$a_{nk}x_k + \cdots + a_{nn}x_n = b_n.$$

Now suppose that $a_{kk} = 0$ so that step 2.1.1 cannot be executed. If $a_{lk} \neq 0$ for some $l > k$, then equations l and k can be interchanged so that the system still has the form (6.9) but with the *new* coefficient $a_{kk} \neq 0$. If such an interchange is not possible, that is, if $a_{lk} = 0$ for all $l \geq k$, then the bottom part of (6.9) (equations k, $k + 1$, ..., n) actually has the form

When the interchange is not possible

$$a_{k,\,k+1}x_{k+1} + a_{k,\,k+2}x_{k+2} + \cdots + a_{k,\,n}x_n = b_k$$
$$a_{k+1,\,k+1}x_{k+1} + a_{k+1,\,k+2}x_{k+2} + \cdots + a_{k+1,\,n}x_n = b_{k+1}$$
$$\vdots \qquad\qquad (6.10)$$
$$a_{n,\,k+1}x_{k+1} + a_{n,\,k+2}x_{k+2} + \cdots + a_{nn}x_n = b_n.$$

That is, the unknown x_k does not appear. But (6.10) is an *overdetermined* system of $(n - k + 1)$ equations in $(n - k)$ unknowns $x_{k+1}, x_{k+2}, \ldots, x_n$, and hence may or may not have a solution depending on the values of b_k, $b_{k+1}, \ldots, b_n$.

To illustrate these ideas, consider the system

$$-x_1 + 2x_2 + 3x_3 + x_4 = 1$$
$$2x_1 - 4x_2 - 5x_3 - x_4 = 0$$
$$-3x_1 + 8x_2 + 8x_3 + x_4 = 2$$
$$x_1 + 2x_2 - 6x_3 + 4x_4 = -1.$$

After the first reduction, the system has the form

$$-x_1 + 2x_2 + 3x_3 + x_4 = 1$$
$$0 + 0 + x_3 + x_4 = 2$$
$$0 + 2x_2 - x_3 - 2x_4 = -1$$
$$0 + 4x_2 - 3x_3 + 5x_4 = 0.$$

In order to continue, the **second and third equations must be interchanged.**

$$-x_1 + 2x_2 + 3x_3 + x_4 = 1$$
$$2x_2 - x_3 - 2x_4 = -1$$
$$x_3 + x_4 = 2$$
$$4x_2 - 3x_3 + 5x_4 = 0.$$

The next reduction step gives

$$-x_1 + 2x_2 + 3x_3 + x_4 = 1$$
$$2x_2 - x_3 - 2x_4 = -1$$
$$x_3 + x_4 = 2$$
$$-x_3 + 9x_4 = 2,$$

and one more reduction produces the upper triangular form. To show where the process breaks down when the equations are singular, consider the system

$$3x_1 + 2x_2 - x_3 + 2x_4 = 1$$
$$3x_1 + 4x_2 + x_3 + x_4 = 3$$
$$-6x_1 - 2x_2 + 4x_3 - 3x_4 = 5$$
$$-3x_1 - 6x_2 - 3x_3 - x_4 = 2.$$

It is not obvious that this system has no solution. However, **after two reduction steps**, the system becomes

$$3x_1 + 2x_2 - x_3 + 2x_4 = 1$$
$$2x_2 + 2x_3 - x_4 = 2$$
$$2x_4 = 5$$
$$-x_4 = 7,$$

which clearly has no solution. Moreover, the elimination process cannot continue. Notice that if the right side of the third equation had been -14, then the last two equations in the reduced system would be $2x_4 = -14$ and $-x_4 = 7$, whose solution is $x_4 = -7$. By using this value in the first two equations with *any* value for x_3, we find a solution to the entire system; hence, there are infinitely many solutions.

The previous discussion can be summarized as follows:

If the system (6.1) is nonsingular, then at each reduction step, it is always possible to interchange equations so that the reduction can continue. If the system (6.1) is singular, then the reduction process (in exact arithmetic) will eventually produce a system of the form (6.10). If (6.10) has no solution, then also (6.1) has no solution. If (6.10) does have a solution, then any value can be assigned to x_k, and (6.1) has an infinite number of solutions.

It is important to realize that the above discussion ignores the effect of rounding error during the reduction. In practice, if the reduction process is applied to a singular system, then rounding errors may cause all of the a_{kk} to be nonzero in step 2.1.1. This will be examined more carefully in Section 6.4.

6.2.4   Partial Pivoting

The rounding error analysis in Section 6.4 will show that the process of interchanging equations is also necessary for controlling the rounding error. In fact, we shall see that complications may arise not only when $a_{kk} = 0$ in step 2.1.1 but even when a_{kk} is small. Thus the complete algorithm for reducing the system to upper triangular form should be essentially Algorithm (6.7) but with an interchange of equations so that each time step 2.1.1 is executed the denominator a_{kk} will be as large as possible. This will cause the multipliers m_{ij} to be less than 1 in absolute value. This interchanging of equations is called **partial pivoting**, and the complete algorithm is called

Gaussian elimination with partial pivoting. As an example of how partial pivoting affects the reduction process, consider the system

<table>
<tr><td>

Example:
Partial
pivoting
is needed
</td><td>

$$(1.000 \times 10^{-5})x_1 + 1.000x_2 = 1.000$$
$$1.000x_1 + 1.000x_2 = 2.000.$$
</td><td>(6.11)</td></tr>
</table>

The exact solution is $x_1 = 1.00001000\ldots, x_2 = .99998999\ldots$. If this system is **reduced with four-decimal-digit arithmetic** *without* partial pivoting, the result is the triangular system

<table>
<tr><td>

Reduced
system with
4-digit
arithmetic
</td><td>

$$1.000 \times 10^{-5}x_1 + 1.000x_2 = 1.000$$
$$-1.000 \times 10^5 x_2 = -1.000 \times 10^5.$$
</td></tr>
</table>

The *exact* solution to this system is the rather strange result $x_2 = 1, x_1 = 0$! However, if the equations are interchanged, then the four-digit arithmetic gives the reduced system

<table>
<tr><td>

Reduced
system after
pivoting
</td><td>

$$1.000x_1 + 1.000x_2 = 2.000$$
$$1.000x_2 = 1.000.$$
</td></tr>
</table>

The solution now is $x_1 = 1.000$, $x_2 = 1.000$. The reason for the large error in the first solution will be explained in Section 6.5.

6.2.5   The Algorithm

An INFL description of the algorithm for Gaussian elimination with partial pivoting is complicated by the fact that, in practice, it is very time-consuming and wasteful to interchange all of the coefficients of the two pivoted equations. Instead the pivoting is accomplished by interchanging the *subscripts* of the coefficients. This technique results in a very efficient but somewhat complicated algorithm. The following program uses an array called "sub" to store the subscripts that denote the equation numbers. Initially, sub(i) has the value i for $i = 1, 2, \ldots, n$. But if equations 3 and 7, for example, are interchanged, then sub(3) will have the value 7 and sub(7) will have the value 3.

Algorithm for Gaussian Elimination with Partial Pivoting

1 Input $\{n,\ a_{ij},\ b_i\}$
2 For $i = 1, 2, \ldots, n$
  2.1 $\text{sub}(i) = i$
3 For $k = 1, 2, \ldots, n - 1$
  3.1 $\max \leftarrow 0$
  3.2 For $i = k, k + 1, \ldots, n$
    3.2.1 $\text{abs} \leftarrow |a_{\text{sub}(i),\,k}|$
    3.2.2 If $\max < \text{abs}$,
      Then $\max = \text{abs}$
      $\text{indx} \leftarrow i$
  3.3 If $\max = 0$, then error exit   (6.12a)
  3.4 $j \leftarrow \text{sub}(k)$
  3.5 $\text{sub}(k) \leftarrow \text{sub}(\text{indx})$
  3.6 $\text{sub}(\text{indx}) \leftarrow j$
  3.7 $\text{pivot} \leftarrow a_{\text{sub}(k),\,k}$
  3.8 For $i = k + 1, k + 2, \ldots, n$
    3.8.1 $m_{\text{sub}(i),\,k} \leftarrow -a_{\text{sub}(i),\,k}/\text{pivot}$
    3.8.2 For $j = k + 1, k + 2, \ldots, n$
      3.8.2.1 $a_{\text{sub}(i),\,j} \leftarrow a_{\text{sub}(i),\,j}$
           $+ m_{\text{sub}(i),\,k}\, a_{\text{sub}(k),\,j}$
    3.8.3 $b_{\text{sub}(i)} \leftarrow b_{\text{sub}(i)} + m_{\text{sub}(i),\,k}\, b_{\text{sub}(k)}$
4 Output $\{n,\ a_{ij},\ b_i,\ m_{ij},\ \text{sub}(i)\}$

This algorithm may seem a little less imposing if steps 3.8 are compared with steps 2.1 of Algorithm (6.7). In the above program we always write $a_{\text{sub}(i),\,j}$ instead of a_{ij}.

The output from this algorithm is the upper triangular system (6.4). In order to compute the solution to these equations, it is necessary to go through the backsolving process. The pivoting that was done during the elimination must, however, also be included in the backsolving. The following is a modification of the backsolving algorithm (6.5) that uses the output from (6.12a), including the pivot information, to solve the system of equations:

Algorithm to Solve the Upper Triangular System
as Computed by (6.12a)

[The input is from the output of (6.12a).]

1 Input $\{n,\ a_{ij},\ b_i,\ \text{sub}(i)\}$
2 $x_n \leftarrow b_{\text{sub}(n)}/a_{\text{sub}(n),\,n}$
3 For $k = n - 1, n - 2, \ldots, 1$
  3.1 $x_k \leftarrow b_{\text{sub}(k)}$ (6.12b)
  3.2 For $i = k + 1, k + 2, \ldots, n$
    3.2.1 $x_k \leftarrow x_k - a_{\text{sub}(k),\,i}\, x_i$
  3.3 $x_k \leftarrow x_k/a_{\text{sub}(k),\,k}$
4 Output $\{x_1, x_2, \ldots, x_n\}$

Notice that the multipliers m_{ik}, which are computed in step 3.8.1 of Algorithm (6.12a), are not used in the solution Algorithm (6.12b). In fact, it appears that there is no reason at all to save and output them since after being used in steps 3.8.2.1 and 3.8.3 [of Algorithm (6.12a)] they are not used again. There are several reasons, however, why it is advisable to save them and include them in the output of Algorithm (6.12a). First of all, they require no additional storage. That is, step 3.8.1, which computes $m_{\text{sub}(i),\,k}$, uses the value of $a_{\text{sub}(i),\,k}$ for the last time. The next steps do some computations that produce a *zero* for the new value of $a_{\text{sub}(i),\,k}$, and this zero need not be stored. Thus the value of $m_{\text{sub}(i),\,k}$ could be stored in place of $a_{\text{sub}(i),\,k}$. This is easily done by writing step 3.8.1 as

Modification to save storage	$a_{\text{sub}(i),\,k} \leftarrow -a_{\text{sub}(i),\,k}/\text{pivot}$

and then using $a_{\text{sub}(i),\,k}$ instead of $m_{\text{sub}(i),\,k}$ in steps 3.8.2.1 and 3.8.3.

Secondly, it will be shown in Section 6.5 that it is often necessary to re-solve the system (6.1) with the same coefficients a_{ij} but with different right-hand sides. Since the multipliers m_{ij} and the coefficients of the upper triangular system do not depend on the right-hand side, they need not be recomputed. The only part of Algorithm (6.12a) that must be repeated is step 3.8.3, which involves the right-hand side. The following algorithm uses the results of Algorithm (6.12a) to solve the system (6.1) with new values c_1, $c_2, \ldots, c_n$ in place of $b_1, b_2, \ldots, b_n$.

Algorithm for Re-solving the Linear System (6.1) with New Values $c_1, c_2, \ldots, c_n$ on the Right-Hand Side

$\Big[$ The input to this algorithm includes the output from Algorithm (6.12a). $\Big]$

  1 Input $\{n, a_{ij}, m_{ij}, \text{sub}(i), c_i\}$
  2 For $k = 1, 2, \ldots, n - 1$
     2.1 $x_k \leftarrow c_{\text{sub}(k)}$
     2.2 $c_{\text{sub}(k)} \leftarrow c_k$
     2.3 For $i = k + 1, k + 2, \ldots, n$
        2.3.1 $c_i \leftarrow c_i + m_{\text{sub}(i),\,k} x_k$ (6.12c)
  3 $x_n \leftarrow c_n / a_{\text{sub}(n),\,n}$
  4 For $k = n - 1, n - 2, \ldots, n$
     4.1 For $i = k + 1, \ldots, n$
        4.1.1 $x_k \leftarrow x_k - a_{\text{sub}(k),\,i} x_i$
     4.2 $x_k \leftarrow x_k / a_{\text{sub}(k),\,k}$
  5 Output $\{x_1, x_2, \ldots, x_n\}$

In this algorithm steps 2.1 and 2.2 interchange the components c_k and $c_{sub(k)}$ just as equations k and $sub(k)$ were interchanged in Algorithm (6.12a). Step 2.3 applies the reduction process to the new right-hand side, and step 4 is the backsolving algorithm (6.12b). The initialization steps 2 and 3.1 of (6.12b) are incorporated into the interchange steps and step 3 of Algorithm (6.12c).

6.2.6   Application to Tridiagonal Systems

Many applications give rise to linear equations with certain special properties that simplify the method of solution. One such property is **bandedness**. This means that the system can be written in a form such as

$$a_{11}x_1 + a_{12}x_2 + a_{13}x_3 \qquad\qquad\qquad\qquad\qquad = b_1$$

$$a_{21}x_1 + a_{22}x_2 + a_{23}x_3 + a_{24}x_4 \qquad\qquad\qquad = b_2$$

$$a_{31}x_1 + a_{32}x_2 + a_{33}x_3 + a_{34}x_4 + a_{35}x_5 \qquad = b_3$$

$$a_{42}x_2 + a_{43}x_3 + a_{44}x_4 + a_{45}x_5 + a_{46}x_6 = b_4$$

$$\ddots \qquad\qquad \vdots$$

$$a_{n,\,n-2}x_{n-2} + \cdots + a_{nn}x_n = b_n.$$

That is, the coefficients a_{ij} are zero whenever $|i - j| > w$ for some w less than n. In the above example, $w = 2$. If $w = 0$, the system is diagonal, and when $w = 1$, it is called **tridiagonal**. In this latter case it is often more convenient to use a somewhat different notation for the coefficients. In fact, we write the system as

$$d_1x_1 + u_1x_2 \qquad\qquad\qquad\qquad = b_1$$

$$l_2x_1 + d_2x_2 \quad + u_2x_3 \qquad\qquad = b_2$$

$$l_3x_2 + d_3x_3 + u_3x_4 = b_3$$

$$\ddots \qquad \vdots$$

$$l_nx_{n-1} + d_nx_n = b_n.$$

Evidently the coefficients can now be stored in three one-dimensional arrays $D(\cdot)$, $U(\cdot)$, and $L(\cdot)$, instead of a full two-dimensional array $A(I, J)$.

Gaussian elimination for a tridiagonal system is simplified by the fact that the system already has nearly triangular form. All that needs to be done is to eliminate the coefficients $l_2, l_3, l_4, \ldots, l_n$. If pivoting is not needed, this elimination can be easily done by a slight modification of the general algorithm (6.7):

Algorithm to Reduce a Tridiagonal System

1   Input $\{n, d_i, u_i, l_i, b_i\}$
2   For $k = 1, 2, \ldots, n - 1$
     2.1 $m_k \leftarrow -l_{k+1}/d_k$
     2.2 $d_{k+1} \leftarrow d_{k+1} + m_k u_k$ (6.12d)
     2.3 $b_{k+1} \leftarrow b_{k+1} + m_k b_k$
3   Output $\{n, d_i, u_i, b_i\}$

The backsolving algorithm can be modified in a similar manner. Clearly, the amount of computation is drastically reduced, and one might expect the accuracy of the computed solution to be slightly improved (see Section 6.4, Exercise 5). Unfortunately, if pivoting is needed, then the algorithm becomes somewhat more complicated. Consider, for example, the system

$$10^{-5}x_1 + x_2 \qquad\quad = 1$$
$$x_1 + x_2 + x_3 = -1$$
$$x_2 + x_3 = 2.$$

If the first two equations are interchanged, then the system has the form

$$x_1 + x_2 + x_3 = -1$$

$\boxed{\text{System after pivoting}}$ $10^{-5}x_1 + x_2 \qquad\quad = 1$

$$x_2 + x_3 = 2.$$

This system is no longer tridiagonal because the (new) coefficient a_{13} is not zero. Generally, if pivoting is included in the algorithm, it is necessary to have an extra array to store the possibly nonzero coefficients $a_{13}, a_{24}, a_{35}, \ldots, a_{n-2, n}$. Even in this case, however, the reduction process is still considerably simpler than for the general algorithm (6.12) (see Exercise 6).

EXERCISES

1   Use Algorithm (6.7) to transform the following systems into upper triangular form:
    (a) $x_1 + 2x_2 - x_3 = 1$ (b) $2x_1 + x_2 + 2x_3 = 0$
           $-x_1 + x_2 + x_3 = 2$ $x_1 - x_2 + x_3 = 1$
           $2x_1 - x_2 + 2x_3 = -1$ $2x_2 - x_3 = 2.$

2   Show that Algorithm (6.7) can be used to transform the system (6.8) into an upper triangular system that can be solved by backsolving.

3 Use Gaussian elimination with partial pivoting and four decimal digit arithmetic to transform the following systems into upper triangular form:

(a) $x_1 + 2x_2 = 1$ (b) $2x_1 - x_2 + x_3 = 1$
      $2x_1 - x_2 = -1$ $2x_1 + 3x_3 = 1$
                                   $x_1 + 2.5x_2 - .5x_3 = 0.5$

(c) $3x_1 + 2x_2 - x_3 = 0$
      $3 \times 10^{-2}x_1 + 2.001 \times 10^{-2}x_2 + .99x_3 = 1$
      $x_1 + 1.667x_2 + .667x_3 = 2.$

4 Consider the (singular) system

$$3x_1 + 2x_2 - x_3 + 2x_4 = 1$$
$$2x_2 + 2x_3 - x_4 = 2$$
$$2x_4 = \alpha$$
$$-x_4 = 1.$$

Show that for certain values of α the system has *no* solution, whereas for other values the system has infinitely many solutions.

5 Show that if (6.1) can be reduced to the form (6.9), where $a_{kk} = a_{k+1,k} = \cdots = a_{nn} = 0$, then the system must be singular.

6 Describe an algorithm for solving a tridiagonal system by Gaussian elimination with partial pivoting. Assume the coefficients of the system are l_i, d_i, u_i, $i = 1, 2, \ldots$, as described in Section 6.2.6. Let c_i, $i = 1, 2, \ldots, n$, be an additional array that is initially zero. When equations k and $k + 1$ are interchanged, then c_k should be given the value u_{k+1}.

6.3[†] Matrix Applications of Gaussian Elimination

A slightly deeper understanding of matrix theory is required for this section than for the remainder of the chapter. Concepts such as determinants, inverses, and permutation matrices should be familiar to the reader. Here we discuss two important matrix applications of the Gaussian elimination algorithms. They will not be used, however, in later sections.

6.3.1 Matrix Computations—LU Decomposition

The basic Gaussian elimination algorithm developed in Section 6.2 can be used to solve the matrix–vector equation (6.2). There are several other problems that have meaning *only* in the context of matrix theory but that can be solved with the help of this same algorithm. We shall consider in this section two such problems, namely, computing the *determinant* of a matrix, and computing the *inverse* of a matrix. In order to describe these computations, it is necessary to express part of the Gaussian elimination algorithm in matrix terms. For this, we first show that operation (iv) of subsection 6.2.1

[†] This section is optional.

applied to the matrix (6.3) is equivalent to multiplying A on the left by a particularly simple matrix. That is, suppose equation k is to be multiplied by the number m_{ik} and then added to equation i, the result to replace equation i. This can be done by multiplying A on the left by the matrix

$$
M_{ik} = \begin{pmatrix} 1 & 0 & & & & \cdots & & 0 \\ 0 & \cdot & & & & & & \cdot \\ & \cdot & \cdot & & & & & \\ & & \cdot & 1 & & & & \cdot \\ \cdot & & & 0 & & & & \\ & & & m_{ik} & \cdot & & & \\ & & & 0 & & \cdot & & \\ & & & \vdots & & & \cdot & \\ 0 & \cdots & & 0 & & \cdots & & 1 \end{pmatrix}. \tag{6.13}
$$

That is, M_{ik} is the identity matrix but with the multiplier m_{ik} in the (i, k) position. For example, with $n = 3$ and $i = 3$, $k = 1$, we have

$$
M_{31} A = \begin{pmatrix} 1 & 0 & 0 \\ 0 & 1 & 0 \\ m_{31} & 0 & 1 \end{pmatrix} \begin{pmatrix} a_{11} & a_{12} & a_{13} \\ a_{21} & a_{22} & a_{23} \\ a_{31} & a_{32} & a_{33} \end{pmatrix}
$$

$$
= \begin{pmatrix} a_{11} & a_{12} & a_{13} \\ a_{21} & a_{22} & a_{23} \\ m_{31} a_{11} + a_{31} & m_{31} a_{12} + a_{32} & m_{31} a_{13} + a_{33} \end{pmatrix}.
$$

Since the reduction Algorithm (6.7) consists of a sequence of operations of the above type, it should be clear that we can describe it as a series of matrix multiplications

$$
M_{n, n-1} \cdots M_{ik} \cdots M_{41} M_{31} M_{21} A = U. \tag{6.14}
$$

Here the matrices M_{ik} are defined by (6.13) with the **multipliers** m_{ik} given by

$$
m_{ik} = \frac{-a_{ik}}{a_{kk}}, \qquad \begin{array}{l} k = 1, 2, \ldots, n-1 \\ i = k+1, k+2, \ldots, n. \end{array}
$$

These multipliers are computed in step 2.1.1 of Algorithm (6.7). The result of all these matrix multiplications is the coefficient matrix of the reduced system (6.4); that is, U in (6.14) is an **upper triangular matrix**

$$
U = \begin{pmatrix} u_{11} & u_{12} & & u_{1n} \\ 0 & u_{22} & & u_{2n} \\ \vdots & & \ddots & \vdots \\ 0 & \cdots & 0 & u_{nn} \end{pmatrix}.
$$

To illustrate this, consider again the case $n = 3$. Then

$$M_{21} = \begin{pmatrix} 1 & 0 & 0 \\ m_{21} & 1 & 0 \\ 0 & 0 & 1 \end{pmatrix}, \qquad m_{21} = -a_{21}/a_{11}$$

and

$$M_{21} A = \begin{pmatrix} a_{11} & a_{12} & a_{13} \\ m_{21} a_{11} + a_{21} & m_{21} a_{12} + a_{22} & m_{21} a_{13} + a_{23} \\ a_{31} & a_{32} & a_{33} \end{pmatrix}$$

$$= \begin{pmatrix} a_{11} & a_{12} & a_{13} \\ 0 & a'_{22} & a'_{23} \\ a'_{31} & a'_{32} & a'_{33} \end{pmatrix}.$$

Here the new $(2, 1)$ element is zero because

$$m_{21} a_{11} + a_{21} = (-a_{21}/a_{11}) a_{11} + a_{21} = -a_{21} + a_{21} = 0.$$

Next we determine

$$M_{31} = \begin{pmatrix} 1 & 0 & 0 \\ 0 & 1 & 0 \\ m_{31} & 0 & 1 \end{pmatrix}$$

and compute the product $M_{31} M_{21} A$. This gives the matrix

$$\begin{pmatrix} a_{11} & a_{12} & a_{13} \\ 0 & a'_{22} & a'_{23} \\ 0 & a''_{32} & a''_{33} \end{pmatrix}.$$

Finally multiplication by

$$M_{32} = \begin{pmatrix} 1 & 0 & 0 \\ 0 & 1 & 0 \\ 0 & m_{32} & 1 \end{pmatrix}$$

gives the upper triangular matrix

$$M_{32} M_{31} M_{21} A = \begin{pmatrix} a_{11} & a_{12} & a_{13} \\ 0 & a'_{22} & a'_{23} \\ 0 & 0 & a'''_{33} \end{pmatrix}.$$

The elements of the above matrix are the values of a_{ij} that are the output of Algorithm (6.7).

An interesting property of the matrix M_{ik} defined by (6.13) is that its inverse is obtained by simply changing the sign of the (i, k) element. That is,

$$
M_{ik}^{-1} = \begin{pmatrix} 1 & & & & & \\ 0 & \cdot & & \cdot & & \\ 0 & & \cdot & 1 & & \\ \cdot & & & \vdots & \cdot & \\ \cdot & & & -m_{ik} & \cdot & \\ \cdot & & & \vdots & & \cdot \\ 0 & \cdots & 0 & \cdots & 0 & 1 \end{pmatrix}.
\tag{6.15}
$$

This is easily verified by simply showing that if M_{ik} is multiplied by the matrix given in (6.15), the result is the identity matrix I. Indeed in the $n = 3$ case, for example, we have for M_{21}:

$$
\begin{pmatrix} 1 & 0 & 0 \\ m_{21} & 1 & 0 \\ 0 & 0 & 1 \end{pmatrix} \begin{pmatrix} 1 & 0 & 0 \\ -m_{21} & 1 & 0 \\ 0 & 0 & 1 \end{pmatrix} = \begin{pmatrix} 1 & 0 & 0 \\ m_{21} - m_{21} & 1 & 0 \\ 0 & 0 & 1 \end{pmatrix} = I.
$$

Thus since M_{ik}^{-1} exists, the formula (6.14) can be written as

$$
A = M_{21}^{-1} M_{31}^{-1} M_{41}^{-1} \cdots M_{n,\,n-1}^{-1} U,
\tag{6.16}
$$

where M_{ik}^{-1} is given by (6.15). Finally this equation can be simplified by using the fact that

$$
M_{21}^{-1} M_{31}^{-1} \cdots M_{n1}^{-1} \cdots M_{n,\,n-1}^{-1} = \begin{pmatrix} 1 & 0 & 0 & \cdots & & 0 \\ -m_{21} & 1 & 0 & \cdots & & 0 \\ -m_{31} & -m_{32} & 1 & 0 & & 0 \\ \vdots & & & \cdot & \cdot & \vdots \\ & & & & & 0 \\ -m_{n1} & \cdots & & -m_{n,\,n-1} & & 1 \end{pmatrix}.
$$

Again we leave the reader to verify the general case and show the details only for the case $n = 3$:

$$
M_{21}^{-1} M_{31}^{-1} M_{32}^{-1} = \begin{pmatrix} 1 & 0 & 0 \\ -m_{21} & 1 & 0 \\ 0 & 0 & 1 \end{pmatrix} \begin{pmatrix} 1 & 0 & 0 \\ 0 & 1 & 0 \\ -m_{31} & 0 & 1 \end{pmatrix} \begin{pmatrix} 1 & 0 & 0 \\ 0 & 1 & 0 \\ 0 & -m_{32} & 1 \end{pmatrix}
$$

$$
= \begin{pmatrix} 1 & 0 & 0 \\ -m_{21} & 1 & 0 \\ -m_{31} & 0 & 1 \end{pmatrix} \begin{pmatrix} 1 & 0 & 0 \\ 0 & 1 & 0 \\ 0 & -m_{32} & 1 \end{pmatrix} = \begin{pmatrix} 1 & 0 & 0 \\ -m_{21} & 1 & 0 \\ -m_{31} & -m_{32} & 1 \end{pmatrix}.
$$

With this, we can write (6.16) as

$$
A = LU,
\tag{6.17}
$$

where L is the **lower triangular matrix**

$$L = \begin{pmatrix} 1 & 0 & 0 & \cdots & 0 \\ -m_{21} & 1 & 0 & & \vdots \\ \vdots & & & \ddots & 0 \\ -m_{n1} & \cdots & -m_{n,\,n-1} & & 1 \end{pmatrix}$$

and U is the **upper triangular matrix** corresponding to the reduced system (6.4).

Triangular matrices L and U whose product equals A are called an *LU decomposition* of A. The matrix interpretation of Gaussian elimination is just the computation of an *LU* decomposition.

As shown in the previous section, for some matrices it is necessary to interchange rows (pivot) in order to complete the Gaussian elimination process. This can also be formulated in matrix terms by using *permutation matrices*. Recall that a permutation matrix is an $n \times n$ matrix that has exactly one 1 in every row and column and zeros elsewhere. Examples of 3×3 permutation matrices are

$$\begin{pmatrix} 1 & 0 & 0 \\ 0 & 1 & 0 \\ 0 & 0 & 1 \end{pmatrix}, \begin{pmatrix} 1 & 0 & 0 \\ 0 & 0 & 1 \\ 0 & 1 & 0 \end{pmatrix}, \begin{pmatrix} 0 & 1 & 0 \\ 0 & 0 & 1 \\ 1 & 0 & 0 \end{pmatrix}, \begin{pmatrix} 0 & 0 & 1 \\ 0 & 1 & 0 \\ 1 & 0 & 0 \end{pmatrix}, \text{ etc.}$$

The ith and jth rows of a matrix A can be interchanged by multiplying A on the left by a special permutation matrix, which we shall denote by P_{ij}. The matrix P_{ij} is constructed from the identity matrix by interchanging rows i and j; that is

Permutation
matrix

$$P_{ij} = \begin{pmatrix} 1 & & & & & & & & & \\ & \ddots & & & & & & & & \\ & & 1 & & & & & & & \\ & & & 0 & \cdots & 1 & & & & \leftarrow \text{row } i\\ & & & & 1 & & & & & \\ & & & \vdots & & \ddots & & \vdots & & \\ & & & & & & 1 & & & \\ & & & 1 & \cdots & & 0 & & & \leftarrow \text{row } j.\\ & & & & & & & 1 & & \\ & & & & & & & & \ddots & \\ & & & & & & & & & 1 \end{pmatrix}$$

$$\text{col.}\qquad\qquad \text{col.}\qquad\qquad\qquad (6.18)$$
$$i \qquad\qquad\qquad\;\; j$$

Thus, in order to interchange rows 1 and 3 of the 3×3 matrix A, we multiply A from the left by

$$P_{13} = \begin{pmatrix} 0 & 0 & 1 \\ 0 & 1 & 0 \\ 1 & 0 & 0 \end{pmatrix}$$

to give

$$\begin{pmatrix} 0 & 0 & 1 \\ 0 & 1 & 0 \\ 1 & 0 & 0 \end{pmatrix} \begin{pmatrix} a_{11} & a_{12} & a_{13} \\ a_{21} & a_{22} & a_{23} \\ a_{31} & a_{32} & a_{33} \end{pmatrix} = \begin{pmatrix} a_{31} & a_{32} & a_{33} \\ a_{21} & a_{22} & a_{23} \\ a_{11} & a_{12} & a_{13} \end{pmatrix}.$$

Gaussian elimination with partial pivoting is a combination of elimination steps and row interchanges. Thus, it can be described in matrix terms as a sequence of multiplications by a combination of matrices of the form (6.13) and of the form (6.18). That is, equation (6.14) becomes

$$M_{n,\,n-1} P_{n-1} M_{n,\,n-2} M_{n-1,\,n-2} P_{n-1} \cdots M_{n,\,1} M_{n-1,\,1} \cdots M_{31} M_{21} P_1 A$$
$$= U. \tag{6.19}$$

Here $P_1, P_2, \ldots, P_{n-1}$ are the permutation matrices (6.18) that perform the row interchanges. The result of this series of multiplications is the upper triangular matrix U. Now (6.19) can be written as

$$A = P_1^{-1} M_{21}^{-1} M_{31}^{-1} \cdots P_{n-1}^{-1} M_{n,\,n-1}^{-1} U = \tilde{L} U. \tag{6.20}$$

Notice that P_k of the form (6.18) has the property that $P_k^{-1} = P_k$. (That is, the *inverse* of interchanging rows i and j is the interchanging of rows i and j!) The matrix $\tilde{L}$ in (6.20) may not be lower triangular; but it can be shown that $\tilde{L} = PL$, where P is a permutation matrix and L is a lower triangular matrix as before (see Exercise 8). Thus (6.20) can be written as

$$A = PLU, \tag{6.20'}$$

where

P is a permutation matrix
L is a lower triangular matrix
U is an upper triangular matrix.

Equation (6.20'), written as $PA = LU$, is of great theoretical importance. It shows that, for any nonsingular system (6.1), it is always possible to write the equations down in some order so that the Gaussian elimination algorithm can be carried out. It can even be shown that the L matrix in (6.20') is just the product of the M_{ij}^{-1} matrices in (6.20). That is, the multipliers m_{ij} and the reduced matrix U, which are obtained by interchanging rows *during* the computation, are exactly the same as would be computed if the interchanging were done *before* the computation had begun.

6.3.2 Computing Determinants

Consider the case when A has an LU decomposition (6.17). Since the determinant of a product of two matrices is equal to the product of the determinants, we have

$$\det(A) = \det(L) \det(U).$$

But the determinant of a *triangular* matrix is just the product of the diagonal elements. Thus we have

$$\det(L) = 1 \cdot 1 \cdot 1 \cdots 1 = 1$$

and

$$\det(U) = u_{11} u_{22} \cdots u_{nn}.$$

Therefore *the determinant of A is just the product of the diagonal elements of the reduced matrix U.*

In the more general case (6.20) when pivoting is used, we must write

$$\det(A) = \det(P_2^{-1}) \det(M_{21}^{-1}) \cdots \det(M_{n,\,n-1}^{-1}) \det(U).$$

Because of (6.15) we still have $\det(M_{ij}^{-1}) = 1$. It can be easily shown that if $P_k \neq I$, then $\det(P_k^{-1}) = -1$. That is, if P_k^{-1} really causes an interchange of rows, then its determinant is -1. Thus

$$\det(A) = \pm \det(u) = \pm u_{11} u_{22} \cdots u_{nn}. \tag{6.21}$$

Here the plus or minus sign depends on whether there is an even or odd number of row interchanges during the reduction.

We summarize this discussion as follows:

> Let A be any $n \times n$ matrix, and let U be the upper triangular matrix that results from applying Gaussian elimination with partial pivoting. Then
>
> $$\det(A) = \pm u_{11} u_{22} \cdots u_{nn}$$
>
> where the plus or minus sign depends upon whether an even or an odd number of row interchanges occurred.

To compare this method of computing $\det(A)$ with the "standard" method of expanding A by minors along some row or column, recall that this latter method requires the computation of the sum of $n!$ terms. Each of these terms consists of the product of n elements of A, for a total operation count of $(n! - 1)A + (n!(n - 1))M$. On the other hand, it will be shown in Section 6.4 that the reduction process without pivoting involves $\frac{1}{3}(n^3 - n)A + \frac{1}{3}(n^3 - n)M + \frac{1}{2}n(n - 1)D$ operations. The product of the diagonal elements of the upper triangular matrix is an additional $(n - 1)$ multiplications, for a total of $(n^3 + 2n - 3)/3$ multiplications. Table 6.2 gives some comparisons between these two methods for several values of n. Obviously, even if division is twice as expensive as multiplication, the Gaussian elimination method is faster for small n as well as for large n.

TABLE 6.2   *Number of Operations to Evaluate* $\det(A)$ *by Gaussian Elimination* (G.E.) *and by Expansion by Minors*

n	2	3	4	10	20
Additions					
G. E.	2	8	20	330	2660
minors	1	5	23	$\sim 3 \times 10^6$	$\sim 2 \times 10^{18}$
Multiplications					
G. E.	3	10	23	339	2679
minors	2	12	72	$\sim 3 \times 10^7$	$\sim 40 \times 10^{18}$
Divisions					
G. E.	1	2	6	45	190
minors	0	0	0	0	0

6.3.3   Computing Inverses

Cramer's method for finding the inverse of a nonsingular $n \times n$ matrix, which is often taught in courses on matrix theory, involves the computation of $n + 1$ determinants. As shown above, even by using Gaussian elimination, the computation of determinants is fairly expensive. A much more efficient method for finding A^{-1} is based on the following observation.

Let e_i denote the ith column of the identity matrix I; that is, let

$$e_i = \begin{pmatrix} 0 \\ \vdots \\ 0 \\ 1 \\ \vdots \\ 0 \end{pmatrix} \leftarrow i\text{th component.}$$

Let b_i denote the ith column of the matrix A^{-1}. Then from the equation

$$AA^{-1} = I$$

it follows that

$$Ab_i = e_i, \qquad i = 1, 2, \ldots, n.$$

Thus the solution to the equation

$$Ax = e_i \tag{6.22}$$

is just the ith column of A^{-1}. Hence, A^{-1} can be determined by solving (6.22) for $i = 1, 2, \ldots, n$. This can be done efficiently with Gaussian elimination by using Algorithms (6.12a)–(6.12c). That is, the following algorithm computes the columns of A^{-1}:

Algorithm to Compute A^{-1}

1   Input $\{n, a_{ij}\}$
2   Use Algorithms (6.12a) and (6.12b)
    to solve $Ax = e_1$. Call the solution d_1
3   For $i = 2, 3, \ldots, n$ (6.23)
    3.1 Use Algorithm (6.12c) to solve
         $Ax = e_i$. Call the solution d_i
4   Output $\{d_1, d_2, \ldots, d_n\}$

It will be shown in Section 6.6 that the reduction Algorithm (6.12a) requires approximately $\frac{1}{3}n^3$ additions and multiplications, whereas Algorithms (6.12b) and (6.12c) use only approximately n^2 operations. Hence, the use of Algorithm (6.12b) and the $(n - 1)$ applications of Algorithm (6.12c) involve altogether only slightly more computations than does the initial reduction algorithm. The entire process is *considerably* better than Cramer's rule.

A word of caution about using inverses is appropriate here. It might be thought that if the system $Ax = b$ is to be solved for several different right-hand sides b, it would be best to compute A^{-1} and then simply multiply A^{-1} by b. Note, however, that it requires $n^2A + n^2M$ operations to compute the product $A^{-1}b$. On the other hand, once the reduction Algorithm (6.12a) has been applied, the system $Ax = b$, for any vector b, can be solved by Algorithm (6.12c) in only slightly more than $n^2A + n^2M$ operations. Thus the work required to find A^{-1} is extra overhead that can be avoided by simply using Algorithms (6.12a)–(6.12c) together.

EXERCISES

1 Find the LU decomposition of the following matrices (no pivoting is necessary):

$$\begin{pmatrix} 2 & 1 \\ -1 & 2 \end{pmatrix}, \quad \begin{pmatrix} 2 & -1 & 0 \\ -1 & 2 & -1 \\ 0 & -1 & 2 \end{pmatrix}, \quad \begin{pmatrix} 3 & 0 & 0 \\ -1 & 2 & 0 \\ 2 & 1 & 2 \end{pmatrix}.$$

2 Give the permutation matrix that, if multiplied on the left, will produce the indicated interchange of rows.

   (a) Interchange rows 1 and 3.
   (b) Interchange rows 1 and 3, then rows 3 and 4.
   (c) Interchange rows 3 and 1, then rows 1 and 4.

3 What happens if the matrix A is multiplied on the *right* by a permutation matrix?

4 Find the determinants of the matrices in Exercise 1.

5 Use Gaussian elimination (with pivoting) to find the determinants of the following matrices:

$$\begin{pmatrix} 0 & 1 \\ 1 & 1 \end{pmatrix}, \quad \begin{pmatrix} 3 & 2 & -1 & 2 \\ -6 & -2 & 4 & -3 \\ -3 & -6 & -3 & -1 \\ 3 & 4 & 1 & 1 \end{pmatrix}, \quad \begin{pmatrix} -1 & 2 & 3 & 1 \\ 2 & -4 & -5 & -1 \\ -3 & 8 & 8 & 1 \\ 1 & 2 & -6 & 4 \end{pmatrix}.$$

6 Use Gaussian elimination to find the inverses of the matrices in Exercise 1.

7 Use Cramer's rule to find the inverse of the matrix

$$\begin{pmatrix} 2 & 1 & -1 \\ 1 & 3 & 2 \\ -1 & 2 & 4 \end{pmatrix}.$$

Then use Gaussian elimination. Compare the amount of work involved. Which is easier?

8 Show that if A is any $n \times n$ nonsingular matrix, then there exists a permutation matrix P so that PA has an LU decomposition; i.e., $PA = LU$.
[*Hint* i: Since A is nonsingular, there exists a permutation matrix P_1 so that

$$P_1 A = \begin{bmatrix} a & r \\ c & B \end{bmatrix},$$

where B is $(n-1) \times (n-1)$ and $a \neq 0$.]
[*Hint* ii: $P_1 A$ can be written as

$$P_1 A = \begin{bmatrix} 1 & 0 \\ (1/a)c & I \end{bmatrix} \begin{bmatrix} a & r \\ 0 & B_2 \end{bmatrix},$$

where $B_2 = B - (1/a)cr$.]
[*Hint* iii: Now use induction; i.e., B_2 is nonsingular, so there exists a permutation matrix P_2 with $P_2 B_2 = L_2 U_2$, and with

$$P_3 = \begin{bmatrix} 1 & 0 \\ 0 & P_2 \end{bmatrix}$$

it follows that

$$P_3 P_1 A = \begin{bmatrix} 1 & 0 \\ (1/a)P_2 C & L_2 \end{bmatrix} \begin{bmatrix} a & r \\ 0 & U_2 \end{bmatrix} = LU.$$

9 Modify Algorithm (6.12a) to compute the determinant of A. (That is, delete the computations involving the right-hand side b_i and keep track of the number of row interchanges.)

6.4 Rounding Error Analysis

The rounding error analysis for the Gaussian elimination algorithm is straightforward but somewhat lengthy. It is important to realize, however, that even this rather complicated algorithm can be analyzed by the same elementary techniques used in earlier chapters. The error analysis is important for two reasons. It shows clearly the need for pivoting, and it confirms the nice accuracy that is generally observed with Gaussian elimination.

6.4.1 The Reduction Process

Let us first examine the rounding errors that occur during the reduction process. For simplicity we shall begin by analyzing Algorithm (6.7). That is, we assume that no pivoting is needed in order to reduce the system to triangular form. The effect of pivoting will be considered later.

To see how much rounding error is involved in each reduction step, let k be some fixed value between 1 and $n - 1$, and let a_{ij}, b_i denote the coefficients of the system just *before* step 2.1 is executed. That is, we are considering the **system just before the kth reduction step**:

$$a_{11}x_1 + a_{12}x_2 + \cdots + a_{1k}x_k + \cdots + a_{1n}x_n = b_1$$

$$a_{22}x_2 + \cdots + a_{2k}x_k + \cdots + a_{2n}x_n = b_2$$

$$\vdots$$

$$a_{kk}x_k + \cdots + a_{kn}x_n = b_k \qquad (6.24)$$

$$\vdots$$

$$a_{nk}x_k + \cdots + a_{nn}x_n = b_n.$$

Now let a'_{ij}, b'_i be the coefficients that are computed during this step. The multiplier m_{ik}, which is given by step 2.1.1 has the **computed value**

$$m_{ik} = \text{fl}[-a_{ik}/a_{kk}] = (-a_{ik}/a_{kk})\langle 1 \rangle, \qquad i = k + 1, \ldots, n. \quad (6.25a)$$

The **computed coefficients** are

$$a'_{ij} = \text{fl}[a_{ij} + m_{ik}a_{kj}] = a_{ij}\langle 1 \rangle + m_{ik}a_{kj}\langle 2 \rangle \qquad (6.25b)$$

$$b'_i = \text{fl}[b_i + m_{ik}b_k] = b_i\langle 1 \rangle + m_{ik}b_k\langle 2 \rangle, \qquad (6.25c)$$

where $i = k + 1, k + 2, \ldots, n$ and $j = k + 1, k + 2, \ldots, n$. The only other computed quantities are the coefficients of x_k in equations $k + 1$, $k + 2, \ldots, n$. These are "computed" to be zero; that is

$$a'_{ik} = 0, \qquad i = k + 1, k + 2, \ldots, n.$$

Thus, during the kth reduction step, the system (6.24) is changed to

$$
\begin{aligned}
a_{11}x_1 + a_{12}x_2 + & & \cdots & & + a_{1n}x_n = b_1 \\
a_{22}x_2 + & & \cdots & & + a_{2n}x_n = b_2 \\
& \ddots & & & \vdots \\
a_{kk}x_k + a_{k,\,k+1}x_{k+1} + & \cdots & + a_{kn}x_n = b_k & & \text{(6.26a)} \\
a'_{k+1,\,k+1}x_{k+1} + & \cdots & + a_{k+1,\,n}x_n = b'_{k+1} & & \\
& & & & \vdots \\
a'_{n,\,k+1}x_{k+1} + & \cdots & + a'_{nn}x_n = b'_n,
\end{aligned}
$$

where a'_{ij} and b'_i are given by equations (6.25a)–(6.25c).

In keeping with the principle of **backward error analysis** we must determine a system of the form (6.24) that the algorithm, with *exact* arithmetic, would reduce to (6.26a). To this end suppose we "undo" the reduction process that gave (6.26a). That is, we multiply equation k in (6.26a) by m_{ik} and then *subtract* it from equation i. The resulting equation i for $i > k$ is

$$
-m_{ik}a_{kk}x_k + (a'_{i,\,k+1} - m_{ik}a_{k,\,k+1})x_{k+1} + \cdots + (a'_{ij} - m_{ik}a_{kj})x_j + \cdots
$$
$$
+ (a'_{i,\,n} - m_{ik}a_{kn})x_n = b'_i - m_{ik}b_k. \qquad \text{(6.26b)}
$$

Now by (6.25a)–(6.25c) we can write the coefficients in (6.26b) as follows:

$$
\begin{aligned}
-m_{ik}a_{kk} &= a_{ik}\langle 1 \rangle & & \text{[from (6.25a)]} \\
a'_{ij} - m_{ik}a_{kj} &= a_{ij}\langle 1 \rangle + m_{ik}a_{kj}\langle 2 \rangle_0 & & \text{[from (6.25b)]} & & \text{(6.26c)} \\
b'_i - m_{ik}b_k &= b_i\langle 1 \rangle + m_{ik}b_k\langle 2 \rangle_0 & & \text{[from (6.25c)].}
\end{aligned}
$$

Thus (6.26b) has the form

$$\bar{a}_{ik}x_k + \bar{a}_{i,\,k+1}x_{k+1} + \cdots + \bar{a}_{ij}x_j + \cdots + \bar{a}_{in}x_n = \bar{b}_i, \qquad \text{(6.27a)}$$

where

$$\bar{a}_{ik} = -m_{ik}a_{kk} = a_{ik}\langle 1 \rangle, \qquad i = k + 1, \ldots, n \qquad \text{(6.27b)}$$

and

$$\bar{a}_{ij} = a_{ij}\langle 1 \rangle + m_{ik}a_{kj}\langle 2 \rangle_0, \qquad i, j = k + 1, \ldots, n \qquad \text{(6.27c)}$$

$$\bar{b}_i = b_i\langle 1 \rangle + m_{ik}b_k\langle 2 \rangle_0, \qquad i = k + 1, \ldots, n. \qquad \text{(6.27d)}$$

What we have shown can be summarized as follows. Just before the kth reduction step, the system of equations has the form (6.24). After the kth reduction step, the system is (6.26a). We have shown that when the same reduction step, but with **exact arithmetic**, is applied to the system

$$a_{11}x_1 + a_{12}x_2 + \qquad\qquad \cdots \qquad\qquad + a_{1n}x_n = b_1$$

$$a_{22}x_2 + \qquad\qquad \cdots \qquad\qquad + a_{2n}x_n = b_2$$

$$\ddots \qquad\qquad\qquad\qquad\qquad \vdots$$

$$a_{kk}x_k + a_{k,k+1}x_{k+1} + \cdots + a_{kn}x_n = b_k \qquad (6.28)$$

$$\bar{a}_{k+1,k}x_k + \bar{a}_{k+1,k+1}x_{k+1} + \cdots + \bar{a}_{k+1,n}x_n = \bar{b}_{k+1}$$

$$\vdots$$

$$\bar{a}_{n,k}x_k + \bar{a}_{n,k+1}x_{k+1} + \cdots + \bar{a}_{nn}x_n = \bar{b}_n$$

the result is again the system (6.26a). This is indeed a backward error analysis for a single reduction step because we have produced a system (6.28) that our algorithm with *exact arithmetic* reduces to the computed form (6.26a). It only remains to compare the system (6.28) with the actual system (6.24). For this we write (6.27b)–(6.27d) as

$$\bar{a}_{ik} - a_{ik} = a_{ik}\langle 1\rangle_0,$$

$$k = k + 1, \ldots, n$$

| Comparison of the coefficients |

$$\bar{a}_{ij} - a_{ij} = a_{ij}\langle 1\rangle_0 + m_{ik}a_{kj}\langle 2\rangle_0,$$

$$i, j = k + 1, \ldots, n$$

$$(6.29)$$

$$\bar{b}_i - b_i = b_i\langle 1\rangle_0 + m_{ik}b_k\langle 2\rangle_0,$$

$$i = k + 1, \ldots, n.$$

To simplify the bounds that we shall obtain, let

| Growth factors |

$$\rho_k = \max_{i,j=k+1,\ldots,n} \{|a_{ij}|, |m_{ik}|\,|a_{kj}|\}$$

$$\beta_k = \max_{i=k+1,\ldots,n} \{|b_i|, |m_{ik}|\,|b_k|\}.$$

(Remember that we are still considering only the kth reduction step.) Then (6.29) with the usual estimate (2.37) gives, for $i = k + 1, \ldots, n$,

| Error bounds |

$$|\bar{a}_{ik} - a_{ik}| \le \rho_k\mu'$$

$$|\bar{a}_{ij} - a_{ij}| \le 3\rho_k\mu', \qquad j = k + 1, \ldots, n \qquad (6.30)$$

$$|\bar{b}_i - b_i| \le 3\beta_k\mu'.$$

To complete the analysis of the entire reduction algorithm we simply observe that each reduction step involves a perturbation of the coefficients of, at most, $3\rho_k \mu'$ (or $3\beta_k \mu'$ for the right-hand side). Since there are $n - 1$ reduction steps, the total perturbation cannot exceed $3(n - 1)\rho_k \mu'$. We summarize this as follows:

Suppose that Algorithm (6.7), with t-digit floating point arithmetic, is applied to the system (6.1). Then the result is an upper triangular system of equations. This *same* upper triangular system can be obtained by applying Algorithm (6.7), with *exact arithmetic*, to a system

$$\bar{a}_{11}x_1 + \bar{a}_{12}x_2 + \cdots + \bar{a}_{1n}x_n = \bar{b}_1$$

$$\bar{a}_{21}x_1 + \bar{a}_{22}x_2 + \cdots + \bar{a}_{2n}x_n = \bar{b}_2$$

$$\vdots$$

$$\bar{a}_{n1}x_1 + \bar{a}_{n2}x_2 + \cdots + \bar{a}_{nn}x_n = \bar{b}_n,$$

(6.31)

where

$$|a_{ij} - \bar{a}_{ij}| \le 3(n - 1)\rho\mu'$$

$$|b_i - \bar{b}_i| \le 3(n - 1)\beta\mu'$$

(6.32)

and

$$\rho = \max_{k = 1, 2, \ldots, n - 1} \rho_k$$

$$\beta = \max_{k = 1, 2, \ldots, n - 1} \beta_k.$$

It is not hard to see that this same analysis also applies to Algorithm (6.12a). Indeed the pivoting that is used in (6.12a) simply causes the reduction steps to be done in a different order. Each reduction step, however, is exactly the same as the one just analyzed above. Of course, the computed *values* will be different, and hence the values of ρ_k and β_k will be different. This will be discussed in the next subsection.

Since the reduction algorithm, with exact arithmetic, does not change the solution to the system of equations, it follows that the computed output from Algorithm (6.7) [or Algorithm (6.12a)] has the same solution as the system (6.31). Thus the back-solving algorithms will actually attempt to compute a solution of (6.31) and not of the given system (6.1). If (6.31) is a reasonably small perturbation of the system (6.1), then we would expect the solutions of the two systems to be close together. This will be discussed in more detail in Section 6.5. For now we only observe that the accuracy of the algorithm is measured by the bounds in (6.32), which in turn depend on the factors ρ and β. The consequence of this will be considered next.

6.4.2   Growth Factors

The bounds (6.32) will be suitably small provided the values of ρ_k and β_k are not large. There are two reasons why these so-called **growth factors** might be very large. The first is that the multipliers m_{ik} might be very large, and the second is that the newly computed coefficients a_{ij} or b_i might be large. The first danger is easily avoided; namely, always use partial pivoting so that the multipliers satisfy $|m_{ik}| \leq 1$. If this is the case, then the definition of ρ_k and β_k becomes

Growth factors with partial pivoting

$$\rho_k = \max_{k+1 \leq i,\, j \leq n} \{|a_{ij}^{(k)}|\}$$

$$\beta_k = \max_{k+1 \leq i \leq n} \{|b_i^{(k)}|\},$$

(6.33)

where $a_{ij}^{(k)}$, $b_i^{(k)}$ denote the coefficients at the kth reduction step. Thus ρ and β in (6.32) simply measure the sizes of the coefficients in the original system, and in all the systems computed during the reduction to upper triangular form.

It can be shown that if the original coefficients are scaled so that $|a_{ij}| \leq 1$ for $i, j = 1, 2, \ldots, n$, then partial pivoting guarantees that $\rho_k \leq 2^{k-1}$. In fact, there exist examples where this bound is actually attained (see Exercise 5). In practice, however, it rarely happens that ρ_k becomes larger than 10, when partial pivoting is used. In order to ensure that ρ_k does not become too large, **complete pivoting** is sometimes recommended. This means that, in addition to interchanging equations, we also renumber the unknowns (i.e., interchange the coefficients of x_i with the coefficients of x_j) in order to maximize the divisor a_{kk} in step 2.1.1 of (6.7). On modern computers, however, the additional expense of testing *all* the coefficients in the unreduced equations to find the largest is usually not worth the slight increase in accuracy. (In fact, there is no *guarantee* that there will be an increase in accuracy.) On computers that have reasonably fast double-precision operations it may be cheaper to do partial pivoting in double-precision than to use complete pivoting. A variation of Gaussian elimination, called the **Crout method**, is also recommended in connection with the use of double precision. In this method the computations can be arranged so that only certain operations need to be done in double precision. The computed results, however, are nearly as accurate as if *all* the operations had been done in double precision.

The description and analysis of the Crout method requires the use of matrix theory and will not be discussed further.

6.4.3 Need for Scaling

Another important application of the backward error bounds (6.32) is to show the importance of proper scaling of the equations. The fact that the bounds on the perturbed coefficients are all of the same *absolute* size suggests that the coefficients themselves should all be of the same absolute size. Consider, for example, the system

| Example: Badly scaled system | $$1.0x_1 + 1.0 \times 10^5 x_2 = 1.0 \times 10^5$$ $$1.0x_1 + 1.0x_2 = 2.0.$$ | (6.34) |

Suppose this system were solved by Gaussian elimination with partial pivoting and with four-decimal-digit arithmetic. Then the estimate (6.32) indicates that we may have actually solved a system whose coefficients differ from those in (6.34) by as much as

| Perturbation of (6.34) predicted by (6.32) | $$3(n-1)\rho\mu' = 3 \times 10^5 (1.06)10^{1-4} \cong 3 \times 10^2.$$ |

A perturbation of this size in the large coefficient (1.0×10^5) is not too unreasonable. However, the other coefficients may also be perturbed by this amount so that the second equation may be *totally* changed. In fact, the computed solution to (6.34) (with four-digit arithmetic) is $x_2 = 1.0000$, $x_1 = 0$. Since the exact solution is $x_1 = 1.0000100\ldots$, $x_2 = .99998999\ldots$, we see that the error analysis (6.31), (6.32) has correctly predicted a large error.

The solutions just discussed may seem familiar because, in fact, (6.34) is just the system (6.11) but with the first equation multiplied by 10^5. The computed solution to (6.34) is the same as the solution we obtained for (6.11) *without* the use of partial pivoting. Thus another possible explanation for the large errors in the computed solution to (6.34) is that poor scaling has circumvented the partial pivoting strategy. Whatever the explanation one chooses, the effect is to create an absolute perturbation (6.32) that results in a very large *relative* perturbation of some coefficients.

In order to make sure that the absolute perturbations (6.32) have the same *relative* effect on all of the coefficients, the coefficients should be appropriately scaled. The goal of any scaling technique is to obtain a system of equations all of whose coefficients are approximately the same size and whose solution is also the solution to the original system. In particular cases it is often easy to see how to scale the equations. However, a general algorithm for doing this would be complicated and expensive. A simple procedure that is often used and that seems to work well in practice is called **equilibration**. In this method elementary operations of type (i) (multiply an equation by a nonzero constant) are performed on each equation so that the largest coefficient in each equation is approximately 1.0 in absolute value. We say "approximately 1.0" because the multiplying constant should be chosen to be a power of the computer's number base. This avoids rounding errors since such multiplications cause changes in the exponents of the coefficients but no changes in the fractional parts. Note that the system (6.11) is equilibrated in this sense, but the coefficients are *not* all of the same size. The partial pivoting strategy seems to take care of the remaining discrepancies in size. The following is an algorithm for equilibration. A similar process, in which the *unknowns* are scaled, is described in Exercise 4.

Algorithm to Equilibrate the System (6.1)

Input consists of the coefficients a_{ij} and b_i. Also needed is a function subprogram $E(x)$ that produces the quantity β^e when x is the floating point number $x = \pm .d_1 d_2 \cdots d_t \times \beta^e$.

1   Input $\{n, a_{ij}, b_i\}$
2   For $i = 1, 2, \ldots, n$
   2.1 $\max \leftarrow |a_{i1}|$
   2.2   For $j = 2, 3, \ldots, n$
      2.2.1   If $|a_{ij}| > \max$,
         then $\max \leftarrow |a_{ij}|$ (6.35)
   2.3   Scale $\leftarrow E(\max)^{-1}$
   2.4   For $j = 1, 2, \ldots, n$
      2.4.1 $a_{ij} \leftarrow a_{ij}$scale
   2.5 $b_i \leftarrow b_i$scale
3   Output $\{a_{ij}, b_i\}$

The above algorithm should be used with care. In particular, the scaled equations should always be examined to make sure that equilibration has not made things *worse*. For example, if the matrix of coefficients

$$A = \begin{bmatrix} 1 & 1 & 10^5 \\ 2 & -1 & 10^5 \\ 1 & 2 & 0 \end{bmatrix}$$

is used as input to this algorithm, the output (with $\beta = 10$) is

$$A = \begin{bmatrix} 10^{-6} & 10^{-6} & .1 \\ 2 \times 10^{-6} & -10^{-6} & .1 \\ .1 & .2 & 0 \end{bmatrix}.$$

Thus the scaling has become *worse* since now there are four coefficients that are of much different sizes from the rest. The real problem with this example is that the unknown x_3 is poorly scaled. (Again we refer the reader to Exercise 4.)

6.4.4 Rounding Error during Backsolving

So far we have only considered the effect of rounding errors during the elimination process. To see how much error is incurred during the **backsolving process**, consider the computational form of (6.5a):

$$\hat{x}_k = \mathrm{fl}[(b_k - a_{k, k+1}\hat{x}_{k+1} - a_{k, k+2}\hat{x}_{k+2} - \cdots - a_{kn}\hat{x}_n)/a_{kk}].$$

Here we use a_{ij}, b_j to denote the coefficients of an upper triangular system. For this analysis we shall ignore the fact that this upper triangular system has actually been *computed* and will contain rounding errors.

The computation of $\hat{x}_k$ is essentially a summation. Hence, a slight modification to (3.3c) gives the result

$$\hat{x}_k = (b_k\langle n - k + 2\rangle - a_{k, k+1}\hat{x}_{k+1}\langle n - k + 2\rangle - \cdots - a_{kn}\hat{x}_n\langle 3\rangle)/a_{kk},$$

for $k = n - 1, n - 2, \ldots, 1$. This follows from (3.3c) by observing that here we have $(n - k + 1)$ terms to be summed, followed by a division. Furthermore, the computation of the products $a_{kj}\hat{x}_j$ contributes one more rounding error to each term. For $k = n$ we have

$$\hat{x}_n = \mathrm{fl}[b_n/a_{nn}] = (b_n/a_{nn})\langle 1\rangle.$$

These formulas for $\hat{x}_k$ and $\hat{x}_n$ can be easily rewritten as the equations:

$$a_{kk}\hat{x}_k + a_{k, k+1}\langle n - k + 2\rangle\hat{x}_{k+1} + \cdots + a_{kn}\langle 3\rangle\hat{x}_n$$
$$= b_k\langle n - k + 2\rangle, \qquad k = 1, 2, \ldots, n - 1$$
$$a_{nn}\hat{x}_n = b_n\langle 1\rangle.$$

If we use the notation

$$\hat{a}_{kj} = \begin{cases} a_{k, k}, & j = k, \\ a_{kj}\langle n - j + 3\rangle, & 1 \le k < j \le n \end{cases}$$
$$\hat{b}_k = \begin{cases} b_k\langle n - k + 2\rangle, & 1 \le k < n \\ b_n\langle 1\rangle, & k = n, \end{cases} \tag{6.36a}$$

then these equations become

System of equations actually solved

$$\hat{a}_{11}\hat{x}_1 + \hat{a}_{12}\hat{x}_2 + \cdots + \hat{a}_{1n}\hat{x}_n = \hat{b}_1$$

$$\hat{a}_{22}\hat{x}_2 + \cdots + \hat{a}_{2n}\hat{x}_n = \hat{b}_2$$

$$\vdots$$

$$\hat{a}_{nn}\hat{x}_n = \hat{b}_n.$$

Hence we have indeed computed the solution to an upper triangular system. However, the system we have solved is not the given system. The difference between the coefficients of these two systems can be estimated by applying (2.37) to (6.36a). The result is

Error estimates

$$|\hat{a}_{kj} - a_{kj}| \le \begin{cases} 0, & j = k \\ |a_{kj}|(n - j + 3)\mu', & 1 \le k < j \le n \end{cases}$$

$$|\hat{b}_k - b_k| \le \begin{cases} (n - k + 2)|b_k|\mu', & 1 \le k < n \\ |b_n|\mu', & k = n. \end{cases}$$

(6.36b)

This shows that the backsolving process is always stable. That is, the *relative* change in the coefficients is bounded by a reasonable multiple (less than $n + 2$) of the unit rounding error μ'. We do not need to worry here about the growth factors ρ and β, except to remember that the upper triangular system that we are solving may be far from the original system (6.1) whose solution is desired.

Finally, note that these backward error analyses are of interest only if it is possible to determine how the solution changes when the coefficients are changed. This will be discussed fully in the next section.

EXERCISES

1   Reduce the following systems to upper triangular form by using four-decimal-digit chopped arithmetic *with* partial pivoting, and then *without* partial pivoting.

(a) $.0001x_1 + 2x_2 = .01$ (b) $x_1 + 2.001x_2 + x_3 = 1$

   $1.001x_1 - x_2 = 1$ $x_1 + 2.000x_2 + 2x_3 = -1$

                                 $x_1 - x_2 - x_3 = 0.$

2   Use four-digit arithmetic to reduce the system (6.34) to upper triangular form, then reduce it with exact arithmetic. Compute the *exact* solutions to both of these upper triangular systems. Now multiply the second equation in (6.34) by 1000 and solve the resulting scaled equations using four-digit arithmetic.

3   Modify the Gaussian elimination Algorithm (6.7) so that the values of ρ and β, which are used in the error estimate (6.33) are also computed. Implement this modified algorithm by a computer program and run it with several sets of equations whose coefficients are approximately 1.0 in magnitude. Can you find such an example for which ρ is *very* large?

4 (a)   Suppose that in the system (6.1) every coefficient of a particular unknown, say x_i, is multiplied by a nonzero constant c. How does this change the solution to the system?

(b)   Write out an algorithm for "column" equilibration in which all coefficients of a particular unknown are multiplied by a constant so that the largest such coefficient is approximately one. (The output must include the scale factors. Why?)

(c)   Apply the algorithm in part (b) to the example following Algorithm (6.35). Discuss the results.

5   Do a backward rounding error analysis for Algorithm (6.12d).

6   Show that when Gaussian elimination is used to solve the system

$$x_1 \qquad\qquad\qquad + x_5 = 1$$

$$x_1 + x_2 \qquad\qquad - x_5 = 0$$

$$-x_1 + x_2 + x_3 \qquad + x_5 = 0$$

$$x_1 - x_2 + x_3 + x_4 - x_5 = 0$$

$$-x_1 + x_2 - x_3 + x_4 + x_5 = 0$$

then the growth factor ρ equals 2^4. Generalize this example to an $n \times n$ system for which the growth factor is 2^{n-1}.

6.5   Error Estimates and Accuracy Improvement

One of the most difficult aspects of numerical computations is estimating the accuracy of a computed result. In this section we examine this problem in the context of solving linear systems. In particular, we describe a method for estimating the accuracy of the solution that also gives a means for improving the computed solution. The problem of ill-conditioned (i.e., unstable) systems is also discussed.

6.5.1   The Residuals

Suppose that $\hat{x}_1, \ldots, \hat{x}_n$ is an approximate solution to the linear system (6.1), where we denote the exact solution by $x_1^*, \ldots, x_n^*$. In order to determine the accuracy of the computed solution, it is necessary to estimate the sizes of the numbers $\varepsilon_1 = |x_1^* - \hat{x}_1|$, $\varepsilon_2 = |x_2^* - \hat{x}_2|$, $\ldots$, $\varepsilon_n = |x_n^* - \hat{x}_n|$. These numbers are just the **errors** in the approximate solution.

The most obvious way to try to estimate these errors is to compute the **residuals**

$$r_1 = b_1 - a_{11}\hat{x}_1 - a_{12}\hat{x}_2 - \cdots - a_{1n}\hat{x}_n$$

$$r_2 = b_2 - a_{21}\hat{x}_1 - a_{22}\hat{x}_2 - \cdots - a_{2n}\hat{x}_n$$

$$\vdots$$

$$r_n = b_n - a_{n1}\hat{x}_1 - a_{n2}\hat{x}_2 - \cdots - a_{nn}\hat{x}_n$$

$$(6.37)$$

and use the absolute values of *these* numbers to indicate the sizes of ε_1, $\varepsilon_2, \ldots, \varepsilon_n$. Unfortunately, this does not work, as is shown by the following example. Consider the system

$$1.000x_1 + 2.000x_2 = 3.000$$
$$.499x_1 + 1.001x_2 = 1.500. \tag{6.38}$$

The exact solution to this system is $x_1^* = x_2^* = 1$, but the residuals, with the "approximate" solution $x_1 = 2.000$, $x_2 = .500$ are $r_1 = 0$, $r_2 = .0015$. Thus, the residuals are considerably smaller than are the errors $\varepsilon_1 = 1.0$, $\varepsilon_2 = .5$.

The difficulty with using the residuals to estimate the accuracy is that, for some linear systems, there are many "approximate" solutions that are not very close to the exact solution but that produce small residuals. This situation is complicated even further by the fact that such systems are also **ill conditioned**. To make this more evident, consider the system

$$1.000x_1 + 2.000x_2 = 3.000$$
$$.500x_1 + 1.002x_2 = 1.500, \tag{6.39}$$

which is the same as the system (6.38) except for two very slight changes in the coefficients. The exact solution to (6.39) is $x_1^* = 3.0$, $x_2^* = 0$, which is rather different from the solution to (6.38).

To analyze this phenomenon more generally, we write (6.37) in the form

$$a_{11}\hat{x}_1 + a_{12}\hat{x}_2 + \cdots + a_{1n}\hat{x}_n = b_1 - r_1$$
$$a_{21}\hat{x}_1 + a_{22}\hat{x}_2 + \cdots + a_{2n}\hat{x}_n = b_2 - r_2$$
$$\vdots$$
$$a_{n1}\hat{x}_1 + a_{n2}\hat{x}_2 + \cdots + a_{nn}\hat{x}_n = b_n - r_n.$$

If we subtract each of these equations from the corresponding equation in (6.1), with the exact solution $x_1^*, \ldots, x_n^*$ for the unknowns in (6.1), we obtain

Relation between errors and residuals

$$a_{11}e_1 + a_{12}e_2 + \cdots + a_{1n}e_n = r_1$$
$$a_{21}e_1 + a_{22}e_2 + \cdots + a_{2n}e_n = r_2$$
$$\vdots \tag{6.40}$$
$$a_{n1}e_1 + a_{n2}e_2 + \cdots + a_{nn}e_n = r_n,$$

where $e_i = x_i^* - \hat{x}_i$, $i = 1, 2, \ldots, n$ and $\varepsilon_i = |e_i|$. If the right-hand sides in (6.40) were *zero*, then the only solution would be the zero solution $e_1 = e_2 = \cdots = e_n = 0$. Thus, if the system were *well* conditioned, then small residuals (i.e., residuals that are close to zero) would imply small errors (i.e., errors that are close to zero). The effect of ill-conditioning here is to allow the possibility of a large solution $e_1, e_2, \ldots, e_n$ to (6.40), even though the right-hand side is small.

6.5.2 Estimating the Condition of a System

In order to make these statements somewhat more quantitative, assume that each equation in (6.1) has been scaled, as defined in Section 6.4. That is, suppose that each equation has been multiplied by a constant so that the largest of the absolute values of the coefficients in the left side of that equation is approximately 1.0. This scaling eliminates the possibility of obtaining a small residual by the trivial process of multiplying an equation in (6.40) by a small number. We want to examine the relationship between the *size* of the right-hand side of (6.40) and the *size* of the solution. Since it is only the *ratio* of these sizes that is really of interest, it is just as well to let the right-hand side be all ones; that is, to let $r_1 = r_2 = \cdots = r_n = 1$. Then if the solution to this system is large *compared to 1.0*, the system must be ill conditioned.

For example, the system (6.38), which is already reasonably scaled, written in the form (6.40) with 1.0, 1.0 for the right-hand side is

$$1.000e_1 + 2.000e_2 = 1.000$$

$$.499e_1 + 1.001e_2 = 1.000.$$

The solution to this is $e_1 = -333.0$, $e_2 = 167.0$, which indicates that the errors in an approximate solution to (6.38) may be more than 100 times larger than the corresponding residuals.

As this example indicates, a reasonable way to use the residuals (6.37) to estimate the accuracy of the approximate solution is to re-solve the system (6.1) with ones on the right side instead of the given values. That is, to compute an (approximate) solution $\bar{x}_1, \bar{x}_2, \ldots, \bar{x}_n$ to

$$a_{11}x_1 + \cdots + a_{1n}x_n = 1.0$$
$$a_{21}x_1 + \cdots + a_{2n}x_n = 1.0$$
$$\vdots$$
$$a_{n1}x_1 + \cdots + a_{nn}x_n = 1.0. \tag{6.41}$$

If $M = \max\{|\bar{x}_1|, |\bar{x}_2|, \ldots, |\bar{x}_n|\}$, then the errors $\varepsilon_1, \ldots, \varepsilon_n$ will be approximately M times as large as the residuals $r_1, r_2, \ldots, r_n$.

So far in this discussion no mention has been made of Gaussian elimination. That is, the entire discussion of error estimation has involved only an approximate solution $\hat{x}_1, \ldots, \hat{x}_n$, which presumably has been computed but not by any particular algorithm. The point at which the Gaussian elimination algorithm becomes important is in the process of solving the system

(6.41) in order to obtain an estimate for the error. In fact, if the approximate solution $\hat{x}_1, \ldots, \hat{x}_n$ has been obtained by using the Gaussian elimination Algorithms (6.12a), (6.12b), then (6.12c) can be used to find quickly the solution to (6.41). We summarize this process as follows:

> To estimate the accuracy of a solution $\hat{x}_1, \ldots, \hat{x}_n$ computed by Algorithms (6.12a) and (6.12b) under the assumption that the system was equilibrated:
> (1) Compute the residuals (6.37), and let $R = \max\{|r_1|, |r_2|, \ldots, |r_n|\}$.
> (2) Use $c_1 = c_2 = \cdots = c_n = 1$ as input to Algorithm (6.12c), with output $z_1, z_2, \ldots, z_n$.
> (3) Let $M = \max\{|z_1|, |z_2|, \ldots, |z_n|\}$. Then the value MR gives an approximation to the maximum error: $\max\{\varepsilon_1, \varepsilon_2, \ldots, \varepsilon_n\}$.

6.5.3 Iterative Improvement

The reason for solving (6.41) is to obtain an approximate relation between the size of the residuals (which can be computed) and the size of the errors (which is not known). Equations (6.40), however, given an *exact* relation between these two sets of numbers. That is, if we were to solve (6.40) for e_1, $e_2, \ldots, e_n$, then these values, together with the approximate solution $\hat{x}_1, \ldots, \hat{x}_n$ used to compute the residuals, determine the exact solution from

$$x_1^* = e_1 + \hat{x}_1, \qquad x_2^* = e_2 + \hat{x}_2, \ldots, x_n^* = e_n + \hat{x}_n. \qquad (6.42)$$

The flaw in this argument, of course, is that (6.40) cannot be solved for the errors any more readily than (6.1) can be solved for the solution directly. However, if Gaussian elimination, that is, (6.12c), is used to obtain an approximate solution $\hat{e}_1, \ldots, \hat{e}_n$ to (6.40), then perhaps

$$\hat{\hat{x}}_1 = \hat{e}_1 + \hat{x}_1, \ldots, \hat{\hat{x}}_n = \hat{e}_n + \hat{x}_n \qquad (6.43)$$

will give a more accurate approximation than $\hat{x}_1, \ldots, \hat{x}_n$. Furthermore, this process can be repeated by using $\hat{\hat{x}}_1, \ldots, \hat{\hat{x}}_n$ to compute new residuals that are again used in (6.40) to determine new corrections. This process is called **iterative improvement**. It is often a very effective way to increase the accuracy of an approximate solution that was obtained by Gaussian elimination. The reason for emphasizing Gaussian elimination in connection with this improvement technique is that the repeated solving of (6.40) with different right-hand sides is only feasible if the entire solution process does not have to be repeated each time. If Algorithms (6.12a) and (6.12b) are used to obtain the approximation $\hat{x}_1, \ldots, \hat{x}_n$, then (6.12c) with $c_1 = r_1, c_2 = r_2, \ldots, c_n = r_n$ can be used to compute $\hat{\hat{x}}_1, \ldots, \hat{\hat{x}}_n$ according to (6.43).

If the coefficients a_{ij} have been equilibrated, as suggested earlier, then the small residuals $r_1, \ldots, r_n$ will be obtained by computing with data (namely, the coefficients) that are much larger than these residuals. Thus subtractive cancellation will probably occur. In order to avoid cancellation, the residuals should be computed in double precision arithmetic. We summarize these points by giving an algorithm for iterative improvement.

Iterative Improvement Algorithm

$$\left[\begin{array}{c} \text{Algorithm for improving the accuracy of a solution to (6.1) that} \\ \text{has been computed by (6.12a) and (6.12b). The input is the} \\ \text{output from those algorithms.} \end{array}\right]$$

   1 Input $\{n, a_{ij}, m_{ij}, \text{sub}(i), x_i\}$
   2 Compute the residuals $r_1, \ldots, r_n$ in double
      precision. Round them to single precision
   3 Use Algorithm (6.12c), with $r_1, r_2, \ldots, r_n$
      as right-hand sides. Let $z_1, z_2, \ldots, z_n$ be   (6.44)
      the output
   4 For $i = 1, 2, \ldots, n$
      4.1 $x_i \leftarrow x_i + z_i$
   5 Repeat steps 2, 3, 4 until $|z_i|$ are
      suitably small

This algorithm can also be used to detect ill-conditioning. If the residuals $r_1, \ldots, r_n$ are small but the corrections $z_1, \ldots, z_n$ are large, then the system is ill conditioned. For well-conditioned systems steps 2, 3, 4 should not have to be repeated more than once or twice.

EXERCISES

1 For each of the following systems, let $\hat{x}_1 = 1.0$, $\hat{x}_2 = 1.0$ be an approximate solution, and (a) find the residuals corresponding to this approximate solution, (b) find the exact solution and hence the error in this approximate solution, (c) find the solution when the right-hand side is 1, 1. Decide, from all this, which of these systems are well conditioned and which are ill conditioned.

$$x_1 + x_2 = 1.9 \qquad\qquad x_1 + x_2 = 1.9 \qquad\qquad x_1 - x_2 = 0$$

$$x_1 - x_2 = .01 \qquad\qquad 2x_1 + 2.01x_2 = 4.01 \qquad\qquad x_1 + x_2 = 1$$

$$x_1 - x_2 = -.01 \qquad\qquad x_1 - x_2 = -.01$$

$$x_1 - 2x_2 = -.99 \qquad\qquad -x_1 + .9x_2 = -.01.$$

2 Write a computer program that will compute the value M defined following equations (6.41). Use this program as part of an error estimate program that accepts as input a system

of equations and an approximate solution and produces as output an estimate for the accuracy of this approximate solution.

3 Modify an existing Gaussian elimination computer program so that the solution to (6.41) is computed while the solution to the given system is being computed.

4 Write a computer program to implement the iterative improvement technique. Test the program with some systems that are ill conditioned as well as some that are well conditioned.

6.6 Efficiency Estimates

The Gaussian elimination method, as described in Section 6.2, is not only accurate but also efficient. Attempts to modify the algorithm often lead to additional computations. In this section we derive efficiency estimates for the various algorithms given in Section 6.2. For comparison we also consider briefly a less efficient form of elimination.

6.6.1 Reduction Process

One of the main reasons for emphasizing the use of Gaussian elimination is that it proves to be a very efficient algorithm. In order to derive efficiency estimates for this method, we consider first the reduction process. To simplify the discussion we shall ignore the partial pivoting procedure and analyze the basic Algorithm (6.7). The partial pivoting Algorithm (6.12a) requires a large number of comparisons in Step 3.2.2 to determine the maximal pivot element. Some computers have special instructions to do these comparisons very quickly. In other computers, however, each comparison is nearly as expensive as a floating point addition. Hence the efficiency of a partial, or complete, pivoting algorithm may depend strongly on the computer and on the compiler's ability to take advantage of special computer hardware.

The floating point operations in Algorithm (6.7) are contained in steps 2.1.1, 2.1.2.1, and 2.1.3. We count them as follows:

step 2.1.1: $(n-1) + (n-2) + \cdots + 1 =$
$$\tfrac{1}{2}n(n-1) \qquad \text{divisions}$$

step 2.1.2.1: $(n-1)(n-1) + (n-2)(n-2) + \cdots + 1 =$
$$\tfrac{1}{6}(n-1)n(2n-1) \qquad \text{additions}$$

step 2.1.2.1: $(n-1)(n-1) + (n-2)(n-2) + \cdots + 1 =$
$$\tfrac{1}{6}(n-1)n(2n-1) \qquad \text{multiplications}$$

step 2.1.3: $(n-1) + (n-2) + \cdots + 1 =$
$$\tfrac{1}{2}n(n-1) \qquad \text{additions}$$

step 2.1.3: $(n-1) + (n-2) + \cdots + 1 =$
$$\tfrac{1}{2}n(n-1) \qquad \text{multiplications.}$$

Thus the total **floating point operation count for the reduction Algorithm** (6.7) is

$$\tfrac{1}{3}(n^3 - n)A + \tfrac{1}{3}(n^3 - n)M + \tfrac{1}{2}(n^2 - n)D.$$

Once the coefficient matrix has been reduced to upper triangular form, Algorithm (6.12c) can be used to apply this same reduction to an arbitrary right-hand side. The only floating point operations in the *reduction* part of (6.12c) are in step 2.3.1, which involves

$$(n-1) + (n-2) + \cdots + 1 = \tfrac{1}{2}(n-1)(n)$$

additions and multiplications. Thus the **total operation count for the reduction of a new right-hand side** is

$$\tfrac{1}{2}(n^2 - n)A + \tfrac{1}{2}(n^2 - n)M.$$

Even for moderate sized n, this is considerably smaller than the operation count for the reduction algorithm.

6.6.2   Backsolving and Iterative Improvement

The backsolving Algorithms (6.4) and (6.12b) are much simpler than the reduction algorithm. Note that partial pivoting has no effect on the number of floating point operations during backsolving. Hence Algorithms (6.4) and (6.12b) will have the same efficiency measures. The operation counts for these algorithms are determined as follows:

step 2:	1	division
step 3.2.1:	$1 + 2 + \cdots + (n-1) = (\tfrac{1}{2})n(n-1)$	additions
step 3.2.1:	$1 + 2 + \cdots + (n-1) = (\tfrac{1}{2})n(n-1)$	multiplications
step 3.3:	$n - 1$	divisions.

Thus the **cost of backsolving** is given by

$$\tfrac{1}{2}(n^2 - n)A + \tfrac{1}{2}(n^2 - n)M + (n)D.$$

Compared with the reduction process, the backsolving is extremely cheap. For example, with $n = 20$, the backsolving requires $190A + 190M + 20D$ operations, as compared with $2660A + 2660M + 190D$ for the reduction process.

The Algorithm (6.12c) for solving the same system but with a different right-hand side has efficiency

$$(n^2 - n)A + (n^2 - n)M + nD.$$

Thus again with $n = 20$, we see that the system can be *re-solved* in only $360A + 360M + 20D$ additional operations. This is only slightly more than 10% of the operations needed to solve the system the first time.

Finally the iterative improvement Algorithm (6.44) requires $n^2A + n^2M$ operations to compute the residuals, followed by the re-solving Algorithm (6.12c), and n additions to add on the corrections. Thus each improvement cycle involves

Cost of iterative improvement	$2n^2A + (2n^2 - n)M + nD$

operations. By comparing this with the efficiency of the reduction algorithm, we see that iterative improvement can be done nearly $\frac{1}{6}n$ times before the cost of the improvement becomes comparable to the cost of the reduction.

6.6.3   Comparison with Other Methods

Several methods similar to Gaussian elimination have at various times been introduced with the aim of reducing the cost of solving linear systems. Most of these methods, however, are *more* expensive to use. To illustrate this, we shall discuss briefly a rather well-known technique called the **Gauss–Jordan method**. This method is again a reduction method in which, at the kth stage, the kth equation is used to eliminate the coefficients of x_k in equations $k + 1$, $k + 2$, ..., n. Then, in addition, the coefficients of x_k in equations $k - 1$, $k - 2$, ..., 1 are also eliminated. After n such steps, the system is in *diagonal* form and can be solved directly without backsolving. Unfortunately, in avoiding the backsolving this method doubles the cost of the elimination. But as we have just seen, the elimination process is roughly $\frac{2}{3}n$ times as expensive as the backsolving. Hence the Gauss–Jordan method is much more costly than the Gauss method as described in Section 6.2.

It is generally felt that Gaussian elimination as defined here is the most efficient method for solving general linear systems that are not too big. By "general" we mean systems that have no special structure, such as bandedness, that might suggest a more efficient algorithm. If the system is so large that not all of the coefficients can be stored in the main memory of the computer, then it may be necessary to consider other methods. In Section 6.8

we shall discuss some iterative methods that are often used for certain very large systems.

EXERCISES

1   Compute the efficiency of Algorithm (6.12a) under the assumption that each comparison in step 3.2.2 is equivalent to one floating point addition.

2   Examine the way in which your computer could do the comparisons in step 3.2.2 of Algorithm (6.12a). Study the instruction set to see if there are special instructions for such a comparison. If possible, obtain a listing of the machine language instructions that your compiler generates to carry out these comparisons.

3   Do Exercise 1 for *complete* pivoting.

4   Derive the efficiency estimates for the Gauss–Jordan method discussed at the end of this section. Compare your result to the efficiency of the Gauss elimination/backsolving method.

5   Derive the efficiency measure for Gaussian elimination applied to a tridiagonal system. Assume that partial pivoting is not used.

6.7[†]   Error Estimates—Matrix Methods

With the aid of some elementary matrix–vector theory, the error estimates of Section 6.5 can be made more precise. We shall develop the concept of a norm and show how it can be used to measure errors.

6.7.1   Measuring Errors

Let X^* denote the **exact solution** to the linear system (6.1). That is,

$$X^* = A^{-1}b,$$

where A is the $n \times n$ matrix given by (6.3). We assume, of course, that A is nonsingular so that its inverse exists. The components of X^* will be denoted by $x_1^*, x_2^*, \ldots, x_n^*$. A computed solution $\hat{x}_1, \ldots, \hat{x}_n$ to (6.1) will be denoted by the vector $\hat{X}$ and the vector E defined by

$$E = \begin{pmatrix} e_1 \\ e_2 \\ \vdots \\ e_n \end{pmatrix} = \begin{pmatrix} x_1^* - \hat{x}_1 \\ x_2^* - \hat{x}_2 \\ \vdots \\ x_n^* - \hat{x}_n \end{pmatrix} = X^* - \hat{X} \tag{6.45}$$

will be called the **error vector**. In order to measure the accuracy of $\hat{X}$ as an approximation to X^*, we must somehow measure the "size" of E. One way

---

[†] This section is optional.

to do this is to compute the **maximum norm** of E as defined by

$$\|E\|_\infty = \max_{1 \le k \le n} |e_k|. \tag{6.46}$$

Clearly, this is a reasonable way to measure how large E is. However, one can easily imagine other measures that are just as reasonable. For example, consider the **sum norm**

$$\|E\|_1 = |e_1| + |e_2| + \cdots + |e_n| \tag{6.47}$$

and the **square norm**

$$\|E\|_2 = (e_1^2 + e_2^2 + \cdots + e_n^2)^{1/2}. \tag{6.48}$$

[The subscripts ∞, 1, 2 in (6.46), (6.47), and (6.48) have theoretical justifications that we shall not discuss.]

In many applications any of these measures can be used to represent the size of a vector. If the particular choice is not important, we shall drop the subscript and simply write $\|E\|$. Some important **properties of these norms** are

$$\|X\| \ge 0 \quad \text{for all vectors} \quad X \quad \text{and}$$
$$\|X\| = 0 \quad \text{only if} \quad X = 0 \tag{6.49a}$$

$$\text{For any number} \quad a, \|aX\| = |a|\,\|X\| \tag{6.49b}$$

$$\text{For any two vectors} \quad X \quad \text{and} \quad Y, \|X + Y\| \le \|X\| + \|Y\|. \tag{6.49c}$$

The fact that $\|\cdot\|_1$ and $\|\cdot\|_\infty$ satisfy properties (6.49) is easily shown. The norm $\|\cdot\|_2$ also satisfies them, but the proof is a little more difficult (see Exercise 5).

In order to use these norms for matrix–vector manipulations, it is necessary also to have similar measurements for *matrices*. Again, there are a variety of ways to measure the size of a matrix; however, *useful* matrix norms should be related somehow to the vector norm that is being used. More precisely, suppose $\|\cdot\|$ denotes any of the vector norms as given by (6.46), (6.47), or (6.48). Then a corresponding matrix norm should satisfy the inequality

$$\|AX\| \le \|A\|\,\|X\| \tag{6.50}$$

for all $n \times n$ matrices A and all n-vectors X. Notice that the first and last quantities in (6.50) are norms of vectors, namely, the vectors AX and X. The second quantity $\|A\|$, however, is a matrix norm. A matrix norm that satisfies (6.50) is said to be **consistent** with the vector norm. It will be seen in the sequel that consistency is important for any kind of error analysis.

It can be shown that a matrix norm which is consistent with the vector norm $\| \cdot \|_\infty$ is the **row-sum norm** defined by

$$\|A\|_\infty = \max_{1 \le i \le n} \sum_{j=1}^{n} |a_{ij}|, \qquad A \text{ is an } n \times n \text{ matrix.} \qquad (6.51)$$

That is, with $\|A\|_\infty$ defined by (6.51) and $\|X\|_\infty$, $\|AX\|_\infty$ defined by (6.46) we have

$$\|AX\|_\infty \le \|A\|_\infty \|X\|_\infty$$

for all $n \times n$ matrices A and all n-vectors X. For example, let

$$A = \begin{pmatrix} -5 & -3 & 1 \\ 4 & 0 & 1 \\ 5 & 1 & -2 \end{pmatrix}$$

$$X = \begin{pmatrix} 2 \\ -5 \\ 1 \end{pmatrix}, \qquad AX = \begin{pmatrix} 6 \\ 9 \\ 3 \end{pmatrix}. \qquad (6.52)$$

Then $\|X\|_\infty = 5$, $\|A\|_\infty = \max\{9, 5, 8\} = 9$, and $\|AX\|_\infty = 9$. Since $9 \le 9 \cdot 5$, the inequality (6.50) holds.

A matrix norm that is consistent with the sum norm (6.47) is the **column-sum norm** defined by

$$\|A\|_1 = \max_{1 \le j \le n} \sum_{i=1}^{n} |a_{ij}|. \qquad (6.53)$$

It can be shown generally that $\|AX\|_1 \le \|A\|_1 \|X\|_1$ for all $n \times n$ matrices A and all $n -$ vectors X. For the example (6.52) we have $\|A\|_1 = 14$, $\|X\|_1 = 8$, and $\|AX\| = 18$, so that the consistency relation does indeed hold.

It is clear from the above example that different norms produce different values for the sizes of vectors and matrices. Since we shall only be interested in qualitative statements, such as "small" errors, "large" solutions, etc., the different *values* produced by different norms are of little consequence. The choice of which norm to use often is made on the basis of computational convenience. Other norms are sometimes used because of their theoretical properties; however, we shall restrict our discussion to the *maximum* and *sum* vector norms, and the corresponding *maximum row-sum* and *maximum column-sum* matrix norms. In situations where there is no possibility for confusion we shall use the notation $\| \cdot \|$ to denote *any* of these norms. However, if matrices and vectors occur in the same expressions, it will be assumed that the corresponding norms are consistent.

Some useful **properties of the matrix norms** (6.51) and (6.53) are

$$\|A\| \ge 0 \qquad \text{for all } A \quad \text{and} \quad \|A\| = 0 \qquad \text{only if } A = 0 \quad (6.54\text{a})$$

$$\|aA\| = |a|\,\|A\| \qquad \text{for any number } a \qquad\qquad\qquad (6.54\text{b})$$

$$\|A + B\| \le \|A\| + \|B\| \qquad \text{for all matrices} \qquad A, B. \qquad (6.54\text{c})$$

In addition, we have the possibility of multiplying matrices, and this leads to another useful property

$$\|AB\| \le \|A\| \cdot \|B\|, \quad \text{all} \quad n \times n \quad \text{matrices } A, B. \tag{6.54d}$$

Proofs of these statements together with a full discussion of norms can be found, for example, in Stewart (1973).

6.7.2   Error Estimates

It was observed in Section 6.5 that the **residual vector**

$$r = b - A\hat{X} \tag{6.55}$$

is often a misleading guide to the accuracy of the approximate solution $\hat{X}$. To see this more clearly, we write (6.55) in the form

$$\hat{X} = A^{-1}b - A^{-1}r.$$

Now use the fact that $A^{-1}b = X^*$ is the exact solution to write this as

$$X^* - \hat{X} = A^{-1}r.$$

That is, the **error vector** (6.45) satisfies

$$E = A^{-1}r.$$

By using either of the norms discussed earlier we find that

$$\|X^* - \hat{X}\| = \|A^{-1}r\| \le \|A^{-1}\| \|r\|. \tag{6.56}$$

[The consistency condition (6.50) has been used here.] Inequality (6.56) is an *absolute* error estimate for the computed solution $\hat{X}$. To obtain a *relative* error estimate, we note that

$$b = Ax^*$$

implies that

$$\|b\| = \|AX^*\| \le \|A\| \|X^*\|.$$

[Again we have used the consistency property (6.50).] Thus

$$1/\|X^*\| \le \|A\|/\|b\|. \tag{6.57}$$

If (6.56) and (6.57) are combined, we obtain

$$\|X^* - \hat{X}\|/\|X^*\| \le \|A\| \|A^{-1}\| \|r\|/\|b\|. \tag{6.58}$$

This is a **bound on the relative error** in $\hat{X}$ as an approximation to the solution vector X^*. The factor $\|r\|/\|b\|$ in this bound is simply the relative residual, which one expects to be small. The other factor

$$\|A\|\,\|A^{-1}\|, \tag{6.59}$$

however, may be quite large. Consider, for example, the system (6.38), which we write in matrix form as $AX = b$ with

$$A = \begin{pmatrix} 1.000 & 2.000 \\ .499 & 1.001 \end{pmatrix}, \qquad b = \begin{pmatrix} 3.000 \\ 1.500 \end{pmatrix}. \tag{6.60}$$

The **exact solution** is

$$X^* = \begin{pmatrix} 1.0 \\ 1.0 \end{pmatrix},$$

but the "**approximate**" solution

$$\hat{X} = \begin{pmatrix} 2.000 \\ .500 \end{pmatrix}$$

gives the **residual**

$$r = b - A\hat{X} = \begin{pmatrix} 3.00 & 3.00 \\ 1.50 & 1.4985 \end{pmatrix} = \begin{pmatrix} 0 \\ .0015 \end{pmatrix}.$$

Now with the maximum vector norm and the row-sum matrix norm (i.e., with $\|\cdot\|_\infty$) we find that

$$\|A\|_\infty = 3, \qquad \|A^{-1}\|_\infty = \left\| \begin{matrix} 333.67 & -666.67 \\ -166.67 & 333.33 \end{matrix} \right\|_\infty = 1000.34. \tag{6.61}$$

Thus the factor (6.59) is 3001.02, and (6.58) implies that the relative error in $\hat{X}$ may be 3000 times larger than the relative residual. This is, in fact, the case since the relative error is

$$\|X^* - \hat{X}\|/\|/\|X^*\| = 1,$$

whereas the relative residual is

$$\|r\|/\|b\| = .0015/3 = .0005.$$

The inequality (6.58) is useful in explaining why a small residual need not imply a small error. However, it has little practical appeal because of the presence of the quantity $\|A^{-1}\|$. Recall from Section 6.3 that the computation of A^{-1} is considerably harder than solving a linear system.

6.7.3   Effect of Changes in the Data

One of the most useful applications of matrix theory to the numerical solution of linear systems is in the understanding of the concept of ill-conditioning. To show very clearly the effect of changes in data on the solution to the system, consider the two systems of equations

Exact system and perturbed system	$AX = b,$ solution X^* $(A + E)X = b,$ solution $\hat{X}.$	(6.62a) (6.62b)

Here we think of E as being a small perturbation of the coefficient matrix A. For now we consider only changes in A; later we shall allow changes in the right-hand side b.

To derive an estimate for $X^* - \hat{X}$ in terms of E, we write (6.62b) as

$$A\hat{X} = b - E\hat{X}$$

or

$$\hat{X} = A^{-1}b - A^{-1}E\hat{X} = X^* - A^{-1}E\hat{X}.$$

Thus by using norms we have

$$
\begin{aligned}
\|\hat{X} - X^*\| = \|-A^{-1}E\hat{X}\| & \\
= \|A^{-1}E\hat{X}\| & \qquad \text{by (6.54b)} \\
\leq \|A^{-1}E\| \, \|\hat{X}\| & \qquad \text{by (6.50)} \\
\leq \|A^{-1}\| \, \|E\| \, \|\hat{X}\| & \qquad \text{by (6.54d)}
\end{aligned}
$$

and hence

$$\|\hat{X} - X^*\|/\|\hat{X}\| \leq \|A^{-1}\| \, \|E\|.$$

This inequality involves a relative error on the left but an absolute error $\|E\|$ on the right. Since E is an error in A, the right side can be relativized by writing it in the form

Error bound: perturbation of A	$\|\hat{X} - X^*\|/\|\hat{X}\| \leq \|A\| \, \|A^{-1}\| \, \|E\|/\|A\|.$	(6.63)

This clearly shows the relative change in the solution that is caused by a certain relative change in the coefficient matrix.

To illustrate this inequality, consider again the system (6.38) and the perturbed system (6.39). That is,

$$A = \begin{pmatrix} 1.000 & 2.000 \\ .499 & 1.001 \end{pmatrix}, \qquad E = \begin{pmatrix} 0 & 0 \\ .001 & -.001 \end{pmatrix}$$

$$X^* = \begin{pmatrix} 1 \\ 1 \end{pmatrix}, \qquad \hat{X} = \begin{pmatrix} 3.0 \\ 2 \end{pmatrix}.$$

Then

$$\|X^* - \hat{X}\|_\infty = \left\| \begin{pmatrix} -2 \\ -1 \end{pmatrix} \right\|_\infty = 2$$

so that

$$\|X^* - \hat{X}\|_\infty / \|\hat{X}\|_\infty = \tfrac{2}{3},$$

whereas

$$\|E\|_\infty / \|A\|_\infty = .002/3 = \tfrac{2}{3} \times 10^{-3}.$$

But the quantity $\|A\|_\infty \|A^{-1}\|_\infty$ has already been determined in (6.61) to be 3001.02, so that the relative change in the solution should be expected to be several thousand times larger than the relative change in the matrix A.

Finally, to examine the effect of changing the right-hand side, consider the two systems

| Perturbation of right-hand side | $AX = b,$ solution X^*
 $AX = b + e,$ solution $\bar{X}.$ | (6.64) |

Then

$$\bar{X} = A^{-1}(b + e) = A^{-1}b + A^{-1}e = X^* + A^{-1}e$$

so that

$$\|\bar{X} - X^*\| = \|A^{-1}e\| \le \|A^{-1}\| \, \|e\|.$$

By using (6.57) we find that

| Error bound: perturbation of b | $\|\bar{X} - X^*\|/\|X^*\| \le \|A\| \, \|A^{-1}\| \, \|e\|/\|b\|.$ | (6.65) |

This represents an estimate similar to (6.63) but relating the relative change in the solution to the relative change in the right-hand side of the system.

6.7.4   Condition Number of Linear Systems

The three estimates (6.58), (6.63), and (6.65) all involve the important factor $\|A\|\,\|A^{-1}\|$. The latter two estimates suggest that if this factor is large, then small changes in the data, A or b, may cause large changes in the solution to the system; that is, the system may be ill-conditioned. For this reason we define the **condition number of a matrix** A to be

$$K(A) = \|A\|\,\|A^{-1}\|. \qquad (6.66)$$

We say that A is *ill-conditioned* if $K(A)$ is large. To make this a little more precise, note that for either of the matrix norms we have by (6.54d) that

$$1 = \|I\| = \|AA^{-1}\| \le \|A\|\,\|A^{-1}\|,$$

where I denotes the identity matrix. Hence for any matrix A, it follows that

$$K(A) \ge 1$$

and if $A = I$ then $K(A) = 1$. Thus in speaking of "large $K(A)$," we mean "large relative to 1.0." The estimates (6.63) and (6.65) show that errors in the data may be multiplied by the factor $K(A)$ in the solution. Estimate (6.58) shows that the error in the solution may be $K(A)$ times as large as the residual.

In much of the previous analysis, it is clear that the sizes of the elements of A^{-1} are of major importance. The condition number, in fact, is just a measure of the size of A^{-1} relative to the size of A. In this regard note that for any nonzero scalar α,

$$K(\alpha A) = K(A).$$

This shows that the condition of A cannot be changed by simply multiplying A or A^{-1} by some appropriate constant.

Finally the definition of $K(A)$ involves the use of a matrix norm, and different norms will give different values. For example, the matrix in the example (6.38) gives, by using (6.61), the values

$$K(A) = 3001.02 \qquad \text{with} \quad \|\cdot\|_{\infty}$$

$$K(A) = 3001.00 \qquad \text{with} \quad \|\cdot\|_{1}.$$

Since we are only interested in whether this value is large or small, the minor variations caused by a particular choice of norm are completely immaterial. Perhaps more important to this discussion is the observation that $K(A)$ is *not* a quantity that is easily computed by using *any* norm. The various inequalities derived above are useful theoretical facts, but the best that can be done in practice is to estimate $K(A)$ by, for example, the method described in Section 6.5.

EXERCISES

1 Compute the norms (6.46), (6.47), and (6.48) of the vectors

$$\begin{pmatrix} 1 \\ 0 \end{pmatrix}, \quad \begin{pmatrix} .01 \\ .001 \\ .0001 \end{pmatrix}, \quad \begin{pmatrix} 10^2 \\ 10^3 \\ 10^4 \end{pmatrix}.$$

2 Compute the norms (6.51) and (6.53) of the matrices

$$\begin{pmatrix} 1 & 2 \\ 3 & 4 \end{pmatrix}, \quad \begin{pmatrix} .001 & .01 \\ .001 & .0001 \end{pmatrix}, \quad \begin{pmatrix} 10^3 & 10^4 \\ 10^3 & 10^4 \end{pmatrix}.$$

3 Prove that (6.49) is true for the norm (6.46).

4 Prove that (6.49) is true for the norm (6.47).

5 Prove that (6.49) is true for the norm (6.48).
   [*Hint*: In order to prove property (6.49c) first show

$$\|X + Y\|_2^2 = \|X\|_2^2 + \|Y\|_2^2 + 2 \sum_{i=1}^{n} x_i y_i.$$

Now use the inequality

$$\sum_{i=1}^{n} x_i y_i \le \|X\|_2 \|Y\|_2.$$

Can you prove *this* inequality?]

6 For each of the systems in Exercise 1 of Section 6.5, compute $K(A)$, using either (or both) of the norms $\| \cdot \|_\infty$, $\| \cdot \|_1$.

7 Determine the cost of computing $K(A)$ using the matrix norm $\| \cdot \|_\infty$.

6.8 Iterative Methods

The Gaussian elimination method is very useful for moderately sized systems of linear equations. Certain applications, however, involve *very* large systems. Indeed, systems of the order of 10,000 equations are not uncommon. The analysis of Section 6.6 shows that to solve this large a system by means of Gaussian elimination would require more than $(\frac{1}{3} \times 10^{12})A + (\frac{1}{3} \times 10^{12})M$ operations. On a computer with an add time of 2×10^{-6} sec and a multiply time of 3×10^{-6} sec, this would take more than $\frac{1}{3} \times 10^{12} \times 2 \times 10^{-6} + \frac{1}{3} \times 10^{12} \times 3 \times 10^{-6}$ sec $= \frac{5}{3} \times 10^6$ sec $\cong 1$ month! In this section we discuss some methods that are often used to solve large systems with a reasonable amount of computer time.

6.8.1   Basic Methods

Associated with **iterative methods** are two problems that do not occur in direct methods such as Gaussian elimination. First it is necessary to have an initial approximation to start the iteration, and second some kind of test is needed to stop the iteration. The convergence theory in Section 6.8.2 shows that for linear systems the choice of **starting values** is not critical. In the algorithms given here we shall include, as part of the input, some starting values for the iteration. In practice, however, these are often chosen to be all ones or all zeros. The **stopping criteria** are somewhat more complicated but again the convergence theory in Sections 6.8.2 and 6.8.3 will be helpful. The algorithms given here do not include stopping conditions except for the case of nonconvergence. Specifically, all of the algorithms include a parameter N_{max} that is the maximum number of iterations allowed. This value should be chosen so that when the iterates are actually converging the stopping criteria will be satisfied long before N_{max} iterates have been computed. Clearly, the value to the used for N_{max} may depend on the stopping criteria. More will be said about stopping criteria and N_{max} in Sections 6.8.3 and 6.8.4.

The simplest iterative method for solving the system (6.1) is derived by solving the first equation for x_1, the second equation for x_2, etc. In general, we write the ith equation in the form

$$x_i = (1/a_{ii})[b_i - a_{i1}x_1 - a_{i2}x_2 - \cdots - a_{i,\,i-1}x_{i-1} - a_{i,\,i+1}x_{i+1} - \cdots - a_{in}x_n].$$

$$(6.67)$$

This suggests that known values for $x_1, \ldots, x_{i-1}, x_{i+1}, \ldots, x_n$ be used on the right side of (6.67) to compute a new value for x_i. Hence, written as an iteration (6.67) has the form

Jacobi method	$$x_i^{(k+1)} = (1/a_{ii})[b_i - a_{i1}x_1^{(k)} - a_{i2}x_2^{(k)}$$ $$- \cdots - a_{i,\,i-1}x_{i-1}^{(k)} - a_{i,\,i+1}x_{i+1}^{(k)}$$ $$- \cdots - a_{in}x_n^{(k)}], \qquad i = 1, 2, \ldots, n. \qquad (6.68)$$

Here the *super*scripts refer to the *iteration* number. In other words, $x_1^{(0)}, x_2^{(0)}, \ldots, x_n^{(0)}$ are the initial values for the unknowns, and $x_1^{(k)}, x_2^{(k)}, \ldots, x_n^{(k)}$ are the values computed during the kth iteration.

Obviously, in order to (6.67) or (6.68) to make sense, a_{ii} must be nonzero for $i = 1, 2, \ldots, n$. If the system is nonsingular, this can always be achieved by interchanging equations (Exercise 13). In most applications for which this algorithm is used, many of the coefficients a_{ij}, $i \neq j$, are zero. This fact should be used in the implementation of the algorithm to avoid useless multiplications of the form $a_{ij} x_j$ when $a_{ij} = 0$, and the resulting addition of zero terms. This, of course, can only be done through careful programming and generally requires that the nonzero coefficients occur in some regular pattern.

A description of this iteration, which is called *Jacobi's method*, is now given.

Algorithm for Jacobi's Method to Solve $AX = b$

1   Input $\{x_1, x_2, \ldots, x_n, a_{ij}, b_i, N_{\max}\}$
2   For $k = 1, 2, \ldots, N_{\max}$
   2.1   For $i = 1, 2, \ldots, n$
      2.1.1 $y_i \leftarrow (1/a_{ii})[b_i - a_{i1} x_1 - a_{i2} x_2 - \cdots - a_{i, i-1} x_{i-1}$
             $- a_{i, i+1} x_{i+1} - \cdots - a_{in} x_n]$
   2.2   If convergence test is satisfied   (6.69)
      2.2.1   Output $\{y_1, y_2, \ldots, y_n\}$ Halt
   2.3   Else
      2.3.1   For $i = 1, 2, \ldots, n$
         2.3.1.1 $x_i \leftarrow y_i$
3   Error termination; no convergence

In this algorithm "old" iterates are denoted by $x_1, x_2, \ldots, x_n$, while the "new" iterates are $y_1, y_2, \ldots, y_n$. If superscripts were used for the iterates, as in (6.68), this would suggest that all of the iterates should be saved. In fact, only the last iterates are used as output, although the next-to-last values may be needed in the convergence test. In an actual implementation of this algorithm, step 2.1.1 should be programmed to take advantage of zero coefficients, as noted earlier.

To illustrate this method consider the system

$$
\begin{aligned}
5x_1 \quad\quad\quad\quad - 3x_4 \; - x_5 &= 2 \\
-x_1 + 4x_2 \quad\quad\quad - x_5 &= 3 \\
2x_3 \; - x_4 \quad\quad\; &= -1 \\
-x_1 \quad\quad\quad + 4x_4 - 2x_5 &= 0 \\
- x_4 + 2x_5 &= -1.
\end{aligned}
\tag{6.70}
$$

With starting values of $x_1 = x_2 = x_3 = x_4 = x_5 = 1$, we find

$$
\begin{aligned}
y_1 &= \tfrac{1}{5}(2 + 3x_4 + x_5) = \tfrac{1}{5}(2 + 3 + 1) = 1.2 \\
y_2 &= \tfrac{1}{4}(3 + x_1 + x_5) = \tfrac{1}{4}(3 + 1 + 1) = 1.25 \\
y_3 &= \tfrac{1}{2}(-1 + x_4) = \tfrac{1}{2}(-1 + 1) = 0 \\
y_4 &= \tfrac{1}{4}(0 + x_1 + 2x_5) = \tfrac{1}{4}(1 + 2) = .75 \\
y_5 &= \tfrac{1}{2}(-1 + x_4) = \tfrac{1}{2}(-1 + 1) = 0.
\end{aligned}
$$

The next iteration gives the values

$$y_1 = .85, \qquad y_2 = 1.05, \qquad y_3 = -1.25, \qquad y_4 = .3, \qquad y_5 = -1.25$$

and the 35th iteration gives the approximate solution

$$x_1 = .086957, \qquad x_2 = .608696, \qquad x_3 = -.652173,$$

$$x_4 = -.304347, \qquad x_5 = -.652173.$$

While computing these iterates, one quickly observes that when y_2, for example, is to be determined, a value for y_1 has just been computed, and one might as well use this value instead of the old iterate x_1. This suggests the **Gauss–Seidel method**, which is derived by writing the (6.67) as an iteration in the form

| Gauss–Seidel method | $$x_i^{(k+1)} = (1/a_{ii})[b_i - a_{i1}x_1^{(k+1)} - a_{i2}x_2^{(k+1)}$$ $$- \cdots - a_{i,i-1}x_{i-1}^{(k+1)}$$ $$- a_{i,i+1}x_{i+1}^{(k)} - \cdots - a_{in}x_n^{(k)}].$$ (6.71) |

The difference between this iteration and (6.68) is that the values of $x_j^{(k+1)}$ are used as soon as they become available. A description of this algorithm is given next.

Algorithm to Solve a Linear System by the Gauss–Seidel Method

1   Input $\{x_1, x_2, \ldots, x_n a_{ij}, b_i, N_{max}\}$
2   For $k = 1, 2, \ldots, N_{max}$
   2.1   For $i = 1, 2, \ldots, n$
      2.1.1 $y_i \leftarrow (1/a_{ii})[b_i - a_{i1}y_1 - \cdots - a_{i,i-1}y_{i-1}$
            $- a_{i,i+1}x_{i+1} - \cdots - a_{in}x_n]$
   2.2   If convergence test is satisfied   (6.72)
      2.2.1 $\{$Output $y_1, y_2, \ldots, y_n\}$ Halt
   2.3   Else
      2.3.1   For $i = 1, 2, \ldots, n$
         2.3.1.1 $x_i \leftarrow y_i$
3   Error termination; no convergence

Notice that in this algorithm, step 2.1.1 employs the newly computed values for $y_1, y_2, \ldots, y_{i-1}$ to determine y_i. If the convergence test uses *only* the values of the current iterate, then the variables $y_1, y_2, \ldots, y_n$ are not needed. That is, step 2.1.1 could be written as

$$x_i \leftarrow (1/a_{ii})[b_i - a_{i1}x_1 - \cdots - a_{i,\,i-1}x_{i-1} - a_{i,\,i+1}x_{i+1} - \cdots - a_{in}x_n]$$

$$(6.73)$$

and step 2.3.1 could be omitted. If this algorithm is applied to the system (6.70) with starting values all equal to 1.0, the following values are determined:

$$y_1 = \tfrac{1}{5}(2 + 3x_4 + x_5) = \tfrac{1}{5}(2 + 3 + 1) = 1.2$$
$$y_2 = \tfrac{1}{4}(3 + y_1 + x_5) = \tfrac{1}{4}(3 + 1.2 + 1) = 1.3$$
$$y_3 = \tfrac{1}{2}(-1 + x_4) = \tfrac{1}{2}(-1 + 1) = 0$$
$$y_4 = \tfrac{1}{4}(0 + y_1 + 2x_5) = \tfrac{1}{4}(1.2 + 2) = .8$$
$$y_5 = \tfrac{1}{2}(-1 + y_4) = \tfrac{1}{2}(-1 + .8) = .1.$$

The next iteration gives

$$y_1 = .86, \qquad y_2 = .94, \qquad y_3 = -.1, \qquad y_4 = .165, \qquad y_5 = -.4175$$

and the 18th iteration gives the approximation solution

$$x_1 = .086957, \qquad x_2 = .608696, \qquad x_3 = -.652173,$$
$$x_4 = -.304347, \qquad x_5 = -.652174.$$

In this example the Gauss–Seidel method has produced the solution in fewer iterations than has the Jacobi method. Indeed one might expect that such is always the case, and therefore there is no point in considering the Jacobi method at all. In fact, there are systems of equations for which the Jacobi method converges while the Gauss–Seidel method does not (Exercise 4). Both methods are important tools.

Before the advent of computers these iterative methods were carried out with the aid of desk calculators. Many people who performed such calculations observed a certain regularity in the changes in the iterates. By taking advantage of this regularity, the convergence could often be accelerated. The resulting methods, called *over-relaxation methods*, are important for solving large linear systems of special types. Their analysis requires strong use of matrix theory and is beyond the scope of this book [see Varga (1963) and Young (1971) for further discussion].

The analysis of any iterative method is simplified by first writing it in the form

$$x_i^{(k+1)} = \sum_{j=1}^{n} c_{ij}x_j^{(k)} + d_i, \qquad i = 1, 2, \ldots, n. \qquad (6.74)$$

Here the c_{ij} are combinations of the a_{ij}, and d_i is related to b_i. For example, the Jacobi method, written as (6.68) has the form (6.74) with

Jacobi method in the form (6.74)

$$c_{ij} = \begin{cases} 0 & \text{if} \quad i = j \\ -a_{ij}/a_{ii} & \text{if} \quad i \neq j \end{cases}, \quad d_i = b_i/a_{ii}. \qquad (6.75)$$

It is not so easy to write the Gauss–Seidel method in the form (6.74); however, it is not difficult to prove by induction that it *can* be written this way. Indeed, equation (6.71) with $i = 1$ already has the form (6.74, $i = 1$) with $c_{11} = 0$, $c_{1j} = -a_{1j}/a_{11}$, $j = 2, 3, \ldots, n$. If we assume that the iterates $x_2^{(k+1)}, \ldots, x_{l-1}^{(k+1)}$ as given by (6.71) have the form (6.74), then equation l in (6.71) can be written as

$$
\begin{aligned}
x_l^{(k+1)} &= \frac{1}{a_{ll}} \left[b_l - \sum_{j=1}^{l-1} a_{lj} x_j^{(k+1)} - \sum_{i=l+1}^{n} a_{li} x_i^{(k)} \right] \\
&= \frac{1}{a_{ll}} \left[b_l - \sum_{j=1}^{l-1} a_{lj} \left(\sum_{i=1}^{n} c_{ji} x_i^{(k)} + d_j \right) - \sum_{i=l+1}^{n} a_{li} x_i^{(k)} \right] \\
&= \sum_{i=1}^{l} \left(\sum_{j=1}^{l-1} \frac{-a_{lj} c_{ji}}{a_{ll}} \right) x_i^{(k)} + \sum_{i=l+1}^{n} \left(\sum_{j=1}^{l-1} \frac{-a_{lj} c_{ji}}{a_{ll}} - \frac{a_{li}}{a_{ll}} \right) x_i^{(k)} \\
&\quad + \frac{1}{a_{ll}} \left(b_l - \sum_{j=1}^{l-1} a_{lj} d_j \right).
\end{aligned} \qquad (6.76)
$$

Thus

Gauss–Seidel method in the form (6.74)

$$x_l^{(k+1)} = \sum_{i=1}^{n} c_{li} x_i^{(k)} + d_l, \qquad (6.77a)$$

where

$$
c_{li} = \begin{cases} -\left(\sum\limits_{j=1}^{l-1} a_{lj} c_{ji} \right) \Big/ a_{ll}, & i = 1, 2, \ldots, l \\[2ex] -\left(\sum\limits_{j=1}^{l-1} a_{lj} c_{ji} + a_{li} \right) \Big/ a_{ll}, & i = l+1, \ldots, n \end{cases} \qquad (6.77b)
$$

$$d_l = \left(b_l - \sum_{j=1}^{l-1} a_{lj} d_j \right) \Big/ a_{ll}. \qquad (6.77c)$$

Even though this defines the coefficients c_{ij} recursively, these equations should never be used to compute the iterates $x_i^{(k+1)}$, $i = 1, 2, \ldots, n$. The formula (6.71) is much simpler to use. The only reason for considering the form (6.74) is to facilitate the discussion of convergence. In that regard note that if $\lim_{k \to \infty} x_i^{(k)} = x_i^*$, $i = 1, 2, \ldots, n$, then by taking limits on both sides of the equation in (6.74), we have

$$x_i^* = \sum_{j=1}^{n} c_{ij} x_j^* + d_i, \qquad i = 1, 2, \ldots, n. \tag{6.78}$$

This fact will be useful in the following convergence theory.

6.8.2   Convergence Theory

A complete discussion of the iterative methods derived above requires the use of certain concepts from matrix theory that are usually covered in more advanced courses. By using only elementary facts, however, it is possible to give several results that will clearly show some of the basic ideas inherent in the study of iterative methods. Consider first of all the general formulation of an iterative method as given by equations (6.74). The desired solution x_1^*, $x_2^*, \ldots, x_n^*$ satisfies (6.78). Next define the **errors at the kth iteration** by $\varepsilon_i^{(k)} = x_i^{(k)} - x_i^*$ for $i = 1, 2, \ldots, n$. Then, by subtracting (6.78) from (6.74), we obtain the equations

$$\varepsilon_i^{(k+1)} = \sum_{j=1}^{n} c_{ij} \varepsilon_j^{(k)}. \tag{6.79}$$

The **convergence** of the iteration (6.74) is equivalent to $\lim_{k \to \infty} \varepsilon_i^{(k)} = 0$, $i = 1$, $2, \ldots, n$, where $\varepsilon_i^{(k)}$ satisfy (6.79). There are two important consequences of this observation. The first is that convergence does not depend on the d_i in (6.74), or equivalently, on the b_i in the original system, since the terms containing these values do not appear in (6.79). The second consequence is that the convergence does not depend upon the size of the initial errors; i.e., it does not depend upon how close the starting values $x_1^{(0)}, \ldots, x_n^{(0)}$ are to the solution. To see this, suppose the starting values have been chosen so that the errors $\varepsilon_i^{(k)}$ go to zero as $k \to \infty$. Now suppose some other set of starting values with initial errors $\delta_i^{(0)}$ is used in the iteration, where $\delta_i^{(0)} = \alpha \varepsilon_i^{(0)}$, $\alpha \neq 0$. Then the errors $\delta_i^{(k)}$ also satisfy equations (6.79) so that

$$\delta_i^{(1)} = \sum_{j=1}^{n} c_{ij} \delta_j^{(0)} = \sum_{j=1}^{n} c_{ij} (\alpha \varepsilon_j^{(0)}) = \alpha \sum_{j=1}^{n} c_{ij} \varepsilon_j^{(0)} = \alpha \varepsilon_i^{(1)}.$$

Similarly, $\delta_i^{(k)} = \alpha \varepsilon_i^{(k)}$ for *any* k so that if $\lim_{k \to \infty} \varepsilon_i^{(k)} = 0$ then also $\lim_{k \to \infty} \delta_i^{(k)} = 0$. Thus, because of the linearity of the iteration (6.74), if the iterates converge for all starting values sufficiently close to the solution, then they converge for *any* starting values. For this reason it is not necessary to be overly concerned with the choice of starting values. Of course, the best available approximations to the solution should always be used as starting values in order to reduce the number of iterations that are actually required.

The simplest convergence theorem is obtained from (6.79) by noting that

$$|\varepsilon_i^{(k+1)}| \le \sum_{j=1}^{n} |c_{ij}| \, |\varepsilon_j^{(k)}|. \tag{6.80}$$

Now, let $\|\varepsilon^{(k)}\|$ be the number

Size of the errors

$$\|\varepsilon^{(k)}\| = \max\{|\varepsilon_1^{(k)}|, \, |\varepsilon_2^{(k)}|, \, \ldots, \, |\varepsilon_n^{(k)}|\}, \tag{6.81}$$

which measures the size[†] of the errors $\varepsilon_1^{(k)}, \ldots, \varepsilon_n^{(k)}$. Then from (6.80) we have

$$|\varepsilon_i^{(k+1)}| \le \sum_{j=1}^{n} |c_{ij}| \, \|\varepsilon^{(k)}\|. \tag{6.82}$$

Next define the number $\|c\|$ by

Size of the coefficients c_{ij}

$$\|c\| = \max_i \sum_{j=1}^{n} |c_{ij}| \tag{6.83}$$

as a measure of the size of the coefficients c_{ij}. Then from (6.82) it follows that

$$|\varepsilon_i^{(k+1)}| \le \|c\| \, \|\varepsilon^{(k)}\|$$

and hence that

$$\|\varepsilon^{(k+1)}\| \le \|c\| \, \|\varepsilon^{(k)}\|. \tag{6.84}$$

Now this entire argument can be repeated to show that

$$\|\varepsilon^{(k)}\| \le \|c\| \, \|\varepsilon^{(k-1)}\|.$$

By combining these two inequalities, we obtain

$$\|\varepsilon^{(k+1)}\| \le \|c\| \, \|\varepsilon^{(k)}\| \le \|c\|^2 \|\varepsilon^{(k-1)}\|.$$

[†] The reader who studied the optional Section 6.7 will recognize the quantities defined by (6.81) and (6.83) to be the ∞-norms of vectors and matrices, respectively.

By repeating this we obtain finally

$$\boxed{\begin{array}{l} \text{Error at} \\ \text{the } k\text{th} \\ \text{iteration} \end{array}} \qquad \|\varepsilon^{(k+1)}\| \le \|c\|^{k+1}\|\varepsilon^{(0)}\|. \tag{6.85}$$

Clearly if $\|c\| < 1$, then the right side tends to zero as $k \to \infty$, and hence $\|\varepsilon^{(k+1)}\| \to 0$. Thus the iterates $x_1^{(k+1)}, \ldots, x_n^{(k+1)}$ converge to the solution, and we have the result:

> If $\|c\|$, as defined by (6.83), satisfies $\|c\| < 1$, then the iteration (6.74) converges for any starting values $x_1^{(0)}, \ldots, x_n^{(0)}$.

The condition $\|c\| < 1$ is a statement about the size of the coefficients c_{ij} in (6.74). There are several ways besides (6.83) to measure the size of these coefficients. Each of these ways leads to a convergence result of the above kind. For example, we could also use the definitions[†]

$$\boxed{\begin{array}{l} \text{Another} \\ \text{measure} \\ \text{of size} \end{array}}$$
$$\begin{aligned} \|\varepsilon^{(k)}\| &= |\varepsilon_1^{(k)}| + |\varepsilon_2^{(k)}| \\ &\quad + \cdots + |\varepsilon_n^{(k)}| = \sum_{i=1}^{n} |\varepsilon_i^{(k)}| \end{aligned} \tag{6.86}$$
$$\begin{aligned} \|c\| &= \max_j \{|c_{1j}| + |c_{2j}| \\ &\quad + \cdots + |c_{nj}|\} = \max_j \sum_{i=1}^{n} |c_{ij}|. \end{aligned} \tag{6.87}$$

Then from (6.80) we have

$$\begin{aligned} \|\varepsilon^{(k+1)}\| &= \sum_{i=1}^{n} |\varepsilon_i^{(k+1)}| \\ &\le \sum_{i=1}^{n} (|c_{i1}||\varepsilon_1^{(k)}| + |c_{i2}||\varepsilon_2^{(k)}| + \cdots + |c_{in}||\varepsilon_n^{(k)}|) \\ &= \sum_{i=1}^{n} |c_{i1}||\varepsilon_1^{(k)}| + \sum_{i=1}^{n} |c_{i2}||\varepsilon_2^{(k)}| + \cdots + \sum_{i=1}^{n} |c_{in}||\varepsilon_n^{(k)}| \\ &\le \|c\|(|\varepsilon_1^{(k)}| + |\varepsilon_2^{(k)}| + \cdots + |\varepsilon_n^{(k)}|) \\ &= \|c\| \|\varepsilon^{(k)}\|. \end{aligned}$$

[†] Again, for those readers who have studied Section 6.7, we note that equations (6.86) and (6.87) define the norms $\|\cdot\|_1$ for vectors and matrices.

This result is the same as (6.84) but with $\|c\|$ and $\|\varepsilon\|$ defined by (6.86) and (6.87). Just as before it follows that

$$\|\varepsilon^{(k+1)}\| \le \|c\|^k \|\varepsilon^{(0)}\|,$$

so that the above convergence result is also true when $\|c\|$ is defined by (6.87).

To see how these results apply to specific methods, consider first the **Jacobi iteration**. The coefficients c_{ij} are given by (6.75) as

$$c_{ij} = \begin{cases} 0, & i = j \\ -a_{ij}/a_{ii}, & i \ne j \end{cases}.$$

Thus with $\|c\|$ defined by (6.83) we have

$$\|c\| = \max_i \left\{ \sum_{j=1}^n |c_{ij}| \right\} = \max_i \left\{ \sum_{j=1, j \ne i}^n |a_{ij}|/|a_{ii}| \right\}.$$

Now $\|c\| < 1$ provided that

$$\sum_{j=1, j \ne i}^n |a_{ij}|/|a_{ii}| < 1 \qquad \text{for} \quad i = 1, 2, \ldots, n.$$

This is equivalent to the condition

$$\sum_{j=1, j \ne i}^n |a_{ij}| < |a_{ii}|, \qquad i = 1, 2, \ldots, n. \tag{6.88}$$

A system of equations in which the coefficients satisfy (6.88) is said to be **strictly diagonally dominant**. This analysis can be summarized in the following form:

> If the system (6.1) is strictly diagonally dominant, then the Jacobi method converges for any starting values $x_1^{(0)}, \ldots, x_n^{(0)}$.

A similar result also holds for the Gauss–Seidel method. To show this, assume the system (6.1) is strictly diagonally dominant and consider the coefficients c_{li}, as defined by (6.77b), for $l = 2, 3, \ldots, n$, with $c_{11} = 0$, $c_{1j} = -a_{1j}/a_{11}, j = 2, 3, \ldots, n$. Then the coefficients $c_{1j}, j = 1, 2, \ldots, n$ are the same as for the Jacobi method, so it follows immediately that

$$\sum_{j=1}^n |c_{1j}| < 1.$$

We shall prove by induction that $\sum_{j=1}^{n} |c_{ij}| < 1$ for all $i \leq n$. For this, assume that $\sum_{i=1}^{n} |c_{ki}| < 1$ for $k < l$. Then by (6.77b)

$$\sum_{i=1}^{n} |c_{li}| = \frac{1}{|a_{ll}|} \left| \sum_{i=1}^{l} \left(\sum_{j=1}^{l-1} a_{lj} c_{ji} \right) + \sum_{i=l+1}^{n} \left(\sum_{j=1}^{l-1} a_{lj} c_{ji} + a_{li} \right) \right|$$

$$\leq \frac{1}{|a_{ll}|} \left\{ \sum_{j=1}^{l-1} \left(\sum_{i=1}^{n} |c_{ji}| \, |a_{lj}| \right) + \sum_{i=l+1}^{n} |a_{li}| \right\}.$$

Now by the induction hypothesis, $\sum_{i=1}^{n} |c_{ji}| < 1$ for $j = 1, 2, \ldots, l-1$, so that

$$\sum_{i=1}^{n} |c_{li}| \leq \frac{1}{|a_{ll}|} \left\{ \sum_{j=1}^{l-1} |a_{lj}| + \sum_{i=l+1}^{n} |a_{li}| \right\} < 1,$$

where this last inequality follows from the strict diagonal dominance of the coefficients a_{ij}. This completes the induction step, and we can conclude that $\|c\| < 1$, where c_{ij} are given by the Gauss–Seidel method, and $\|c\|$ is defined by (6.83). This gives the result:

If the system (6.1) is strictly diagonally dominant, then the Gauss–Seidel method converges for any starting values $x_1^{(0)}, \ldots, x_n^{(0)}$.

As an example of how these results are applied, consider the system

$$
\begin{aligned}
(2 + h)x_1 - x_2 &&&= b_1 \\
-x_1 + (2 + h)x_2 - x_3 &&&= b_2 \\
-x_2 &+ (2 + h)x_3 - x_4 &&= b_3 \\
&&&\vdots \\
&- x_{n-1} + (2 + h)x_n &&= b_n,
\end{aligned}
\tag{6.89}
$$

where $h > 0$ is a small parameter. Such a system arises as a result of solving certain *differential equations*. Clearly the coefficients here are strictly diagonally dominant so that the Jacobi and Gauss–Seidel methods will converge.

It is important to realize that the theorems proved here are rather "weak" theorems, in the sense that the iteration might converge even though the system is not diagonally dominant. For example, the system of equations given in Exercise 4 is not diagonally dominant nor do the Jacobi iteration coefficients satisfy Theorem 6.1. Nevertheless the Jacobi method does converge for *any* starting values. The reason for this is that there is another, more complicated way of measuring the size of the coefficients c_{ij} that does give a value less than 1.

6.8.3   Rounding Errors and Stopping Criteria

In order to examine the effect of rounding errors on the Algorithms of Section 6.8.1, consider first the **Jacobi method**, in which the computed iterates are

$$\hat{x}_i^{(k+1)} = \text{fl}[(b_i - a_{i1}\hat{x}_1^{(k)} - \cdots - a_{in}\hat{x}_n^{(k)})/a_{ii}], \quad i = 1, 2, \ldots, n, \quad (6.90)$$

where the term $a_{ii}\hat{x}_i^{(k)}$ is omitted from the right side. With the notation of Section 2.5.3, this can be written as

$$\hat{x}_i^{(k+1)} = (b_i\langle n+1\rangle - a_{i1}\hat{x}_1^{(k)}\langle n+1\rangle - a_{i2}\hat{x}_2^{(k)}\langle n\rangle - \cdots - a_{in}\hat{x}_n^{(k)}\langle 3\rangle)/a_{ii}.$$

That is, the computation of $\hat{x}_i^{(k+1)}$ is just a summation of n terms in which most terms are computed products followed by a division. Thus the above expression is (3.3c) but with two additional rounding errors to compensate for the multiplications $a_{ij}\hat{x}_j^{(k)}$ and the division by a_{ii}. We rewrite this as

$$\hat{x}_i^{(k+1)} = (b_i - a_{i1}\hat{x}_1^{(k)} - \cdots - a_{in}\hat{x}_n^{(k)})/a_{ii} + E_i^{(k)}, \quad (6.91a)$$

where

$$E_i^{(k)} = (b_i\langle n+1\rangle_0 - a_{i1}\hat{x}_1^{(k)}\langle n+1\rangle_0 - \cdots - a_{in}\hat{x}_n^{(k)}\langle 3\rangle_0)/a_{ii}. \quad (6.91b)$$

Thus from (2.37) we have

Bound on the rounding error	$\begin{aligned}\lvert E_i^{(k)}\rvert \leq \{&(n+1)\lvert b_i\rvert + (n+1)\lvert a_{i1}\hat{x}_1^{(k)}\rvert + n\lvert a_{i2}\hat{x}_2^{(k)}\rvert \\ &+ \cdots + 3\lvert a_{in}\hat{x}_n^{(k)}\rvert\}\mu'/\lvert a_{ii}\rvert.\end{aligned}$ (6.92)

If $\lim_{k\to\infty} \hat{x}_i^{(k)} = \hat{x}_i^*$, then taking limits in (6.91) we find that

$$\hat{x}_i^* = (b_i - a_{i1}\hat{x}_1^* - \cdots - a_{in}\hat{x}_n^*)/a_{ii} + E_i^*. \quad (6.93)$$

It is not clear that E_i^* will have the form given by (6.91b), but certainly the bound (6.92) will hold. That is, E_i^* in (6.93) will satisfy

$$\lvert E_i^*\rvert \leq \frac{(n+1)\lvert b_i\rvert + (n+1)\lvert a_{i1}\hat{x}_1^*\rvert + \cdots + 3\lvert a_{in}\hat{x}_n^*\rvert}{\lvert a_{ii}\rvert}\mu'. \quad (6.94a)$$

From (6.93) we have, by multiplying by a_{ii},

$$a_{ii}\hat{x}_i^* = b_i - a_{i1}\hat{x}_1^* - \cdots - a_{in}\hat{x}_n^* + a_{ii}E_i^*$$

or

$$(a_{i1}\hat{x}_1^* + a_{i2}\hat{x}_2^* + \cdots + a_{in}\hat{x}_n^* - b_i) = a_{ii}E_i^*.$$

The left side of this is just the residual that results from substituting the limit of the computed values into the equations. If we denote this residual by R_i, then $R_i = a_{ii} E_i^*$, and from (6.94a) we find that

Bound on the residual	$\begin{aligned}|R_i| \leq &((n+1)|b_i| + (n+1)|a_{i1} \hat{x}_1^*| \\ &+ n|a_{i2}\hat{x}_2^*| + \cdots + 3|a_{in}\hat{x}_n^*|)\mu'.\end{aligned}$ (6.94b)

Here we again note that the term $a_{ii} \hat{x}_i^*$ does not occur in this expression.

Stated another way (6.93) shows that the limit values $\hat{x}_i^*$ are the solution to the system of equations

$$a_{i1} x_1 + a_{i2} x_2 + \cdots + a_{in} x_n = b_i + R_i, \qquad i = 1, 2, \ldots, n, \qquad (6.95)$$

where $R_1, \ldots, R_n$ are bounded by (6.94b). Thus if the original system (6.1) is ill-conditioned, then the limit $\hat{x}_i^*$ may be quite far from the exact solution x_i^*.

With the notation (6.81) and (6.83) the bound (6.94b) can be simplified, but increased somewhat, to give

Simple bound on the residuals	$\begin{aligned}|R_i| \leq &[(n + 1)|b_i| + ((n + 1)|a_{i1}| \\ &+ n|a_{i2}| + \cdots + 3|a_{in}|)\|\hat{x}^*\|]\mu' \\ \leq &(n + 1)(|b_i| + \|A\| \|\hat{x}^*\|)\mu'.\end{aligned}$ (6.96)

In deriving these estimates we have assumed for simplicity that all of the arithmetic operations indicated in (6.90) were actually performed and possibly contributed to the total error. It was stated in Section 6.8.1, however, that iterative methods are often applied to systems in which many of the coefficients are zero. In this case the errors will be much smaller than indicated by (6.96). In fact, suppose that in the ith equation there are only m_i nonzero coefficients. Then instead of there being n subtraction and multiplication errors, there will be only m_i such errors. The bound (6.96) can be replaced by

$$|R_i| \leq (m_i + 1)(|b_i| + \|A\| \|\hat{x}^*\|)\mu' \qquad (6.97)$$

in which m_i is substantially smaller than n.

To illustrate these bounds consider the equations (6.89). It is easily seen that $\|A\| = 4 + h$, and m_i is either 2 or 3. Thus the residuals satisfy

$$|R_i| \leq 4(|b_i| + (4 + h)\|\hat{x}^*\|)\mu'.$$

With $b_i = 1, i = 1, 2, \ldots, n, \|\hat{x}^*\| \leq 1$ so that

$$|R_i| \leq (4 + 16h)\mu'.$$

This indicates that, since this system turns out to be well-conditioned, the limit of the computed values will be an accurate approximation to the solution.

The bound given in (6.97) is a *computable* quantity if $\|\hat{x}^{(k)}\|$ is used in place of $\|\hat{x}^*\|$ and suggests a criteria for stopping the iteration. Since the residuals can never be guaranteed to be substantially less than this bound, it is reasonable to stop iterating when the residuals approach this quantity. Finally, it is worth noting that the residual at the kth iteration, namely,

$$R_i^{(k)} = b_i - a_{i1}\hat{x}_1^{(k)} - \cdots - a_{ii}\hat{x}_i^{(k)} - \cdots - a_{in}\hat{x}_n^{(k)}, \tag{6.98}$$

can be easily computed. In fact

$$R_i^{(k)} = \frac{[b_i - a_{i1}\hat{x}_1^{(k)} - \cdots - a_{i,\,i-1}\hat{x}_{i-1}^{(k)} - a_{i,\,i+1}\hat{x}_{i+1}^{(k)} - \cdots - a_{in}\hat{x}_n^{(k)}]}{a_{ii}} a_{ii}$$

$$- a_{ii}\hat{x}_i^{(k)}$$

$$= \hat{x}_i^{(k+1)}a_{ii} - a_{ii}\hat{x}_i^{(k)}$$

so that

$$R_i^{(k)} = a_{ii}(\hat{x}_i^{(k+1)} - \hat{x}_i^{(k)}).$$

Hence the **iteration should be stopped** when

Stopping criteria	$\displaystyle |\hat{x}_i^{(k+1)} - \hat{x}_i^{(k)}| \le \frac{(m_i + 1)(|b_i| + \|A\|\,\|\hat{x}^{(k)}\|)}{|a_{ii}|},$

$$i = 1, 2, \ldots, n. \tag{6.99}$$

The analysis of the Gauss–Seidel method is substantially the same and is left as an exercise.

6.8.4 Efficiency—Rate of Convergence and Sparseness

The efficiency of the algorithms discussed in this section depend upon two things: how fast the iterates converge and how many floating point operations are needed to compute each iterate. There is an additional consideration, however, that should be taken into account. All of the numerical problems analyzed in previous chapters have required relatively small amounts of data—certainly small enough to be stored easily in the fast memory of any reasonably sized computer. For large linear systems, however, the amount of data is sufficiently great so that often the storage and retrieval of this data is more time-consuming than the actual numerical computation. This is especially true if extensive use must be made of auxiliary storage devices, e.g., tapes, drums, or disks. For example, if a system of 1000 linear equations is to be solved by Gaussian elimination, the storage of the coefficients requires 1,000,000 memory locations. Since many computer systems do not allow that much data storage in the fast memory, only part of the coefficients can be held there; the rest must be stored on tape, etc. However, during the kth elimination step, all the equations below the kth one must be transformed; hence, all of these coefficients must be loaded into the fast memory at some time during this elimination step.

Very often the large linear systems that occur in applications have the property that most of the coefficients are zero. Such systems are said to be **sparse**. It is not uncommon to encounter sparse systems in which less than 1% of the coefficients are nonzero. By using special data-handling techniques, it is possible to store only the nonzero coefficients, thereby making it possible to store rather large *sparse* systems. Unfortunately, unless special techniques are used, elimination methods such as Gaussian elimination quickly create so many nonzero elements that sparsity is destroyed. Iterative methods, on the other hand, always work with the original system of equations. Hence, if the system is sparse, then the algorithm does not change this fact.

Thus when comparing the efficiency of these methods with only the floating point operations as a measure, the iterative methods sometimes appear to be rather bad. When the effort required for data handling is included in this measure, the iterative methods often look much better.

If the ith equation has m_i nonzero coefficients, then both the Jacobi and Gauss–Seidel methods require $(m_i)A + (m_i)M + (1)D$ operations to compute $x_i^{(k+1)}$ from the previous iterates. Hence each iteration requires a total of

Efficiency measure for Jacobi and Gauss–Seidel

$$\left(\sum_{i=1}^{n} m_i \right) A + \left(\sum_{i=1}^{n} m_i \right) M + (n)D \qquad (6.100)$$

floating point operations. To make this more precise, suppose that $m_i = (.01)n$, i.e., that in each equation only 1% of the coefficients are nonzero. Then (6.100) is

$$(.01n^2)A + (.01n^2)M + (n)D.$$

N iterations will require

Efficiency of N iterations for sparse systems

$$(.01n^2N)A + (.01n^2N)M + (nN)D \qquad (6.101)$$

operations. As a comparison, we note that (6.101) with $N \cong 33n$ is approximately the same as the efficiency measure for Gaussian elimination. Thus the Jacobi or Gauss–Seidel methods applied to such a sparse matrix require no more floating point operations than does Gaussian elimination provided these methods converge in $33n$ iterations.

A reasonable estimate for the number of iterations needed to obtain the approximate solution is rather hard to obtain. The basis for such an estimate is the inequality (6.85), which we write in the form

$$\|\varepsilon^{(k)}\| \le \|c\|^k \|\varepsilon_0\|. \tag{6.102}$$

This relates the error at the kth iteration to the error in the initial values. It is clear from (6.102) that the smaller $\|c\|$ is the faster the errors $\|\varepsilon^{(k)}\|$ go to zero. For example, if $\|c\| = \frac{1}{2}$, and $\|\varepsilon_0\| = 1$, then $\|\varepsilon^{(k)}\| \le (\frac{1}{2})^k$, so that after about 25 iterations the error will be less than 10^{-7}. (That is, $(\frac{1}{2})^{25} \le 10^{-7}$). On the other hand, if $\|c\| = .8$, then 70 iterations would be required to make $\|\varepsilon^{(k)}\| \le 10^{-7}$. The value of $\|c\|$ is sometimes called the *convergence factor* for the iteration (6.74). The convergence results in Section 6.8.2 only required that $\|c\| < 1$. It is clear, however, that if $\|c\|$ is very close to one then it may take a *very* large number of iterations to obtain a sufficiently accurate result. The success of the overrelaxation methods mentioned earlier results from the ability in certain cases to make $\|c\|$ *very* small so that convergence is very rapid.

If $\|\varepsilon^{(0)}\|$ and $\|c\|$ are known (or can be estimated), then it is possible to use (6.102) to estimate the number of iterations required to reduce the error to some prescribed amount δ. In fact, k should satisfy the inequality

$$\|c\|^k \|\varepsilon^{(0)}\| \le \delta.$$

By taking logarithms of both sides, we obtain

$$k \log\|c\| + \log\|\varepsilon^{(0)}\| \le \log \delta. \tag{6.103}$$

For convergence the condition $\|c\| < 1$ is needed so that $\log\|c\| < 0$, and thus (6.103) becomes

$$k \ge \frac{\log \delta - \log\|\varepsilon^{(0)}\|}{\log\|c\|}. \tag{6.104}$$

The difficulties with using this estimate are twofold. First of all, iterative methods often are not easily written in the "successive substitution" form (6.74) that was used to obtain (6.102). Secondly, it is not clear how best to measure the size of the coefficients c_{ij}, that is, how to define $\|c\|$ to obtain the best estimate. For example, the iteration

$$x_1^{(k+1)} = 1.0 - .4x_2^{(k)} + .1x_3^{(k)}$$
$$x_2^{(k+1)} = -2.5 - .3x_1^{(k)} + .001x_3^{(k)}$$
$$x_3^{(k+1)} = 3.0 - .01x_1^{(k)} - .4x_2^{(k)}$$

has $\|c\| = .5$ with (6.83), but with (6.87) we have $\|c\| = .8$. The first of these, with the estimate (6.104), indicates that 25 iterations are needed in order to reduce an initial error of one to less than 10^{-7}. The second value for $\|c\|$ implies that 70 iterations are required.

In general, the value of $N_{\max}$ to be used in iterative algorithms should be an overestimate. Techniques such as those leading to (6.104) are often useful in particular situations, but general procedures for obtaining accurate estimates do not exist.

EXERCISES

1   For the case $n = 3$, show in detail that the Gauss–Seidel method has the form of (6.74). [That is, verify (6.76) and (6.77) for the $n = 3$ case.]

2   Solve the following system by the Gauss–Seidel and Jacobi methods:

$$4x_1 - x_2 \qquad\; = 1$$
$$-x_1 + 4x_2 - x_3 = 0$$
$$- x_2 + 4x_3 = 1.$$

3   Write a program to implement the Gauss–Seidel and Jacobi methods. Use $N_{\max} = 100$ and test for convergence by computing $e = \max_{i=1,\,\dots,\,n} |x_i - y_i|$ and stopping the iteration when $e \le 10^{-4}$. Test your program on several systems, including (6.89), with various right-hand sides.

4   For the system of equations

$$x_1 + 2x_2 - 2x_3 = 1$$
$$x_1 + x_2 + x_3 = 1$$
$$2x_1 + 2x_2 + x_3 = 1$$

show that the Jacobi method converges while the Gauss–Seidel method does not. [Use $x_1 = x_2 = x_3 = 1$ as the starting values.]

5   For the equations given in Exercise 4 write out the error equations (6.79) for both the Jacobi and the Gauss–Seidel methods. Use these to show that the Jacobi method converges but the Gauss–Seidel method does not.

6   For which of the following systems can you *guarantee* the convergence of the Jacobi method? For those systems that are in doubt, write out the error equations (6.79) and see if

you can decide whether the iterates will converge.

(a)　　　　$2x_1 - x_2 = 0$　　　　　　(b)　$x_1 + 2x_2 = 1$

　　　　$-x_1 + 2x_2 - x_3 = 1$　　　　　　$x_1 + x_2 + x_3 = 0$

　　　　　　$-x_2 + 2x_3 = 0$　　　　　　　$2x_2 + x_3 = 1$

(c)　$4x_1 + 4x_2 + x_3 = 1$　　　　(d)　$2x_1 - x_2 = -100$

　　　$x_1 + 8x_2 + x_3 = 0$　　　　　$-x_1 + 2x_2 - x_3 = 50$

　　　　　$x_2 + 3x_3 = -1$　　　　　　$-x_2 + 2x_3 = 25.$

7　For each of the following iterations determine $\|C\|$ using both (6.83) and (6.87) and tell which of the iterations converge:

(a)　$x_1^{(k+1)} = .5x_1^{(k)} + .1x_2^{(k)}$　　　　(b)　$x_1^{(k+1)} = .1x_2^{(k)} - .2x_3^{(k)}$

　　$x_2^{(k+1)} = .8x_1^{(k)} - .1x_2^{(k)}$　　　　　$x_2^{(k+1)} = -.3x_1^{(k)} + .1x_3^{(k)}$

　　　　　　　　　　　　　　　　　$x_3^{(k+1)} = .5x_1^{(k)} - .6x_2^{(k)}$

(c)　$x_1^{(k+1)} = -\frac{1}{2}x_2^{(k)}$　　　　　(d)　$x_1^{(k+1)} = x_2^{(k)} + x_3^{(k)}$

　　$x_2^{(k+1)} = -\frac{1}{2}x_1^{(k)} - \frac{1}{2}x_3^{(k)}$　　　　$x_2^{(k+1)} = -.1x_1^{(k)} + x_2^{(k)}$

　　$x_3^{(k+1)} = -\frac{1}{2}x_2^{(k)}$　　　　　　$x_3^{(k+1)} = x_1^{(k)} + x_2^{(k)}.$

8　The essential property of the quantities $\|C\|$ and $\|\varepsilon\|$, as defined by (6.81) and (6.83) or by (6.86) and (6.87), is that for $\varepsilon_1^{(k+1)}, \ldots, \varepsilon_n^{(k+1)}$ given by (6.79), it follows that $\|\varepsilon^{(k+1)}\| \le \|C\|\,\|\varepsilon^{(k)}\|$. Prove that this inequality is *not* valid for $\|\varepsilon\|$ defined by (6.81) and $\|C\|$ defined by (6.87).

9　Let $\hat{x}_i^*, i = 1, 2, \ldots, n$ be the limit of the computed values from the Gauss–Seidel method. Show that these computed values are the solution to a system of the form (6.95) and find a bound on the quantities $R_i, i = 1, \ldots, n$. Compare this bound with the bound (6.96).

10　Rewrite Algorithm (6.69) so that the test for convergence (step 2.2) makes use of the inequality (6.99).

11　Derive a formula for computing the residuals (6.98) from the Gauss–Seidel iterates $\hat{x}_i^{(k)}$.

12　For the system (6.89) with $n = 1000$ determine the number of operations needed for each iteration of the Jacobi method. How many iterations are possible in 10 sec on your computer?

13　Prove that if A is nonsingular, then it is possible to interchange rows so that all diagonal elements are nonzero.

Suggestions for Further Reading

　　Gaussian elimination and several variants are discussed in detail by Stewart (1973). FORTRAN programs are given in Forsythe et al. (1977). A complete rounding error analysis is done in Wilkinson (1963) and in Forsythe and Moler (1967). Iterative methods are studied in depth by Young (1971) and Varga (1962).

Chapter 7

NONLINEAR EQUATIONS

7.1 Basic Methods

In this section the most commonly used methods for solving nonlinear equations are described. All of these methods are iterative in nature, and hence we begin with some remarks about iterations in general.

7.1.1 Iterative Algorithms

The problem to be studied in this chapter is that of finding one or more solutions to an equation of the form

$$f(x) = 0, \tag{7.1}$$

where f is some nonlinear function of the real variable x. In all but the simplest cases, such as when f is a linear or quadratic function, the equation cannot be solved directly and some numerical method must be found to obtain an approximation to the solution. Most such methods have the form of an **iteration** where a starting value x_0 is somehow determined, and then the method produces a *sequence of iterates* $x_1, x_2, x_3, \ldots$. Under certain conditions the iterates will converge to a solution; that is, $\lim_{k \to \infty} x_k = x^*$, where $f(x^*) = 0$. Of course, in practice only a finite number of iterates x_1, $x_2, \ldots, x_N$ are actually computed. An important question is to determine how many iterates to compute in order to obtain a sufficiently accurate approximation without unnecessary computation. All iterative algorithms must include one or more conditions that, when satisfied, cause the algorithm to stop computing further iterates. These **stopping conditions** will be studied in detail throughout this chapter, but a few preliminary observations are appropriate. First of all, there are two possible computational interpretations of the problem stated as equation (7.1). The first is that of computing a value $\hat{x}$ that should be very close to x^*, where $f(x^*) = 0$; that is, $\hat{x}$ should be close to the *exact* solution of (7.1). The second interpretation is that the desired value $\hat{x}$ should make $|f(\hat{x})|$ very small. As will be shown later, these two conditions are often quite independent of one another. The physical problem that produced equation (7.1) is sometimes the only guide as to which interpretation is to be assumed. If the second interpretation is appropriate, then the stopping criteria is very simple—stop when $|f(x_N)|$ is sufficiently small. If the first interpretation is indicated, then the stopping criteria is complicated by the fact that the size of $|x^* - x_N|$ cannot be examined as long as x^* is not known. Often, the best that can be done in this case is to try to use *analytic* techniques to relate $|x^* - x_N|$ to the difference $|x_N - x_{N-1}|$ that *can* be easily examined. (In fact, for reasons that should be obvious by now, the *relative* difference $|x_N - x_{N-1}|/|x_N|$ should be examined.)

In addition, it may happen that the iteration does not converge at all, and hence such convergence criterion may *never* be satisfied. To prevent the algorithm from computing an infinite number of iterates, it is necessary to count the number of iterates that have been computed and to stop when this number exceeds a certain maximum value. Thus algorithms for solving (7.1) may involve at least three **stopping parameters** ε_1, ε_2, N_{max}. The effect of these is to stop the algorithm after computing x_N, where either $|x_N - x_{N-1}|/|x_N| \leq \varepsilon_1$, or $|f(x_N)| \leq \varepsilon_2$, or $N \geq N_{max}$. The problem of determining reasonable values for ε_1 and ε_2 will be discussed in Section 7.3. The choice of a value for N_{max} will depend on how *fast* the iterates are expected to converge (if they do converge) and will be studied in Section 7.4.

7.1.2   Successive Substitution

Many of the commonly used iterative methods have the form

Successive substitution

$$x_{k+1} = g(x_k), \qquad k = 0, 1, \ldots, \tag{7.2}$$

in which each iterate is obtained from the previous one by a simple function evaluation. Such a method is called a **method of successive substitution**. The function g in (7.2) is called the *iteration function*.

For example, suppose the equation to be solved is

$$x + e^x - 2 = 0. \tag{7.3}$$

Then a possible iteration of the form (7.2) is

$$x_{k+1} = 2 - e^{x_k}, \qquad k = 0, 1, \ldots . \tag{7.4}$$

Notice that if $x_0, x_1, x_2, \ldots$ as defined by (7.4) converge to a limit value x^*, then by taking limits on both sides of (7.4) we have $x^* = 2 - e^{x^*}$; hence x^* is a solution to (7.3).

One way to find an iteration of the form (7.2) is to write the given equation (7.1) in the form

<table>
<tr><td>Fixed
point
equation</td><td>$$x = g(x). \tag{7.5}$$</td></tr>
</table>

This immediately suggests the iteration (7.2). Furthermore, if x^* is the limit of the sequence given by (7.2) and if g is continuous at x^*, then by taking the limit as $k \to \infty$, we see that

$$x^* = g(x^*). \tag{7.6}$$

A value x^* that satisfies (7.6) is called a **fixed point** of the function g (since g leaves the point fixed). Thus the problem of finding a "zero point" of a function f can often be solved by finding a fixed point of a related function g, using successive substitution. Any equation (7.1) can be put into one or more fixed point forms; the simplest is to let $g(x) = f(x) + x$. However, it is not always the case that the iteration (7.2) derived from the fixed point equation converges. In fact, it may take a bit of ingenuity to derive a convergent iteration. As an example, consider again (7.3), which has a solution in the interval $[0, 1]$. The iteration (7.4) is obtained by writing (7.3) in the form

$$x = 2 - e^x. \tag{7.7}$$

Now with $x_0 = 0$, (7.4) gives the iterates

<table>
<tr><td rowspan="4">Iterates
given by
(7.4)</td><td>$x_1 = 1$</td><td>$x_5 = 1.92094$</td></tr>
<tr><td>$x_2 = -.71828$</td><td>$x_6 = -4.82709$</td></tr>
<tr><td>$x_3 = 1.51242$</td><td>etc.,</td></tr>
<tr><td>$x_4 = -2.53761$</td><td></td></tr>
</table>

which do *not* converge to the solution. If, however, the equation is rewritten as $e^x = 2 - x$, then by taking logarithms we find that

$$x = \ln(2 - x).$$

TABLE 7.1   *Iterates from (7.8)*

$x_1 = .69315$	$x_{15} = .44336$
$x_2 = .26767$	$x_{16} = .44252$
$x_3 = .54948$	$x_{17} = .44306$
$x_4 = .37196$	$x_{18} = .44272$
$x_5 = .48755$	$x_{19} = .44296$
$\vdots$	$x_{20} = .44279$
$x_{12} = .44079$	$x_{21} = .44286$
$x_{13} = .44414$	$x_{22} = .44285$
$x_{14} = .44208$	$x_{23} = .44286$

This results in the **iterative method**

$$x_{k+1} = \ln(2 - x_k), \qquad k = 0, 1, \ldots . \tag{7.8}$$

More specifically, with $x_0 = 0$ the iterates (7.8) are as shown in Table 7.1. These appear to be converging to a value close to .44286. The problem of determining which iterations converge will be studied in the next section.

7.1.3   Newton's Method

Often the *form* of the function f suggests a way of rewriting the equation (7.1) in fixed point form, as in the example (7.3), which led to (7.4). If f is differentiable, then a fixed point form, which has several desirable features, is derived by replacing the function $f(x)$ by a *linear* function $l(x)$ that is tangent to f at the starting point x_0. That is, let $l(x)$ be defined by

$$l(x) = f(x_0) + f'(x_0)(x - x_0).$$

If we solve the equation $l(x) = 0$ (instead of $f(x) = 0$), we find a new value

$$x_1 = x_0 - f(x_0)/f'(x_0).$$

Repetition of this process then leads to the iteration

$$x_{k+1} = x_k - f(x_k)/f'(x_k), \qquad k = 0, 1, \ldots . \tag{7.9}$$

This is called **Newton's method** (see Fig. 7.1). It has the form (7.2), with iteration function

Iteration function for Newton's method

$$g(x) = x - f(x)/f'(x). \tag{7.10}$$

If this method is applied to (7.3), for which $f(x) = x + e^x - 2$, then (7.9) becomes

$$x_{k+1} = x_k - (x_k + e^{x_k} - 2)/(1 + e^{x_k}).$$

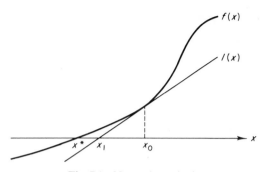

Fig. 7.1 Newton's method.

With $x_0 = 0$ we have $x_1 = .5$, $x_2 = .44385$, $x_3 = .44286$. By comparing these with the iterates in (7.8), we see that in this case Newton's method has produced the solution in fewer iteration steps than (7.8). But, of course, this came at the expense of a somewhat more complicated formula for determining each iterate.

An algorithm for using Newton's method to solve (7.1) has the form:

Newton's Method Algorithm

> The input must include subroutines for evaluating $f(x)$ and $f'(x)$
> for any value of x. The starting value x_0, as well as the three
> stopping parameters N_{max}, ε_1, ε_2 must also be supplied.

1 For $N = 0, 1, \ldots, N_{max}$
   1.1 $f_0 \leftarrow f(x_0)$
   1.2 If $|f_0| \le \varepsilon_2$
      Then
        1.2.1 Output $\{x_0, \text{"}f(x)\text{ is small"}\}$
        1.2.2 Halt
      Else
        1.2.3 $f'_0 \leftarrow f'(x_0)$
        1.2.4 $x_1 \leftarrow x_0 - (f_0/f'_0)$
        1.2.5 If $|x_1 - x_0| \le \varepsilon_1 |x_1|$
           Then
             1.2.5.1 Output $\{x_1, \text{"}|x_N - x_{N-1}|\text{ is small"}\}$
             1.2.5.2 Halt
           Else
             1.2.5.3 $x_0 \leftarrow x_1$
2 Error message, "No convergence after N_{max} iterations"
3 Halt

In the above algorithm just the current iterate is saved. The index N is only used to count the number of iterations.

7.1.4 Secant Method

One major difficulty with Newton's method is the need to evaluate the derivative at each step. If $f'(x)$ is difficult (i.e., expensive) to compute, it may be advisable to use the **modified Newton method**:

$$x_{k+1} = x_k - f(x_k)/f'(x_0), \qquad k = 0, 1, \ldots . \tag{7.11}$$

That is, f' is evaluated only at x_0.

Another modification to Newton's method, which avoids the derivative entirely, is to replace the derivative by the **divided difference**

$$[f(x_k) - f(x_{k-1})]/(x_k - x_{k-1}).$$

Then the iteration becomes

$$x_{k+1} = x_k - f(x_k) \Big/ \left(\frac{f(x_k) - f(x_{k-1})}{x_k - x_{k-1}} \right), \qquad k = 1, 2, 3, \ldots, \tag{7.12a}$$

which can be written more simply as

$$x_{k+1} = \frac{x_k f(x_{k-1}) - x_{k-1} f(x_k)}{f(x_{k-1}) - f(x_k)}, \qquad k = 1, 2, 3, \ldots . \tag{7.12b}$$

This is called the **secant method** because x_{k+1} is just the intersection of the *secant* line through $f(x_k)$ and $f(x_{k-1})$ with the x axis (see Fig. 7.2). An

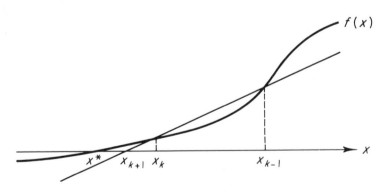

Fig. 7.2 Secant method.

important observation about the secant method is that the iteration (7.12a, b) is *not* of the form (7.2). Instead, it has the form

$$x_{k+1} = g(x_k, x_{k-1}) \tag{7.13}$$

in which x_{k+1} is determined by *both* x_k and x_{k-1}. Such an iteration is called a **two-step method**. The analysis of such methods is considerably more complicated than the simpler type (7.2). Also, it is clear that in order to use this method, two starting points x_0, x_1 must be provided; the iteration then determines x_2, x_3, ... If this method is applied to equation (7.3) with $x_0 = 0$, $x_1 = 1$, we have

$$x_2 = \frac{x_1 f(x_0) - x_0 f(x_1)}{f(x_0) - f(x_1)} = .36788$$

$$x_3 = \frac{x_2 f(x_1) - x_1 f(x_2)}{f(x_1) - f(x_2)} = .43006$$

and $x_4 = .443147$, $x_5 = .44286$. This indicates that, at least for this equation, the secant method produces the answer more slowly than does Newton's method but more quickly than (7.8).

An algorithm for the secant method should be based on formula (7.12a) rather than (7.12b). The reason for this is that subtractive cancellation in (7.12b) is likely to occur when x_k and x_{k-1} are near the solution. On the other hand, if (7.12a) is used as written or in the form

$$x_{k+1} = x_k - \frac{f(x_k)(x_k - x_{k-1})}{f(x_k) - f(x_{k-1})}, \tag{7.12c}$$

then overflow may result. In the following algorithm the possibilities of cancellation and overflow are reduced by writing (7.12a) as

Safe and accurate form of secant method	$$x_{k+1} = x_k - \left[(x_{k-1} - x_k)\left(\frac{f(x_k)}{f(x_{k-1})}\right) \right] \bigg/ \left(1 - \frac{f(x_k)}{f(x_{k-1})}\right). \tag{7.12d}$$

Formula (7.12d) gives a **safe and accurate way** to compute x_{k+1} provided that $|f(x_k)| < |f(x_{k-1})|$. This condition is easily guaranteed by interchanging, when necessary, the values of x_k and x_{k-1}. This interchange may cause a set of iterates different from those defined by (7.12) to be computed. For example, if $x_0 = 1$, $x_1 = 0$, $f(x_0) = .5$, $f(x_1) = 1$, then $x_2 = 2$. Suppose now that $f(x_2) = .25$. Then (7.12) gives

$$x_3 = \frac{x_2 f(x_1) - x_1 f(x_2)}{f(x_1) - f(x_2)} = \frac{2 \cdot 1 - 1 \cdot (.25)}{1 - .25} = 2.333.$$

If the interchange rule is followed, then x_1 becomes 1, and $f(x_1)$ is .5, so that

$$x_3 = \frac{x_2 f(x_1) - x_1 f(x_2)}{f(x_1) - f(x_2)} = \frac{2(.5) - 1 \cdot (.25)}{.5 - .25} = 3.0.$$

Note, however, that when the iterates are close to a solution and are converging, then indeed one would expect $|f(x_k)|$ to be smaller than $|f(x_{k-1})|$, so that no interchange is necessary.

Secant Method Algorithm

> The input must include a subroutine for evaluating $f(x)$. Two starting points x_0, x_1 are needed, as well as the three stopping parameters $N_{\max}$, ε_1, ε_2.

1 $f_0 \leftarrow f(x_0)$
2 $f_1 \leftarrow f(x_1)$
3 If $|f_1| > |f_0|$
   Then
      3.1 Interchange x_0 and x_1
      3.2 Interchange f_0 and f_1
4 For $k = 0, 1, \ldots, N_{\max}$
      4.1 If $|f_1| < \varepsilon_2$
         Then
         4.1.1 Output $\{x_1, \text{“}|f(x)| < \varepsilon_2\text{”}\}$
         4.1.2 Halt
      4.2 $s \leftarrow f_1/f_0$
      4.3 $p \leftarrow (x_0 - x_1)s$
      4.4 $q \leftarrow 1 - s$
      4.5 $x_2 \leftarrow x_1 - p/q$
      4.6 If $|x_1 - x_2| < \varepsilon_1 |x_2|$
         Then
         4.6.1 Output $\{x_2, \text{“}|x_{k-1} - x_k| < \varepsilon_1 |x_k|\text{”}\}$
         4.6.2 Halt
      4.7 $f_2 \leftarrow f(x_2)$
      4.8 If $|f_2| > |f_1|$
         Then
         4.8.1 $x_0 \leftarrow x_2$
         4.8.2 $f_0 \leftarrow f_2$
         Else
         4.8.3 $x_0 \leftarrow x_1$
         4.8.4 $f_0 \leftarrow f_1$
         4.8.5 $x_1 \leftarrow x_2$
         4.8.6 $f_1 \leftarrow f_2$
5 Error message, "No convergence after $N_{\max}$ iterations"
6 Halt

Here again at each iteration only the two iterates needed for computing the next iterate are saved. After the first step, the "interchange" of iterates, as discussed earlier, is combined with the replacement of old iterates in step 4.8.

7.1.5 Interval Methods

All of the methods that have been discussed so far in this chapter determine a sequence of numbers that one hopes will converge to a solution x^*. There are also a variety of methods that try to determine a sequence of *intervals* that converge to a solution. That is, each interval in the sequence should contain a particular root x^*, and the lengths of the intervals should decrease to zero. An important advantage of interval methods is that rigorous bounds on the error $|x_k - x^*|$ are readily available. Indeed, if $[x_k, x_{k-1}]$ is an interval that is known to contain a solution x^*, then $|x_k - x^*| \leq |x_k - x_{k-1}|$. Also, with an appropriate starting interval, most interval methods are *guaranteed* to converge.

The major disadvantage to interval methods is that usually a starting interval that contains a solution must be found. One way to make certain that a continuous function f has a zero in an interval $[x_0, x_1]$ is to show that

$$f(x_0)f(x_1) < 0. \tag{7.14}$$

Indeed, if (7.14) holds, then $f(x_0)$ and $f(x_1)$ must have opposite signs. If f is continuous on $[x_0, x_1]$, f must take all values between $f(x_0)$ and $f(x_1)$. In particular, f must take the value 0 at some x^* between x_0 and x_1. If f has a *double* zero at a point x^* [that is, $f(x^*) = 0$ and $f'(x^*) = 0$], or if f has two zeros close together (see Fig. 7.3), then the condition (7.14) will *not* hold, even though the interval $[x_0, x_1]$ contains a solution.

For all of the interval methods discussed in this section, it will be assumed that the starting points x_0, x_1 satisfy the inequality (7.14). It should be clearly understood that this assumption limits the applicability of these methods.

The simplest interval method is the **bisection method**, in which the initial interval $[x_0, x_1]$ is bisected into two subintervals $[x_0, x_2]$ and $[x_2, x_1]$ with

$$x_2 = \tfrac{1}{2}(x_0 + x_1).$$

The process is repeated on whichever of the subintervals satisfies the inequality (7.14). That is, if $f(x_0)f(x_2) < 0$, then continue the bisection on the interval $[x_0, x_2]$. Otherwise, use the interval $[x_2, x_1]$. Clearly, the length of the interval is reduced by half each time so the method is guaranteed to converge.

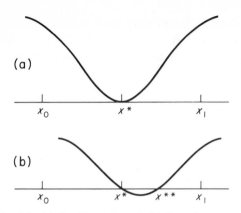

Fig. 7.3 (a) Double zero at x^*. (b) Two close zeros.

Bisection Algorithm

The input consists of two points x_0, x_1 for which $f(x_0)f(x_1) < 0$. Only two stopping parameters ε_1, ε_2 are needed since convergence is guaranteed. A subroutine to evaluate $f(x)$ for any x in the initial interval is also required.

1   $f_0 \leftarrow f(x_0)$, $f_1 \leftarrow f(x_1)$
2   If $f_0 f_1 > 0$ then *Error Exit*
3   While $|f_0| > \varepsilon_2$ and $|f_1| > \varepsilon_2$
    3.1   If $|x_0 - x_1| < \varepsilon_1|x_1|$
      Then
          3.1.1   Output $\{x_0, x_1\}$
          3.1.2   Halt
    3.2 $x_2 \leftarrow .5(x_0 + x_1)$
    3.3 $f_2 \leftarrow f(x_2)$
    3.4   If $f_2 f_0 < 0$
      Then   (7.15)
          3.4.1 $x_1 \leftarrow x_2$
          3.4.2 $f_1 \leftarrow f_2$
      Else
          3.4.3 $x_0 \leftarrow x_2$
          3.4.4 $f_0 \leftarrow f_2$
4   If $|f_0| \leq \varepsilon_2$
      Then
    4.1   Output $\{x_0\}$, Halt
      Else
    4.2   Output $\{x_1\}$, Halt

Steps 3.4 of this algorithm replace x_0 or x_1 with the new point x_2 according to the rule

| Replacement rule | if $f(x_0)f(x_2) < 0$, replace x_1 by x_2
 if $f(x_0)f(x_2) > 0$, replace x_0 by x_2. | (7.16) |

If the **bisection method is applied to the example (7.3)** with $x_0 = 0$, $x_1 = 1$, we find the following sequence of intervals:

$[0, 1]$, $[0, .5]$, $[.25, .5]$, $[.375, .5]$,

$[.4375, .5]$, $[.4375, .46875]$, $[.4375, .453125]$,

$[.4375, .44531]$, $[.44141, 44531]$,

$[.44141, .44336]$, $[.44238, .44336]$,

$[.44238, .44287]$, $[.44263, .44287]$,

$[.44275, .44287]$, $[.44281, .44287]$,

$[.44284, .44287]$, $[.44284, .44286]$, $[.44285, .44286]$.

For this example the method required 17 iterations to obtain an interval of length 10^{-5}. In fact, this will be true no matter *what* equation is being solved. That is, given an initial interval of length 1, the bisection method applied to *any* function f will require 17 steps to reduce the interval to one of length less than 10^{-5} (see Section 7.4.2). Because of this fixed and rather slow convergence, the bisection method is not highly recommended.

The slowness of the bisection method is understandable in view of the fact that the method makes little use of the function f itself. Indeed only the *sign* of the function at various points is used. In order to obtain intervals that decrease more rapidly, the method of **false position** (*regula falsi*) is often suggested. In this method, given the interval $[x_0, x_1]$ as before, we compute the "secant point"

$$x_2 = x_1 - f(x_1) \bigg/ \left[\frac{f(x_1) - f(x_0)}{x_1 - x_0} \right] \tag{7.17}$$

just as in (7.12a–d). It is easily seen geometrically (see Fig. 7.4) that if f is continuous then the secant point x_2 often lies in the interval $[x_0, x_1]$. If x_2 is

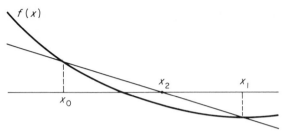

Fig. 7.4

used to replace x_0 or x_1 according to the rule (7.16), then the new interval will still contain the solution x^*. A drawback with the false position method is that it may not "converge" in the sense described above. That is, the lengths of the intervals may not decrease to zero. For example, if the method is applied to the example (7.3) with $x_0 = 0$, $x_1 = 1$, the following intervals are found:

False position method applied to (7.3)	[0, 1], [.36788, 1], [.43006, 1], [.44067, 1], [.44248, 1], [.44279, 1], [.44283, 1], [.44285, 1].

In this example the right endpoint never changes; that is, the secant point is *always* to the left of the zero point. Nevertheless, by examining these intervals, one can easily conclude that there is a solution near .44285.

The example given above shows that the false position method sometimes requires more steps to obtain a solution than does the secant method. In fact, it will be shown in Section 7.4 that this will usually be the case when one endpoint of the false position method stays fixed. On the other hand, the secant method does not have the guaranteed convergence property of the false position method. An interesting combination of the false position, secant, and bisection methods has been developed by Dekker and modified by Brent. The **Dekker–Brent algorithm** is rather complicated, but the main features of it can be illustrated by the following simplified version. Suppose that, at the kth step, we have computed a set of three points $\{x_k, x_{k-1}, y_k\}$, where

x_k is the newest point
x_{k-1} is the previous point
y_k is the most recent point for which $f(x_k)f(y_k) < 0$.

Thus the interval between x_k and y_k contains a zero point x^*. The next set of points will be $\{x_{k+1}, x_k, y_{k+1}\}$, where x_{k+1} is the **secant point** (7.12c), that is,

$$x_{k+1} = x_k - f(x_k) \Big/ \left(\frac{f(x_k) - f(x_{k-1})}{x_k - x_{k-1}} \right) \qquad (7.18a)$$

provided that this point lies between x_k and y_k and that $y_k \neq y_{k-2}$. If $y_k = y_{k-2}$ or (7.18a) gives a value that is not between x_k and y_k, then x_{k+1} is taken to be the **bisection point**; that is,

$$x_{k+1} = \tfrac{1}{2}(x_k + y_k). \qquad (7.18b)$$

The value of y_{k+1} is determined according to

$$y_{k+1} = \begin{cases} x_k & \text{if } f(x_k)f(x_{k+1}) < 0 \\ y_k & \text{otherwise.} \end{cases} \qquad (7.18c)$$

The formula (7.18c) guarantees that the interval between x_{k+1} and y_{k+1} is the smallest interval computed up to now which is known to contain a solution. The value to be used for x_{k+1}, as described above, is just the same as in the secant method unless this point is outside the interval *or* the endpoint y_k has become stuck. By reverting to the bisection method whenever y_k has not changed, we guarantee that eventually y_k will change and indeed the interval between x_k and y_k will converge to a solution x^*. The iteration is started with two points x_0, x_1 that satisfy (7.14), y_1 is set equal to x_0, and $y_{-1} = y_1 + 1$ so that initially $y_k \neq y_{k-2}$.

TABLE 7.2 *Dekker–Brent Iterates*

k	x_{k-1}	x_k	y_k
0	0	1	0
1	1	.36788	1
2	.36788	.43006	1
3	.43006	.44315	.43006
4	.44315	.44286	.44315
5	.44286	.44285	.44286

This algorithm applied to example (7.3) gives the values in Table 7.2. For this example bisection was never used. That is, the values of x_k are the same as those given by the secant algorithm. Note, however, that here we actually find an *interval* [.44285, .44286] that contains the solution. A more precise description of the algorithm is now given.

Simplified Dekker–Brent Algorithm

$$\left[\begin{array}{l}\text{The input consists of two points } x_0, x_1 \text{ for which} f(x_0)f(x_1) < 0. \\ \text{Two stopping parameters } \varepsilon_1, \varepsilon_2 \text{ are needed, as is a subroutine to} \\ \hspace{3cm}\text{evaluate } f(x).\end{array}\right]$$

1   $y_1 \leftarrow x_0, \ y_{-1} \leftarrow y_0 \leftarrow y_1 + 1$
2   $f_0 \leftarrow f(x_0), f_1 \leftarrow f(x_1)$
3   While $|x_1 - y_1| > \varepsilon_1 |x_1|$ and $|f(x_1)| > \varepsilon_2$
   3.1   If $y_1 \neq y_{-1}$
      Then
        3.1.1 $d \leftarrow f_1(x_1 - x_0)/(f_1 - f_0)$
        3.1.2   If $\text{sign}(d) \neq \text{sign}(x_1 - y_1)$ or $|d| > |x_1 - y_1|$
           Then
             3.1.2.1 $d \leftarrow .5(x_1 - y_1)$
      Else   (7.19)
        3.1.3 $d \leftarrow .5(x_1 - y_1)$
   3.2 $x_0 \leftarrow x_1, f_0 \leftarrow f_1$
   3.3 $x_1 \leftarrow x_1 - d, f_1 \leftarrow f(x_1)$
   3.4 $y_{-1} \leftarrow y_0, \ y_0 \leftarrow y_1$
   3.5   If $f_0 f_1 < 0$
      Then
        3.5.1 $y_1 \leftarrow x_0$
   3.6   Continue
4   Output $\{x_0, x_1, f_0, f_1\}$
5   Halt

As usual, we do not save all of the iterates—only those that are needed for computing the next values. The tests in step 3.1.2 guarantee that the secant point $x_1 - d$, with d given by step 3.1.1, lies between x_1 and y_1.

The algorithm actually described by Dekker and Brent involves a more complicated criteria for deciding when to use the bisection point instead of the secant point. The simplified version given above will do unnecessary bisections for some functions. Also we have not considered the possibility that d, computed by step 3.1.1, will be so small that $\text{fl}(x_1 - d) = x_1$. This situation is also taken care of by the more complicated algorithm. Finally note that the interchange technique, as used in the secant algorithm, could also be used here to avoid overflow and/or subtractive cancellation when computing the secant point.

EXERCISES

1   Write each of the following equations in at least two different fixed point forms [i.e., in the form (7.5)]:

   (a) $x + e^x = 0$ (b) $\sqrt{x} + x^2 - 5 = 0$

(c) $x + \sin x - 1 = 5$ (d) $(1/x) + e^x = 0$.

2   Compute the first three Newton method iterates x_1, x_2, x_3, starting with $x_0 = 1$, for each of the following equations:

(a) $x^2 - e^x = 0$ (b) $\sin x + \cos x = 0$

(c) $x^2 - 2 = 0$ (d) $x \log x + 1 = 0$.

3   For each of the equations in Exercise 2, compute the first four secant method iterates, x_2, x_3, x_4, starting with $x_0 = 0$, $x_1 = 1$.

4   Show that if the *regula falsi* method is applied to the equation $x^2 - 2 = 0$ with $x_0 = 2$, $x_1 = 0$, then x_0 stays fixed, and x_1 is gradually increased until $x_1 \simeq \sqrt{2}$.

5   Prove that x_{k+1} as defined by (7.12) is the intersection with the x axis of the secant line through $(x_k, f(x_k))$ and $(x_{k-1}, f(x_{k-1}))$.

6   Let f be a convex increasing function; that is, $f'(x) > 0$ and $f''(x) > 0$. Let $f(x^*) = 0$ and suppose x_{k-1} and x_k are greater than x^*. Show that the secant point x_{k+1}, given by (7.12), is also greater than x^*. Conclude that, for such functions, the false position method always produces a sequence of intervals with one endpoint fixed.

7   Write a program to implement the Dekker–Brent algorithm. Test it on the functions given in Exercise 2. Also test it on the following more difficult functions:

(a) $f(x) = \begin{cases} 0 & \text{if } x = 0 \\ x - \exp(-x^{-2}) & \text{if } x \neq 0 \end{cases}$

(b) $f(x) = x^9$ (take $x_0 = -1$, $x_1 = 1$)

(c) $f(x) = x(x - 1)^5$ (take $x_0 = -.5$, $x_1 = .99$).

8   Modify the Dekker–Brent algorithm to interchange x_0 and x_1 so that $|f_0| < |f_1|$. Then use (7.12d) to compute the next secant point.

7.2   Convergence Results

Equation (7.3), which was used to illustrate the various methods described in the preceding section, was carefully chosen so that the iterates would converge. However, in practice, many of these methods may fail to converge. In this section we shall develop several results that help in predicting when an iterative method will converge.

7.2.1   Successive Substitution—Fixed Points

Consider, first of all, the successive substitution iteration (7.2). If the iterates x_0, x_1, ... converge to x^*, then as noted earlier x^* is a fixed point of g. That is, $x^* = g(x^*)$, and hence

$$|x_{k+1} - x^*| = |g(x_k) - g(x^*)|. \tag{7.20a}$$

Now, if g is differentiable, then the mean value theorem (see the Appendix) insures that there is some point ξ between x_k and x^* for which

$$g(x_k) - g(x^*) = g'(\xi)(x_k - x^*).\qquad(7.20\text{b})$$

By combining equations (7.20a) and (7.20b) we obtain

$$|x_{k+1} - x^*| = |g'(\xi)|\,|x_k - x^*|,\qquad(7.20\text{c})$$

where ξ is some value between x_k and x^*. Thus x_{k+1} will be closer to x^* than was x_k precisely when $|g'(\xi)|$ is less than 1.0. This suggests that if near x^* the *derivative* of the iteration function is small and, more specifically, less than 1 in magnitude, then we can expect the iteration to converge. Conversely, if near x^* the derivative is large, then the iterates will diverge. To make this more precise, we state the following "**convergence rule**":

> The iteration $x_{k+1} = g(x_k)$, $k = 0, 1, \ldots$, when g is continuous and differentiable, will converge to a fixed point x^* provided that x_0 is in an interval I that has the following properties:
> (i) I contains x^*.
> (ii) If x is in I, then $g(x)$ is also in I.
> (iii) There is some $\alpha < 1$ such that $|g'(x)| \le \alpha$ for all x in I.

Condition (ii) of this convergence rule implies that $x_1 = g(x_0)$ will be in I. Hence, also $x_2 = g(x_1)$ is in I and, in fact, all of the iterates $x_0, x_1, \ldots$ will be in the interval I.

In order to apply this rule, it often suffices to evaluate g at a few points to determine an interval that contains a solution, then estimate the derivative of g in this interval. Consider, for example, the equation

$$2x - \cos x = 0.\qquad(7.21)$$

This has a fixed point form

$$x = \tfrac{1}{2}\cos x$$

so that an iteration function is $g(x) = \tfrac{1}{2}\cos x$. The function $f(x) = 2x - \cos x$ can be easily evaluated at 0 and $\pi/2$:

$$f(0) = -\cos 0 = -1 < 0$$

$$f(\pi/2) = \pi > 0.$$

Thus, since f is continuous, there must be at least one value x^* in the interval $[0, \pi/2]$ at which $f(x^*) = 0$, and $x^* = g(x^*)$. Now, $g'(x) = -\tfrac{1}{2}\sin x$ so that $|g'(x)| \le \tfrac{1}{2}$ for *all* x. Furthermore, $0 \le g(x) \le \tfrac{1}{2} \le \pi/2$ for any x with $0 \le x \le \pi/2$. Hence, the interval $I = [0, \pi/2]$ satisfies the conditions of the convergence rule, and the iteration $x_{k+1} = \tfrac{1}{2}\cos x_k$, $x_0 = 0$, will converge to a solution x^* in I.

7.2.2 Local Convergence—Newton's Method

The convergence rule really involves two distinct conditions. First of all, $|g'(x)|$ must be small for all x in some interval containing the solution, and second, the iterates must all remain inside that interval. Suppose now that $g'(x)$ is continuous, and somehow we are able to show that g' at x^* is small. Then, by continuity, $|g'(x)|$ will be small for all x near x^*. Furthermore, if x_k is sufficiently close to x^*, then, according to (7.20c), x_{k+1} will be even closer. That is, there will be a small interval I containing x^* that satisfies the conditions of the convergence rule. This discussion can be summarized as follows:

> Suppose that g is continuous and has a continuous derivative at all points near a fixed point x^*. If $|g'(x^*)| < 1$, then there exists an interval I containing x^* such that if $x_0 \in I$, then $x_{k+1} = g(x_k), k = 0, 1, \ldots$, are in I, and $\lim_{k \to \infty} x_k = x^*$.

This result is called a **local convergence theorem** because it says that "locally," that is, for x_0 near to x^*, the iteration will converge.

An important application of this principle is the case of Newton's method. That is, suppose $f(x^*) = 0$, where f, f', and f'' exist and are continuous near x^*. Furthermore, suppose $f'(x^*) \neq 0$. If g is given by (7.10), then straightforward differentiation gives

$$g'(x) = 1 - f'(x)/f'(x) + f(x)f''(x)/[f'(x)]^2.$$

Hence, for any x where $f'(x) \neq 0$, we have

$$g'(x) = f(x)f''(x)/[f'(x)]^2. \tag{7.22}$$

Now, since $f(x^*) = 0$ and $f'(x^*) \neq 0$, it follows that $g'(x^*) = 0$. Thus certainly $|g'(x^*)| < 1$ and the local convergence property holds. That is:

> If Newton's method is applied to a function f that is twice differentiable near a zero point x^*, where $f'(x^*) \neq 0$, then the iterates will converge if x_0 is close enough to x^*.

Of course, if we know nothing about the zero point x^*, then we may have no idea how to choose a proper x_0. Nevertheless, it is comforting to know that a method is *guaranteed* to converge if we pick the right starting point. The assumption that $f'(x^*) \neq 0$ is not essential here. It can be shown, for example (see Exercise 5), that if $f'(x^*) = 0$ but $f''(x^*) \neq 0$, then $g'(x^*) = \frac{1}{2}$, and the result still holds. [The situation in which $f(x^*) = f'(x^*) = 0$ involves numerical difficulties that will be discussed in Section 7.3.]

It will be shown in Section 7.4 that when Newton's method converges to a zero point x^* for which $f'(x^*) \neq 0$, then it converges *very* rapidly. This fact, combined with the local convergence property, suggests the following procedure: Use a method that is known to converge, such as bisection or false position, to find a point that is reasonably close to x^*. Now, starting with this point as x_0, use Newton's method to continue on toward x^*. Such a procedure is difficult to implement as an algorithm but may be extremely useful in an interactive computation.

7.2.3 Monotone Convergence—Newton's Method

Sometimes the convergence of an iterative method can be established more easily by applying geometric arguments instead of analytic results such as those used above. Newton's method, for example, has the simple geometric interpretation shown in Fig. 7.1. That is, x_{k+1} is the intersection of the line tangent to f at x_k with the x axis. Furthermore, it is easy to see that for some functions (see Fig. 7.5), the iterates satisfy the inequalities

$$x_0 \geq x_1 \geq x_k > x_{k+1} > \cdots > x^*. \tag{7.23}$$

Whenever (7.23) holds, convergence is assured by the **monotonicity** of the iterates. To make this more precise, suppose that f is continuous and twice

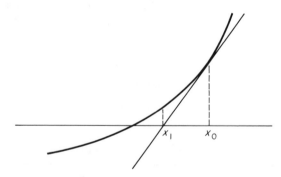

Fig. 7.5 Newton's method.

differentiable in an interval I that contains a solution x^*. Furthermore, assume that f has a graph that is similar to the graph in Fig. 7.5, in the sense that

(i) f is strictly increasing; that is, $x < y$ implies that $f(x) < f(y)$ for all x, y in I, or, equivalently, $f'(x) > 0$ for all x in I;
(ii) f is convex, that is, $f''(x) > 0$ for all x in I.

Now, let x_0 be any point in I that is greater than x^*. Then, by property (i),

$$x_0 > x^* \quad \text{implies} \quad f(x_0) > f(x^*) = 0$$

so that $f(x_0) > 0$. From this and the fact that $f'(x_0) > 0$, it follows that

$$x_1 = x_0 - f(x_0)/f'(x_0) < x_0.$$

Furthermore, with $g(x) = x - f(x)/f'(x)$, (7.20b) gives

$$x_1 - x^* = g(x_0) - g(x^*) = g'(\xi)(x_0 - x^*),$$

where ξ is some value between x_0 and x^*. But (7.22), together with the convexity of f [that is, $f''(x) \geq 0$] implies $g'(\xi) \geq 0$. This, combined with $x_0 - x^* > 0$, gives $x_1 - x^* > 0$. Hence, we have shown that

$$x^* < x_1 < x_0.$$

By repeating this argument, (7.23) is established. It follows easily that $\lim_{k \to \infty} x_k$ exists and is a solution to (7.1). Since f is strictly increasing, x^* is the *only* solution, and we have the result:

Let f be a function that is twice differentiable on some interval I containing a zero point x^*. If f is strictly increasing and convex on I, then for any x_0 in I with $x_0 \geq x^*$, the Newton iterates are monotone decreasing and converge to x^*.

7.2.4   Applications of Monotone Convergence

An important application of the monotone convergence of Newton's method is to the function

$$f(x) = x^2 - a,$$

where a is any positive number. The positive solution to $f(x) = 0$ is just the positive square root of a. The Newton iteration for this function has the simple form

| Newton iterates for $\sqrt{a}$ | $x_{k+1} = \frac{1}{2}[x_k + (a/x_k)], \qquad k = 0, 1, \dots .$ | (7.24) |

The function f here is strictly increasing and convex on the interval $0 < x < \infty$; hence if $x_0 > \sqrt{a}$, then the Newton iterates will be monotone decreasing and will converge to $\sqrt{a}$. In fact, Fig. 7.6 indicates that even if

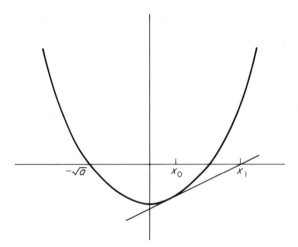

Fig. 7.6   Graph of $f(x) = x^2 - a$.

$x_0 < \sqrt{a}$, but $x_0 > 0$, then $x_1 > \sqrt{a}$ and iterates x_1, x_2, ... will be mono-tone decreasing. That is, Newton's method, applied to this function, will converge to $\sqrt{a}$ with any positive value for x_0. Many computer subroutines for computing $\sqrt{a}$ use precisely this technique.

There are many variations on the conditions given above for monotone convergence of Newton's method. For example, in Fig. 7.6, if we pick an $x_0 < -\sqrt{a}$, then the iterates will be monotone *increasing* and will converge to $-\sqrt{a}$. The curve in Fig. 7.6 is strictly decreasing for $-\infty < x < 0$ but is still convex. Rather than remember all possible such conditions, it is simpler to sketch the graph of the function and then use the geometric interpretation of Newton's method to determine the proper convergence result.

In a similar manner, it is possible to show the convergence of the secant method for special types of functions (see Exercise 6).

As a final comment on the above results, it is perhaps worthwhile to stress the fact that for many of the methods treated here, there are *two* functions that enter into the discussion: the function f whose zero point is sought, and the iteration function g. Some of these results apply to f, some to g. For example, the convergence rule gives conditions on g to ensure convergence of the iteration (7.2), whereas convexity is a condition on f that ensures convergence of Newton's method.

EXERCISES

1 By evaluating the function g at a few points, determine an interval that contains a fixed point.

(a) $g(x) = e^x$ (b) $g(x) = x^3 - 2x^2 + 1$

(c) $g(x) = x \cos x + 1$ (d) $g(x) = -\log x$.

2   Show that the following iterations are locally convergent: (First determine an interval that contains a fixed point of the iteration function.)

(a) $x_{k+1} = \frac{1}{4} e^{x_k}$ (b) $x_{k+1} = \frac{1}{2} \cos x_k$

(c) $x_{k+1} = x_k^3 - 2x_k^2 + \frac{3}{2} x_k + \frac{1}{2}$ (d) $x_{k+1} = e^{.25 x_k}$

3   Sketch the graphs of the following functions and determine an x_0 so that Newton's method will converge monotonically:

(a) $f(x) = x^3 - 2$ (b) $f(x) = e^x - 5$

(c) $f(x) = \cos x - .587765$ (d) $f(x) = (1/x) - 5$.

4   Prove the following strengthened form of the convergence rule. Let g be a continuous function, defined on an interval I, and suppose statements (ii) and (iii) of the convergence rule hold. Then g has a unique fixed point in I, and the iteration (7.2) converges to it, for any x_0 in I.

5   Show that Newton's method is locally convergent even when $f'(x^*) = 0$, but $f''(x^*) \neq 0$. [*Hint*:   Use L'Hospital's rule to show $g'(x^*) = \frac{1}{2}$.]

6   Show graphically that the secant method always converges when f is a convex increasing function on the interval $[x_0, x_1]$, where x_0 and x_1 are the starting points for the iteration, and $f(x_1)f(x_0) < 0$.

7.3   Stability and Effects of Rounding Errors

There are two places where rounding error is involved in the iterative methods described in Section 7.1. The first is in the evaluation of the function f (or the iteration function g, which may include f'), and the second is in computing the new iterate from these function values and the previous iterates. In this section we study how these errors affect the methods.

7.3.1   Evaluating the Function—Stability

The effect of rounding errors that occur during the function evaluations can be understood best by recalling some facts from Chapter 3. There we saw that the computed value of a function $f(x)$, where x is a floating point number, is the *exact* value of another function $\hat{f}(x)$. The amount that $\hat{f}$ differs from f depends on the size of the rounding errors and the algorithm for evaluating f. If f is not a polynomial or more generally not a rational function, then f will first have to be approximated by an appropriate such function before it can be evaluated (approximately). We shall write the relation between f and $\hat{f}$ more precisely as

$$\hat{f}(x) = \mathrm{fl}[f(x)], \tag{7.25}$$

where x is any floating point number for which f is defined. The precise form of $\hat{f}$ can be derived by looking at the rational approximation (if any) used for evaluating f and then doing a rounding error analysis on this approximating function. For the current discussion we are more interested in the fact that any numerical method for solving (7.1) will *actually* attempt to solve the equation

| Equation that
is actually
solved | $$\hat{f}(x) = 0, \tag{7.26}$$ |

where $\hat{f}$ is given by (7.25). Thus the very least effect that rounding (and truncation) errors have on the iterative methods of Section 7.1 is that the methods will try to solve the wrong equation.

To examine the relation between equation (7.26) and the given equation (7.1), we assume that $\hat{f}$ is related to f by

$$\hat{f}(x) = f(x) + g(x)\mu. \tag{7.27}$$

Here, as usual, μ is the unit rounding error for the computer. If f is a polynomial or rational function, then so is g; otherwise g will also include the truncation error that results from approximating f by a function that can be evaluated directly. Next we define the **multiplicity** of a solution x^* to be m whenever

$$f(x^*) = f'(x^*) = \cdots = f^{(m-1)}(x^*) = 0, \qquad f^{(m)}(x^*) \neq 0.$$

[We assume, of course, that f is sufficiently differentiable at x^*.] In case $m = 1$, we call x^* a *simple* root, if $m = 2$, then x^* is called a *double* root, and generally if $m > 1$, we call x^* a *multiple* root. By using the implicit function theorem (see the Appendix and recall Section 1.2.2), we can show that if x^* is a simple root of (7.1) and if μ is sufficiently small, then (7.26) has a solution $x^*(\mu)$ that is related to x^* by (approximately)

| Perturbation
of simple
root | $$x^* - x^*(\mu) \cong [g(x^*)/f'(x^*)]\mu. \tag{7.28a}$$ |

If x^* is a *double* root, then the relation is given by

| Perturbation
of double
root | $$x^* - x^*(\mu) \cong \left[\frac{2g(x^*)}{f''(x^*)} \mu \right]^{1/2}. \tag{7.28b}$$ |

If x^* is a root of multiplicity greater than two, these estimates become more complicated but have somewhat the same form.

From (7.28a) we see that if $f'(x^*)$ is not too small, then (7.26) has a solution that is quite close to the desired solution x^*. Hence in this case (7.1) is stable or well conditioned. If $f'(x^*)$ is small or zero, then (7.28) implies that x^* is ill conditioned. Indeed if, for example, x^* is a double root and $g(x^*) \cong f''(x^*) \cong .5$, $\mu = 10^{-8}$, then (7.28b) shows that $x^*(\mu)$ may differ from x^* by as much as 10^{-4}. That is, if $x^* \cong 1$, then the solution we find to (7.26) may

Fig. 7.7   A function with two roots.

agree with x^* to only *four* figures. The case where $f'(x^*)$ is small but nonzero occurs frequently when f has two or more roots very close together (see Fig. 7.7). We summarize these observations as follows:

A solution x^* to (7.1) for which $f'(x^*)$ is not close to zero is well conditioned. If, however, x^* is a *multiple* root or if there are other roots *very close* to x^*, then (7.1) is an ill-conditioned problem.

The relations (7.28) are useful for helping us to understand ill-conditioning. However, in practice, the situation may be more complicated than these formulas indicate. In particular, it may happen that $\hat{f}$ has *many* zeros near x^*. Consider, for example, the function

$$f(x) = x^4 - 115x^3 + 1575x^2 - 7625x + 12500. \qquad (7.29)$$

This function has zeros only at $x = 5$ and $x = 100$ and is negative in

TABLE 7.3   *Evaluation of (7.29) in the Floating Point System $F(2, 27, -128, 127)$*

x	$\hat{f}(x)$	x	$\hat{f}(x)$
4.99	$.36621094 \times 10^{-3}$	5.004	$.12207031 \times 10^{-3}$
4.991	0	5.005	$-.24414063 \times 10^{-3}$
4.994	$.12207031 \times 10^{-3}$	5.008	$.12207031 \times 10^{-3}$
4.995	0	5.009	$-.12207031 \times 10^{-3}$
4.999	$.12207031 \times 10^{-3}$	5.012	0
5.0	0	5.014	$-.44828125 \times 10^{-3}$
5.003	$-.24414063 \times 10^{-3}$	5.02	$-.85449219 \times 10^{-3}$

between. Table 7.3 gives some computed values of this function, evaluated by Horner's method on a UNIVAC 1108 computer (that is, with $\beta = 2$, $t = 27$). From the changes of sign, it appears that $\hat{f}(x)$ has *several* zero points near $x = 5.0$ (see Fig. 7.8). The fluctuations of $\hat{f}$ near zeros of f have a

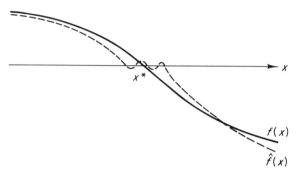

Fig. 7.8   The graph of $f(x)$ and of $\hat{f}(x) = \text{fl}[f(x)]$ near a zero point x^*.

particularly noticeable effect on interval methods. These effects will be discussed next.

7.3.2   Effect of Rounding on Interval Methods

An important step in the interval methods discussed in Section 7.1.5 is the examination of the sign of the function at the new point. That is, if $[x_0, x_1]$ is the current interval which contains x^*, and x_2 is the new point, then the sign of $f(x_2)$ must be compared with the sign of $f(x_0)$ or $f(x_1)$. This is usually done by the replacement rule (7.16). Now, if the *computed* value of $f(x_2)$, that is $\hat{f}(x_2)$ has the wrong sign, then the replacement rule will replace the wrong point. Once this has occurred, the interval $[x_0, x_1]$ will no longer contain a solution of (7.1). Table 7.4 illustrates this phenomena. Here the bisection method is being used to find the root $x^* = 5.0$ of the function f given by (7.29). As noted earlier, this function has roots at 5.0 and 100.0 and is negative in between. The computed value of f at $x = 5.0039063$ is *positive*, however, so that the final interval given in the table does not contain a root.

To understand this effect more generally, we introduce an **uncertainty interval** I^* for a solution x^* of (7.1). I^* should have the property that if x is in I^*, then the computed value of f at x may have the wrong sign. That is, for x in I^*, we are "uncertain" as to whether the sign of $f(x)$ equals the sign of $\hat{f}(x)$. For the example (7.29), Table 7.3 shows that we may let $I^* = [4.99, 5.012]$. In general, we cannot determine an uncertainty interval

TABLE 7.4 *Bisection Method Attempting to Find the Root x* = 5.0 of the Function (7.29). (Computations Done with β = 2, t = 27)*

x_0	x_1	$\hat{f}(\frac{1}{2}(x_0 + x_1))$
4.875	5.25	$-.23 \times 10^{-1}$
4.875	5.0625	$.30 \times 10^{-2}$
4.96875	5.0625	$-.10 \times 10^{-2}$
4.96875	5.0156250	$.10 \times 10^{-2}$
4.9921875	5.0156250	$.10 \times 10^{-2}$
5.0039063	5.0156250	0

very accurately; indeed, we have not even *defined* this interval very accurately. However, this concept greatly simplifies the following discussion.

The size of the uncertainty interval I^* at a solution x^* depends on the computer accuracy, the algorithm that is used to evaluate f, and the stability of the solution x^*. According to formulas (7.28), for example, if f is twice differentiable, and x^* is a double root, then with $\hat{f}$ given by (7.27), the uncertainty interval at x^* will have length at least as big as

$$ r = \left\{ \frac{2 \, |g(x^*)|}{|f''(x^*)|} \mu \right\}^{1/2}. \tag{7.30} $$

The root $x^* = 5$ of the function f given by (7.29) is extremely ill-conditioned. For this example, we have $f(x^*) = f'(x^*) = f''(x^*)$, so that x^* is a *triple* root and the interval I^* is even larger than would be predicted by (7.30). On the other hand, if x^* is a well-conditioned root, and if the algorithm used to evaluate f is stable (that is, subtractive cancellation does not occur), then I^* will be quite small. Indeed, we would expect I^* to be not much longer than μ.

Next, we consider how the existence of uncertainty intervals affects the interval algorithms of Section 7.1.5. For this, let $[x_0, x_1]$ be an initial interval that contains the root x^* that we are seeking, and let I^* be the uncertainty interval for x^*. At each stage of the algorithm we compute a new point x_2 in $[x_0, x_1]$ and then use the replacement rule (7.16) to replace either x_0 or x_1 by x_2. If the computed sign of f at x_2 is correct, then the new interval will again contain x^*. However, if x_2 is in I^*, then the sign of $\hat{f}$ at x_2 may be wrong. In this case the wrong endpoint is replaced, and the new interval no longer contains x^*. Consider, as in Fig. 7.9b, the case where the new interval $[x_0, x_1]$ has x_1 in I^*; and let $\bar{x}_2$ denote the *next* new point. There are two cases to consider:

(i) $\bar{x}_2$ is not in I^*;
(ii) $\bar{x}_2$ is in I^*.

In case (i) the computed sign of f at $\bar{x}_2$ is correct, so the replacement rule

Fig. 7.9   (a) $[x_0, x_1]$ contains x^*; $\hat{f}(x_2)$ has wrong sign. (b) New interval $[x_0, x_1]$ does not contain x^*.

correctly replaces x_0 with $\bar{x}_2$. The new interval $[x_0, x_1]$ still omits x^*, but the endpoint x_1 is in I^*. In case (ii) the replacement rule may replace *either* x_0 or x_1 with x_2. If it replaces x_0, then the new $[x_0, x_1]$ is entirely inside I^*; whereas if x_1 is replaced, then $[x_0, x_1]$ is only partly inside I^*. But in any case the new interval $[x_0, x_1]$ has one or both endpoints in I^*. By repeating this argument, we prove the following result:

> At any stage of an interval algorithm, either the interval $[x_0, x_1]$ contains x^*, or else one or both endpoints x_0, x_1 lie in I^*.

The above discussion shows that once the replacement rule makes a mistake, then the interval $[x_0, x_1]$ will not contain x^*. In fact, further calculations will lead away from x^*, although no further than the length of I^*. Certainly there is no point in continuing the iteration once the root x^* has been skipped over. Generally, the iteration should probably be stopped whenever $|x_0 - x_1|$ is smaller than the length of I^*. If $|x_0 - x_1|$ is *very* small, then it is unlikely that $[x_0, x_1]$ actually contains a root of f.

7.3.3   Effect of Rounding on Successive Substitution

The successive substitution methods, as expressed by (7.2), can be written computationally as

$$\hat{x}_{k+1} = \mathrm{fl}[g(\hat{x}_k)], \qquad k = 0, 1, 2, \ldots, \tag{7.31a}$$

where $\hat{x}_0$ is a floating point starting value. This can be written as

$$\hat{x}_{k+1} = g(\hat{x}_k) + \delta_k, \qquad k = 0, 1, \ldots, \tag{7.31b}$$

where δ_k is the (absolute) error in evaluating g at $\hat{x}_k$; that is,

$$\delta_k = \mathrm{fl}[g(\hat{x}_k)] - g(\hat{x}_k). \tag{7.31c}$$

To see how near these computed iterates $\hat{x}_1, \hat{x}_2, \ldots$ come to the solution x^*, assume that g satisfies the conditions of the convergence rule in Section 7.2.1.

Then by imitating the argument used in equations (7.20) we can use (7.31b) to write

$$|x^* - \hat{x}_k| = |g(x^*) - g(\hat{x}_{k-1}) - \delta_{k-1}|$$

$$\leq |g(x^*) - g(\hat{x}_{k-1})| + |\delta_{k-1}|. \qquad (7.32)$$

Now assume that $|\delta_k| \leq \delta$ for all k, and use the mean value theorem to write $|g(x^*) - g(\hat{x}_{k-1})| = |g'(\xi_k)||x^* - \hat{x}_{k-1}|$, where ξ_k is some value between x^* and $\hat{x}_{k-1}$. Then (7.32) gives

$$|x^* - \hat{x}_k| \leq |g'(\xi_k)||x^* - \hat{x}_{k-1}| + \delta$$

$$\leq \alpha|x^* - \hat{x}_{k-1}| + \delta, \qquad (7.33a)$$

where α is the bound on $|g'(x)|$ as specified in the convergence rule. By repeating this argument with k replaced by $k - 1$, we find

$$|x^* - \hat{x}_{k-1}| \leq \alpha|x^* - \hat{x}_{k-2}| + \delta. \qquad (7.33b)$$

If (7.33b) is substituted into (7.33a), the result is

$$|x^* - \hat{x}_k| \leq \alpha^2|x^* - \hat{x}_{k-2}| + \alpha\delta + \delta. \qquad (7.33c)$$

A repetition of this argument gives finally

$$|x^* - \hat{x}_k| \leq \alpha^k|x^* - \hat{x}_0| + [\alpha^{k-1} + \alpha^{k-2} + \cdots + \alpha + 1]\delta. \qquad (7.33d)$$

Since $0 < \alpha < 1$, the term in brackets satisfies

$$\alpha^{k-1} + \alpha^{k-2} + \cdots + \alpha + 1 \leq 1 + \alpha + \cdots + \alpha^{k-1} + \alpha^k + \cdots$$

$$\leq 1/(1 - \alpha).$$

Thus (7.33d) implies

Error bound for computed iterates

$$|x^* - \hat{x}_k| \leq \alpha^k|x^* - \hat{x}_0| + \delta/(1 - \alpha) \qquad (7.34)$$

Observe the two parts of (7.34). The first term goes to zero as $k \to \infty$ because $\alpha < 1$. The second term, however, is a constant that depends on the rounding error δ and the bound α on the derivative of g. If α is very close to 1, then this second term is very large, and $\hat{x}_k$ may *never* get very near to x^*. To illustrate this, consider the iteration

$$x_{k+1} = g(x_k), \qquad g(x) = .5x^3 - 1.5x^2 + .5001x + 1.4999. \qquad (7.35)$$

This function g has a fixed point at $x^* = 1$. But $|g'(1)| = .9999$, so that the constant α must be at least .9999. Table 7.5 gives some computed values of

TABLE 7.5 *Computed Values of Iterates Given by* (7.35)

k	x_k	k	x_k
0	.9999	111	1.0000100
1	1.0001	112	.99999002
2	.99990004	113	1.0000100
3	1.0000999	114	.99999002
4	.99990016	115	1.0000100
110	.99999001		

the iterates with $x_0 = .9999$. If, on the other hand, Newton's method is applied to the function $f(x) = .5x^3 - 1.5x^2 - .4999x + 1.4999$ to find this same value $x^* = 1.0$, then the constant α will be very small. With $x_0 = .9999$ one Newton iteration gives $x_1 = 1.0000005$. This example illustrates the fact that a slowly convergent method is bad not only because it may take a very long time to get close to the solution, but also because it may *never* get sufficiently close.

7.3.4 Stopping Criteria

The inequality (7.34) suggests a reasonable criteria for stopping the iteration (7.2). By applying (7.34) to $|x^* - \hat{x}_k|$ and to $|x^* - \hat{x}_{k+1}|$, we find that

$$|\hat{x}_{k+1} - \hat{x}_k| = |\hat{x}_{k+1} - x^* + x^* - \hat{x}_k| \le |\hat{x}_{k+1} - x^*| + |x^* - \hat{x}_k|$$

$$\le \alpha^{k+1}|x_0 - x^*| + \delta/(1-\alpha) + \alpha^k|x_0 - x^*| + \delta/(1-\alpha).$$

Hence,

$$|\hat{x}_{k+1} - \hat{x}_k| \le 2\delta/(1-\alpha) + \alpha^k(1+\alpha)|x^* - x_0|, \qquad (7.36)$$

which implies that the difference between successive iterates may possibly *always* be at least as large as $2\delta/(1-\alpha)$. Thus if this quantity can be estimated, then the stopping parameter ε_1 should be chosen accordingly in order to avoid useless computations. If something is known about $g'(x)$ for x near x^*, then α can be estimated. Furthermore, δ can be obtained from a rounding error analysis of g. For the example (7.35), whose iterates are given in Table 7.5, we indicated earlier that $\alpha \ge .9999$. A simple error analysis of g gives $\delta \simeq 3 \times 10^{-8}$ for a floating point system with $\beta = 2, t = 27$. Hence the first term in (7.36) will be approximately 6×10^{-4}. Thus the iterates might *never* agree to more than four or five figures.

The preceding discussion of rounding error was applied to the problem of how to choose the stopping parameter ε_1. That is, how to get the best

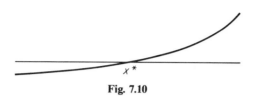

Fig. 7.10

possible accuracy, in the sense of making $|\hat{x}_k - x^*|$ small without unnecessary computation. As observed earlier, however, in many applications the real goal is to find some x for which $|f(x)|$ is small, and it may not matter how close this x is to the exact zero point. For example, if $f(x)$ is a measure of the error in some experiment, then a suitably small error is just as acceptable as zero error. More importantly, for certain functions f, such as the one illustrated by Fig. 7.10, it may be very easy to make $|f(\hat{x}_k)|$ small, but difficult or impossible to get $\hat{x}_k$ close to x^*. On the other hand, a function such as shown in Fig. 7.11 has the property that $|f(\hat{x}_k)|$ is small only if $\hat{x}_k$ is

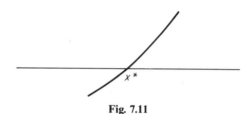

Fig. 7.11

very close to x^*. In fact, rounding error may always keep $\hat{x}_k$ far enough from x^* that $|f(\hat{x}_k)|$ will *never* be very small. Thus in selecting the stopping parameters $\varepsilon_1, \varepsilon_2$ one must consider the following.

(a) What kind of approximation is desired; i.e., should $|x_k - x^*|$ be small or should $|f(x_k)|$ be small?

(b) What kind of function is being solved; i.e., is it like Fig. 7.10 or Fig. 7.11?

(c) Will the rounding error cause too much difficulty in obtaining the desired kind of approximation? If so, one may be willing to settle for something less.

EXERCISES

1 For each of the following functions, write a program to evaluate the function near the zero point $x^* = 1$ and determine the uncertainty interval I^*.
  (a) $f(x) = (\cos \pi x)[1 - \sin (\frac{1}{2}\pi x)/^2$
  (b) $f(x) = x^4 - 112x^3 + 1221x^2 - 2110x + 1000.$

2 Analyze the effect of rounding error on Newton's method. Assume that $f'(x^*) \neq 0$, and that $f(x_k)$ and $f'(x_k)$ can be evaluated with absolute errors δ_k, ε_k, respectively.

3 Sketch the graph of each of the following functions to determine whether it is of the form shown in Fig. 7.10 or Fig. 7.11:
   (a) $f(x) = x^5 - 2$
   (b) $f(x) = x^4 - 8x^3 + 24x^2 - 32x + 16$
   (c) $f(x) = 100 + \ln x$.

4 For the function $f(x) = x^3 - 2x^2 + 1001x - 1000$, how close does x have to be to the zero point $x^* = 1.0$ in order for $|f(x)|$ to be less than 10^{-8}? With eight-digit decimal arithmetic and the bisection method is it possible to find an x so that $|f(x)| \leq 10^{-8}$?

5 The function

$$f(x) = 8 \sin x + 4x^2 - 4\pi x + \pi^2 - 8 = 0$$

has a root of multiplicity three at $x = \pi/2$.
   (a) Verify this fact.
   (b) Write a program to evaluate f at points near $x = \pi/2$. Use the values

$$4\pi \simeq 12.56638, \qquad \pi^2 - 8 \simeq 1.869604.$$

   (c) Estimate the uncertainty interval around the root $x^* = \pi/2$.

6 Use any of the methods of Section 7.1 to find the zero point $x^* = \pi/2$ of the function in Exercise 5. How close does your method come to the root? Explain your results.

7.4 Efficiency—Rates of Convergence

The number of floating point operations that are required by an iterative algorithm depends on the number of iterates that are computed. Thus, in general, it is impossible to estimate the efficiency of the algorithms in Section 7.1 with much accuracy. Nevertheless, rough estimates are useful for comparing different iterative algorithms. These rough estimates are derived by first estimating how many iterates must be computed.

7.4.1 Determining Efficiency

The efficiency with which an iterative method finds an approximate solution to (7.1) is determined by two things:

(i) How much work is involved in computing each iterate?
(ii) How fast do the iterates converge to the solution?

The first of these can be estimated in a straightforward manner by simply counting the number of floating point arithmetic operations. It is assumed, of course, that the operations involved in evaluating the function f, and its derivative in the case of Newton's method, are included in this count. In fact, one major objection to the use of Newton's method is that if $f'(x)$ is very difficult to evaluate then the work involved in determining each iterate may be completely unreasonable.

The answer to the question (ii) cannot be given quite so precisely because the speed with which the iterates converge may depend on the starting value x_0, as well as on the rounding error. Nevertheless, there are some general qualitative statements that can be made concerning iterative methods that at least allow a reasonable comparison to be made between various methods.

7.4.2 Rates of Convergence

Consider first of all the successive substitution method (7.2). The convergence rule says that if

$$|g'(x)| \leq \alpha < 1 \tag{7.37}$$

for all x in a suitable interval; then

Linear convergence

$$|x_k - x^*| \leq \alpha |x_{k-1} - x^*|; \tag{7.38}$$

and hence

$$|x_k - x^*| \leq \alpha^k |x_0 - x^*|. \tag{7.39}$$

Thus the quantity α can be used as a measure of how fast the iteration will converge—the smaller α is, the faster the convergence. We make this more general by saying that if $x_0, x_1, x_2, \ldots$ is *any* sequence that converges to x^*, then the **convergence rate is linear**, *with convergence constant* α, if (7.38) holds for all x_k. Thus if the iteration function g satisfies (7.37), then the successive substitution method (7.2) produces a sequence that converges linearly, provided x_0 is sufficiently close to x^*. Next recall that for Newton's method if f is twice differentiable and $f'(x^*) \neq 0$, then the iteration function has the property that $g'(x^*) = 0$ [see equation (7.22)]. Thus if the sequence $x_0, x_1, x_2, \ldots$ is converging to x^*, then

$$|x_k - x^*| = |g(x_{k-1}) - g(x^*)| = |g'(\xi_{k-1})| \, |x_{k-1} - x^*|,$$

where ξ_{k-1} is between x_{k-1} and x^*. If we let $\alpha_{k-1} = |g'(\xi_{k-1})|$, then $\alpha_{k-1} \to 0$ as $k \to \infty$ since $g'(\xi_{k-1}) \to g'(x^*)$. Hence

Super linear convergence

$$|x_k - x^*| \leq \alpha_{k-1} |x_{k-1} - x^*|,$$
$$\alpha_{k-1} \to 0 \quad \text{as} \quad k \to \infty. \tag{7.40}$$

Clearly this is a stronger statement than (7.38), which led to (7.39). A sequence $x_0, x_1, \ldots$ that converges to x^* and that satisfies (7.40) is said to be **superlinearly convergent**.

Newton's method was used to motivate (7.40), but in fact an even stronger statement can be made. If we expand the iteration function $g(x) = x - f(x)/f'(x)$ in a Taylor's series at x^* [still assuming that $f'(x^*) \neq 0$], we find that

$$g(x) = g(x^*) + g'(x^*)(x - x^*) + \tfrac{1}{2}g''(\xi)(x - x^*),$$

where ξ lies between x and x^*. From the facts that $g(x^*) = x^*$ and $g'(x^*) = 0$ this gives

$$g(x) - x^* = \tfrac{1}{2}g''(\xi)(x - x^*)^2,$$

and hence in terms of the iterates,

Quadratic convergence	$|x_{k+1} - x^*| \leq \alpha |x_k - x^*|^2,$	(7.41)

where α is some upper bound on $|\tfrac{1}{2}g''(\xi)|$. Now if the iterates are converging to x^*, then eventually $|x_k - x^*|$ is small, and $|x_k - x^*|^2$ is even smaller. More precisely if $\alpha = 1$ and $|x_k - x^*|$ is of the order of 0.1, then $|x_{k+1} - x^*|$ will be of the order of .01, $|x_{k+2} - x^*|$ will be approximately .0001, and $|x_{k+3} - x^*|$ will be .00000001. Thus three iterations have reduced the error from 10^{-1} to 10^{-8}. A sequence that converges to x^* and satisfies (7.41) for some α is said to be **quadratically convergent**. Clearly the argument that led to (7.41) will work for *any* iteration (7.2) for which $g'(x^*) = 0$; hence any such iteration will be quadratically convergent.

The **secant-type methods**, including the false position and Dekker–Brent methods, may be analyzed together because they are all *two-step* methods in which the new point x_{k+1} is determined from two previous points x_k, x_{k-1} by the secant approximation (7.12). We can also characterize x_{k+1} as the solution to the equation $l_k(x) = 0$, where l_k is the *linear* function that interpolates f at the two points x_k and x_{k-1}. According to formulas (4.12) and (4.13), if f is twice differentiable, we can write

$$f(x) = l_k(x) + \tfrac{1}{2}(x - x_k)(x - x_{k-1})f''(\xi_k), \qquad (7.42)$$

where ξ_k is some point in an interval containing x, x_k, and x_{k-1}. Furthermore, since $l_k(x_{k+1}) = 0$, we can express $l_k(x)$ as

$$l_k(x) = (x - x_{k+1})[f(x_k) - f(x_{k-1})]/(x_k - x_{k-1}). \qquad (7.43)$$

Now (7.42) implies that

$$0 = f(x^*) = l_k(x^*) + \tfrac{1}{2}(x^* - x_k)(x^* - x_{k-1})f''(\xi_k), \qquad (7.44a)$$

and (7.43) with the mean value theorem gives

$$l_k(x^*) = (x^* - x_{k+1})f'(\eta_k), \qquad (7.44b)$$

where η_k is some point between x_k and x_{k-1}. By substituting (7.44b) into (7.44a) and rearranging terms, we find that

$$(x_{k+1} - x^*) = -\tfrac{1}{2}(x_k - x^*)(x_{k-1} - x^*)f''(\xi_k)/f'(\eta_k) \qquad (7.44c)$$

provided $f'(\eta_k) \neq 0$. Thus if $f'(x) \neq 0$ for x near to x^*, then we have

| Error bound for secant methods | $$\|x_{k+1} - x^*\| \leq K\|x_{k-1} - x^*\|\,\|x_k - x^*\|,$$ | (7.45) |

where K is an upper bound on the factor

$$\tfrac{1}{2}\|f''(\xi_k)/f'(\eta_k)\|.$$

Now if the sequence $x_1, x_2, \ldots$ has been computed by the secant or Dekker–Brent algorithms, and *if* the sequence converges, then (7.45) with $\alpha_k = K\|x_{k-1} - x^*\|$ is just (7.40). That is, these methods are superlinearly convergent. We can also use (7.45) to study the false position method. Now however we must recall that one of the points may become stuck; that is, x_{k-1} may equal x_{k-j} for large j. In this case we cannot conclude that $\alpha_k = K\|x_{k-1} - x^*\| \to 0$ as k increases. However, *if* x_{k-1} is close enough to x^*, then certainly (7.45) implies

$$\|x_{k+1} - x^*\| \leq \alpha_k\|x_k - x^*\|, \qquad \text{where} \quad \alpha_k < 1.$$

Thus the false position method is at least *linearly convergent* with convergence constant $\alpha \leq K\|x_{k-1} - x^*\|$, which may in fact tend to zero.

Finally for the bisection method, we have the simple estimate $\|x_{k+1} - x^*\| \leq \tfrac{1}{2}\|x_k - x^*\|$. This follows immediately from the fact that the interval is reduced by $\tfrac{1}{2}$ each step. Thus the bisection method also converges linearly. We summarize these rate of convergence results in Table 7.6.

7.4.3   Comparing Iterative Methods

To illustrate how Table 7.6 can be used to select an iterative method, consider the equation $f(x) = 0$, where

$$f(x) = x^3 - 1.3x^2 - 14x + 18.2.$$

It is easily seen, by evaluating f at $x = 1$ and $x = 2$, that there is a zero in the interval $[1, 2]$. The iteration function $g(x) = \tfrac{1}{14}(x^3 - 1.3x^2 + 18.2)$ satisfies $\|g'(x)\| \leq \tfrac{1}{2}$ in this interval. Hence the rate of convergence for the iteration

$$x_{k+1} = \tfrac{1}{14}(x_k^3 - 1.3x_k^2 + 18.2) = .07142857x_k^3 - 18.2x_k^2 + 254.8$$

TABLE 7.6   *Rates of Convergence of the Methods in Section 7.1*

Method	Rate of convergence
Successive substitution	Linear with constant α if $\lvert g'(x)\rvert \leq \alpha < 1$ for x near x^*.
Newton's method	Quadratic if x^* is simple. Linear if x^* has multiplicity $m > 1$.
Secant method and Dekker–Brent algorithm	Superlinear if x^* is simple. Linear if x^* has multiplicity $m > 1$.
False position	Superlinear if both endpoints change, and x^* is simple. Linear if one endpoint does not change, or if x^* has multiplicity $m > 1$.
Bisection method	Linear with constant $\alpha = \frac{1}{2}$.

is linear with constant $\alpha = \frac{1}{2}$. (Note that according to Table 7.6 this is the same as the rate of convergence of the bisection method.) If we start with $x_0 = 1.5$, then the initial error satisfies $\lvert x^* - x_0\rvert \leq .5$, so the error at the kth step will satisfy $\lvert x^* - x_k\rvert \leq 10^{-8}$, provided $\alpha^k\lvert x^* - x_0\rvert \leq 10^{-8}$; that is, $(\frac{1}{2})^k.5 \leq 10^{-8}$. By taking logarithms we find that

$$(k + 1)\log(.5) \leq -8 \log 10$$

so

$$k \geq (8/.30103) - 1 \simeq 25.5754.$$

Hence theoretically it may take as many as 26 iterations to reduce the error to 10^{-8}. Each iteration requires three multiplications and two additions, (we use Horner's method, of course) for a total of $52A + 78M$ operations. Now consider Newton's method. Since $f'(x) = 3x^2 - 2.6x - 14 < 0$ for all x in $[1, 2]$, we are assured that convergence is quadratic. If the constant in (7.41) is about 1, then the initial error of .5 is reduced to 10^{-8} in five iterations. Each iteration requires six additions, four multiplications, and one division for a total of $30A + 20M + 5D$ operations during five iterations. Note that we must *assume* here that these iterations actually do converge. This is not obvious in the case of Newton's method. The secant and Dekker–Brent algorithms will both converge superlinearly. It is not so easy to estimate how many iterations will be needed by these methods. We would expect them to be slower than Newton's method but faster than the successive substitution method. Thus an estimate of 10 to 15 iterations is probably reasonable. Note that each iteration takes only $3A + 2M$ operations to evaluate f, but $3A + 1M + 1D$ operations are needed to determine x_{k+1} from (7.12c), for a total of $6A + 3M + 1D$ per iteration. This is about the same as for Newton's method.

The rate of convergence of the bisection method is independent of the equation we are solving. Thus for *any* equation we can determine *a priori* how many iterations are required to reduce an initial interval of length L_i to a final interval of length L_f. In fact

the 1st iteration gives an interval of length $\frac{1}{2}L_i$;
the 2nd iteration gives an interval of length $\frac{1}{2}^2 L_i$;
the 3rd iteration gives an interval of length $\frac{1}{2}^3 L_i$;
etc.

Thus the interval will have length less than L_f after N iterations, provided that N satisfies the equation

$$\tfrac{1}{2}^N L_i \le L_f.$$

By taking logarithms we find that

$$N \log \tfrac{1}{2} \le \log L_f - \log L_i$$

or

$$N \ge \frac{\log L_f - \log L_i}{\log \frac{1}{2}} \simeq \frac{\log L_f - \log L_i}{.30103}.$$

For example, to reduce the interval $[1, 2]$ to an interval of length 10^{-8} requires $N \ge (0 - (-8))/.30103 = 26.57554$ iterations.

In addition to estimating the efficiency of the algorithm, the rate of convergence of an iterative method is also useful for determining a reasonable value for the stopping parameter N_{max}. If, for example, the method is linearly convergent, with $\alpha = \frac{1}{2}$ and initial error of approximately $\frac{1}{2}$, then as just shown it may take as many as 26 iterations to reduce the error to 10^{-8}. Thus a value of $N_{max} = 30$ would be reasonable. On the other hand, Newton's method for this situation should require only about five iterations, so N_{max} should certainly be no larger than 10.

7.4.4 Accelerating Convergence

Iterations that converge linearly may converge too slowly to be of practical use. For example, Table 7.5 gives an iterative method in which more than 100 iterations were needed just to gain one figure of accuracy. Even the simple example (7.8) for solving (7.3) required 23 iterations to get a four figure approximation. With the help of (7.20b), however, it is often possible to accelerate the convergence of an iterative method.

To motivate this acceleration procedure, assume that the iteration

$x_{k+1} = g(x_k)$, $k = 0, 1, \ldots$, satisfies

| Assumption | $x_{k+1} - x^* = \alpha(x_k - x^*)$, $\quad k = 0, 1, \ldots,$ | (7.46) |

where $|\alpha| < 1$. That is, we assume a slightly stronger form of (7.38). Notice that (7.20b) implies that, for some ξ_k between x_k and x^*,

| Real situation | $x_{k+1} - x^* = g(x_k) - g(x^*) = g'(\xi_k)(x_k - x^*).$ | (7.47) |

Thus if $g'(\xi_k)$ does not change with k, then (7.46) is the same as (7.47) with $\alpha = g'(\xi_k)$. For convergence we must have $|\alpha| < 1$. Now in (7.46) there are two unknowns—α and x^*. But we can also write

$$x_k - x^* = \alpha(x_{k-1} - x^*), \tag{7.46'}$$

which gives a second equation in these same two unknowns. We now solve these two equations for x^* to obtain

$$x^* = (x_{k-1}x_{k+1} - x_k^2)/(x_{k+1} - 2x_k + x_{k-1}). \tag{7.48}$$

Thus if we compute three successive iterates and combine them according to (7.48), then the solution is found. Of course, this is all based on the assumption that (7.46) holds. In practice we only have (7.47), and thus (7.48) will not be true. However, if x_k is fairly close to x^*, then ξ_k in (7.47) will be close to x^*, and hence $g'(\xi_k) \simeq g'(x^*)$. That is, our assumption (7.46) is not exactly true but may be nearly so. At least, it is reasonable to expect that x_{k+1}^* defined by

| Improved value | $x_{k+1}^* = (x_{k-1}x_{k+1} - x_k^2)/(x_{k+1} - 2x_k + x_{k-1})$ |

will be closer to x^* than is x_{k+1}. To illustrate this fact, consider the values x_{12}, x_{13}, x_{14} given in Table 7.1. These combine to give the value

$$x_{14}^* = \frac{(.44079)(.44208) - (.44414)^2}{.44208 - 2(.44414) + .44079} = .44286\ldots,$$

which is as accurate as x_{21}.

This procedure, called *Aitken's method*, is a very effective technique. It should only be applied however when it is clear that the iteration *is converging* and that the rate of convergence is linear. To see how the Aitken technique can be used repeatedly, consider the following algorithm, sometimes known as *Steffensen's method*:

Linearly Convergent Iteration with Aitken Acceleration

> The input consists of a starting value x_0, the usual stopping parameters ε_1, ε_2, $N_{\max}$, the function f, and the iteration function $g(x)$.

    1   For $N = 0, 1, 2, \ldots, N_{\max}$

        1.1 $x_1 \leftarrow g(x_0)$

        1.2 $x_2 \leftarrow g(x_1)$

        1.3 $x^* \leftarrow (x_0 x_2 - x_1^2)/(x_2 - 2x_1 + x_0)$

        1.4   If $|f(x^*)| \le \varepsilon_2$

            Then

            1.4.1.   Output $\{x^*, \text{“}|f(x^*)| \text{ is small”}\}$ (7.49)

            1.4.2.   Halt

        1.5   If $|x_1 - x_2| \le \varepsilon_1 |x_1|$

            Then

            1.5.1.   Output $\{x^*, \text{“}|x_N - x_{N-1}| \text{ is small”}\}$

            1.5.2.   Halt

        1.6 $x_0 \leftarrow x^*$

    2   Error message, "No convergence after $N_{\max}$ iterations"

    3   Halt

A **word of warning** concerning this algorithm is appropriate. If x_0 is so far from the solution that x_1 and x_2 are not related by (7.38), then x^* computed in step 1.1.3 may be worse than x_1 or x_2.

EXERCISES

1   For each of the following functions determine how much work is involved in one step of the secant method and one step of Newton's method. (Assume that to evaluate a function such as $\sin x$, e^x, etc. requires five additions and five multiplications.)

    (a) $f(x) = x^4 + 2x^3 - 1.5x + 7$ (b) $f(x) = e^x \sin x + 1$

    (c) $f(x) = x^2 \sin x + 2x$ (d) $f(x) = x^3 e^x + x^2 + 1$.

2   For each of the following iteration functions, estimate the convergence constant α for finding a fixed point near 1.0 and give a reasonable value for the stopping parameter $N_{\max}$:

    (a) $g(x) = \frac{1}{4} e^x$

    (b) $g(x) = x^3 - 2x^2 + \frac{3}{2}x + \frac{1}{2}$.

3   Suppose the iteration $x_{k+1} = g(x_k)$ is superlinearly convergent, with convergence constants $\alpha_k \le 1/k$, $k = 1, 2, 3, \ldots$. How many iterations are needed to reduce an initial error of .5 to an error less than 10^{-8}? (Compare this result with the linearly convergent example described in the text.)

4 Suppose you are asked to solve the equation

$$x^3 - 1.6x^2 - .75x + .5 = 0.$$

By evaluating the function at a few points you determine that there is a solution between 0 and 1. Compare the efficiency of the bisection method, Newton's method, and the iteration

$$x_{k+1} = x_k^3 - 1.6x_k^2 + .25x_k + .5$$

$$x_0 = 1.$$

5 Solve the equation given in Exercise 4 by using the iteration that is written out in Exercise 4. Next use this same iteration with Aitken acceleration and compare your results.

7.5 Roots of Polynomials

In this section we consider a special case of the problem to which this chapter is devoted; namely, the problem of solving $f(x) = 0$ when f is a polynomial. The best general-purpose methods for finding roots of polynomials make strong use of complex variable theory. Here we shall consider only some simple and limited methods, together with several important numerical aspects of the polynomial root problem. Elementary properties of complex numbers will be used throughout.

7.5.1 Special Properties of Polynomials—Horner's Method Revisited

In this section we shall use $p(x)$ to denote a polynomial of degree n with *real* coefficients $a_0, a_1, \ldots, a_{n-1}$; that is,

$$p(x) = x^n + a_{n-1}x^{n-1} + \cdots + a_1 x + a_0. \qquad (7.50)$$

Note that we have *scaled* the polynomial by dividing each coefficient by the coefficient of x^n. The solutions to the equation

$$p(x) = 0 \qquad (7.51)$$

are called the *roots* of p. There are two reasons for considering (7.51) as a different problem from (7.1): The solutions may be complex, and we are usually interested in finding more than one root. These facts make the problem more difficult to solve. On the other hand, the special nature of polynomials makes it possible to apply some very strong mathematical tools toward solving (7.51). Most of these tools are beyond the scope of this book; hence we shall consider only some rather simple results with a few indications of deeper possibilities.

Recall from Chapter 3 that Horner's method [Algorithm (3.18)] is an efficient and accurate way to evaluate an arbitrary polynomial. Here we shall derive the Horner algorithm in a slightly different way that leads to other useful applications. This derivation is based on the formula

Factorization of p	$$p(x) = (x - \hat{x})q(x) + r,$$	(7.52)

where $\hat{x}$ is a fixed (real or complex) number, and $q(x)$ is a polynomial of degree $n - 1$; that is

$$q(x) = x^{n-1} + b_{n-2}x^{n-2} + \cdots + b_1 x + b_0. \tag{7.53}$$

The quantity r in (7.52) is also a real or complex number. The algorithm we shall now derive gives a constructive proof of the existence of the factorization (7.52) for *any* value of $\hat{x}$. We first note that if $\hat{x}$ is substituted for x in (7.52), then we obtain

$$p(\hat{x}) = r. \tag{7.54}$$

Thus if we can compute r in (7.52), we shall have the value of p at $\hat{x}$. To obtain the coefficients $b_0, b_1, \ldots, b_{n-2}$ and the number r, equate coefficients of like powers of x in (7.52). This gives

$$
\begin{aligned}
a_{n-1} &= b_{n-2} - \hat{x} &&\text{(coefficients of } x^{n-1}) \\
a_{n-2} &= b_{n-3} - \hat{x}b_{n-2} &&\text{(coefficients of } x^{n-2}) \\
&\;\;\vdots \\
a_{k+1} &= b_k - \hat{x}b_{k+1} &&\text{(coefficients of } x^{k+1}) &&\text{(7.55a)} \\
&\;\;\vdots \\
a_1 &= b_0 - \hat{x}b_1 &&\text{(coefficients of } x) \\
a_0 &= r - \hat{x}b_0 &&\text{(constant terms).}
\end{aligned}
$$

If we now rewrite these equations to obtain the b_k from the a_k, we find that

$$
\begin{aligned}
b_{n-2} &= a_{n-1} + \hat{x} \\
b_{n-3} &= a_{n-2} + \hat{x}b_{n-2} \\
&\;\;\vdots \\
b_k &= a_{k+1} + \hat{x}b_{k+1} \qquad\qquad (7.55b) \\
&\;\;\vdots \\
b_0 &= a_1 + \hat{x}b_1 \\
r &= a_0 + \hat{x}b_0.
\end{aligned}
$$

This is a **recursive method** for computing the coefficients of q and the value of

p at $\hat{x}$. That is, the value of b_{n-2} that is computed by the first equation must be used in the second equation to get b_{n-3}, and so on. An algorithm for this recursion is now given.

Algorithm for Computing the Factorization (7.52), with q Defined by (7.53) and r Satisfying (7.54)

$$
\begin{aligned}
&1 \quad b_{n-2} \leftarrow a_{n-1} + \hat{x} \\
&2 \quad \text{For } k = n-2, n-3, \ldots, 1 \\
&2.1 \quad b_{k-1} \leftarrow a_k + \hat{x}b_k \\
&3 \quad r \leftarrow a_0 + \hat{x}b_0
\end{aligned}
\tag{7.56}
$$

Except for the fact that intermediate results $(a_k + \hat{x}b_k)$ are saved, this algorithm is the same as the Horner Algorithm (3.18) for evaluating p at $\hat{x}$. The coefficients b_k will be needed for the techniques discussed in Section 7.5.3. They are also of interest because they give the *derivative* of p at $\hat{x}$. That is, if we differentiate (7.52), we find

$$p'(x) = (x - \hat{x})q'(x) + q(x)$$

so that

$$p'(\hat{x}) = q(\hat{x}). \tag{7.57}$$

Thus we can evaluate p' at $\hat{x}$ by simply using Horner's method to evaluate the polynomial q whose coefficients are computed by (7.56). The computation of the *coefficients* of q are combined with the *evaluation* of q in the following algorithm:

Algorithm for Computing the Factorization (7.52) and Evaluating $q(\hat{x}) = p'(\hat{x})$

$$
\begin{aligned}
&1 \quad b_{n-2} \leftarrow a_{n-1} + \hat{x} \\
&2 \quad pp \leftarrow b_{n-2} + \hat{x} \\
&3 \quad \text{For } k = n-2, n-3, \ldots, 1 \\
&3.1 \quad b_{k-1} \leftarrow a_k + \hat{x}b_k \\
&3.2 \quad pp \leftarrow b_{k-1} + \hat{x}pp \\
&4 \quad r \leftarrow a_0 + \hat{x}b_0
\end{aligned}
\tag{7.58a}
$$

Steps 2 and 3.2 are just the Horner's method evaluation of $q(\hat{x})$. The coefficients of q are computed in the previous steps. If these coefficients are not going to be used later, then they need not be saved. Hence an algorithm for just evaluating p and p' at $\hat{x}$ is the following.

Algorithm for Evaluating $p(\hat{x})$ and $p'(\hat{x})$

$$
\begin{aligned}
&1 \quad p \leftarrow a_{n-1} + \hat{x} \\
&2 \quad pp \leftarrow p + \hat{x} \\
&3 \quad \text{For } k = n - 2, n - 3, \ldots, 1 \\
&\qquad 3.1 \quad p \leftarrow a_k + \hat{x}p \\
&\qquad 3.2 \quad pp \leftarrow p + \hat{x}pp \\
&4 \quad p \leftarrow a_0 + \hat{x}p
\end{aligned}
\tag{7.58b}
$$

In order to find *complex* roots of a polynomial, it is sometimes necessary to evaluate p at a complex value of $\hat{x}$. The Horner Algorithm (3.18) or Algorithms (7.56), (7.58a), or (7.58b) can be used for this purpose. For complex $\hat{x}$, however, most of the arithmetic will have to be *complex* arithmetic. This increases the number of (real) floating point computations by nearly a factor of *four* (see Exercise 2). A more efficient technique is based on a generalization of the factorization (7.52), namely,

$$
p(x) = (x - \hat{x})(x - \bar{\hat{x}})q(x) + sx + t. \tag{7.59}
$$

Here $\hat{x} = u + iv$, where u and v are real, and $\bar{\hat{x}} = u - iv$. The subsequent discussion will show that such a factorization always exists and, moreover, s, t, and the coefficients of q are all *real*. Furthermore, by substituting $\hat{x}$ for x in (7.59) we find

$\boxed{\begin{array}{l} p(\hat{x}) \text{ for} \\ \text{complex } \hat{x} \end{array}}$ $\qquad\qquad p(\hat{x}) = s\hat{x} + t. \tag{7.60}$

Thus if s and t can be computed in real arithmetic, then p can be evaluated at complex $\hat{x}$ with only one real–complex multiplication, and one real–complex addition.

To derive an algorithm for computing q, s, and t in (7.59), note first of all that

$\boxed{\begin{array}{l} \text{A real} \\ \text{quadratic} \\ \text{factor} \end{array}}$ $\begin{aligned} (x - \hat{x})(x - \bar{\hat{x}}) &= (x - (u + iv))(x - (u - iv)) \\ &= x^2 - 2ux + (u^2 + v^2) \\ &= x^2 - \alpha x + \beta \end{aligned} \tag{7.61}$

when $\alpha = 2u$ and $\beta = u^2 + v^2$ are real. If we next write q as

$$
q(x) = x^{n-2} + b_{n-3}x^{n-3} + \cdots + b_1 x + b_0,
$$

then (7.59) becomes

$$
x^n + a_{n-1}x^{n-1} + a_{n-2}x^{n-2} + \cdots + a_1 x + a_0
$$
$$
= (x^2 - \alpha x + \beta)(x^{n-2} + b_{n-3}x^{n-3} + \cdots + b_1 x + b_0) + sx + t.
$$

As before, we equate coefficients of like powers of x to obtain

$$a_{n-1} = -\alpha + b_{n-3} \qquad \text{(coefficients of } x^{n-1})$$

$$a_{n-2} = \beta - \alpha b_{n-3} + b_{n-4} \qquad \text{(coefficients of } x^{n-2})$$

$$a_{n-3} = \beta b_{n-3} - \alpha b_{n-4} + b_{n-5} \qquad \text{(coefficients of } x^{n-3})$$

$$\vdots$$

$$a_1 = \beta b_1 - \alpha b_0 + s \qquad \text{(coefficients of } x)$$

$$a_0 = \beta b_0 + t \qquad \text{(constant term).}$$

(7.62)

If these equations are solved for the b_k, we find

$$b_{n-3} = a_{n-1} + \alpha$$

$$b_{n-4} = a_{n-2} + \alpha b_{n-3} - \beta$$

$$b_{n-5} = a_{n-3} + \alpha b_{n-4} - \beta b_{n-3}$$

$$\vdots$$

$$b_k = a_{k+2} + \alpha b_{k+1} - \beta b_{k+2}$$

$$\vdots$$

$$b_0 = a_2 + \alpha b_1 - \beta b_2$$

(7.63a)

and the last two equations in (7.62) give

$$s = a_1 + \alpha b_0 - \beta b_1$$

$$t = a_0 - \beta b_0.$$

(7.63b)

The recursion (7.63a, b) proves that the factorization (7.59) exists. Moreover, since all of the arithmetic in (7.63) is real, it follows that all of the numbers computed by the recursion are real. Thus the following algorithm computes the factorization (7.59) and evaluates $p(x)$ by using only real arithmetic.

Algorithm to Compute the Real Coefficients
$b_{n-3}, b_{n-4}, \ldots, b_0$ and Real s and t to Satisfy (7.60)

$$[\text{Here } \hat{x} = u + iv.]$$

1 $\alpha \leftarrow 2u, \ \beta \leftarrow u^2 + v^2$
2 $b_{n-3} \leftarrow a_{n-1} + \alpha$
3 $b_{n-4} \leftarrow a_{n-2} + \alpha b_{n-3} - \beta$

(7.64)

   (*continued*)

4   For $k = n - 5, n - 6, \ldots, 0$
  4.1 $b_k \leftarrow a_{k+2} + \alpha b_{k+1} - \beta b_{k+2}$
5   $s \leftarrow a_1 + \alpha b_0 - \beta b_1$
6   $t \leftarrow a_0 - \beta b_0$
7   $\eta \leftarrow su + t, \ \zeta \leftarrow sv$

The **complex value** of $p(\hat{x})$ is just $\eta + i\zeta$, where η and ζ are the real values computed in step 7. The work involved in this algorithm is $2nA + (2n + 1)M$. This is better, by almost half, than simply using a complex version of (7.56) or (3.18).

Just as (7.52) was used to obtain $p'(\hat{x})$ efficiently for real $\hat{x}$, we can use (7.59) to compute $p'(\hat{x})$ for complex $\hat{x}$. That is, if (7.59) is differentiated, the result is

$$p'(x) = (x - \hat{x})q(x) + (x - \bar{\hat{x}})q(x) + (x - \hat{x})(x - \bar{\hat{x}})q'(x) + s.$$

By substituting $\hat{x}$ for x, we find

$$p'(\hat{x}) = (\hat{x} - \bar{\hat{x}})q(\hat{x}) + s$$

or, since $\hat{x} - \bar{\hat{x}} = 2vi$, we have

$$p'(\hat{x}) = 2vq(\hat{x})i + s. \tag{7.65}$$

An algorithm to use (7.64) and (7.65) to evaluate p and p' at a complex value $\hat{x}$ is given next.

Algorithm to Evaluate p and p' at a Complex Value
$$\hat{x} = u + iv$$

1   $\alpha \leftarrow 2u, \ \beta \leftarrow u^2 + v^2$
2   $b_{n-3} \leftarrow a_{n-1} + \alpha$
3   $c_{n-5} \leftarrow b_{n-3} + \alpha$
4   $b_{n-4} \leftarrow a_{n-2} + \alpha b_{n-3} - \beta$
5   $c_{n-6} \leftarrow b_{n-4} + \alpha c_{n-5} - \beta$
6   For $k = n - 5, n - 6, \ldots, 2$
  6.1 $b_k \leftarrow a_{k+2} + \alpha b_{k+1} - \beta b_{k+2}$ (7.66)
  6.2 $c_{k-2} \leftarrow b_k + \alpha c_{k-1} - \beta c_k$
7   $b_1 \leftarrow a_3 + \alpha b_2 - \beta b_3$
8   $b_0 \leftarrow a_2 + \alpha b_1 - \beta b_2$
9   $s \leftarrow a_1 + \alpha b_0 - \beta b_1, \ t \leftarrow a_0 - \beta b_0$
10 $s' \leftarrow b_1 + \alpha c_0 - \beta c_1, \ t' \leftarrow b_0 - \beta c_0$
11 $\eta \leftarrow su + t, \ \zeta \leftarrow sv$
12 $\eta' \leftarrow s - 2v^2 s', \ \zeta' \leftarrow 2v(s'u + t')$

The (complex) values of $p(\hat{x})$ and $p'(\hat{x})$ are now given by

$$p(\hat{x}) = \eta + i\zeta, \qquad p'(\hat{x}) = \eta' + i\zeta'.$$

The algorithm requires $(4n - 4)A + (4n - 1)M$ real operations.

We conclude this subsection with two theoretical facts about (7.51) that are important for the subsequent discussions. The first, called the fundamental theorem of algebra, states that (7.51) always has a solution, which may, of course, be complex. If x_1 is such a solution, then (7.52) and (7.54), with $\hat{x} = x_1$ implies that $p(x) = (x - x_1)q(x)$, where $q(x)$ is a polynomial of degree $n - 1$. If we apply the fundamental theorem and (7.52), (7.54) to q, we find $q(x) = (x - x_2)q_2(x)$ for some complex number x_2, and q_2 a polynomial of degree $n - 2$. By repeating this process, we obtain finally a **complete factorization of** p:

$$p(x) = (x - x_1)(x - x_2) \cdots (x - x_n). \tag{7.67}$$

The values $x_1, x_2, \ldots, x_n$ need not be distinct. The number of times a particular value x_i appears in (7.67) is just the **multiplicity** of x_i. It is not hard to see (Exercise 12) that if x_i appears m times in (7.67), then

$$p(x_i) = p'(x_i) = \cdots = p^{(m-1)}(x_i) = 0. \tag{7.68}$$

The second fact about (7.51) that will be useful is that if $\hat{x}$ is a *complex* solution, then $\bar{\hat{x}}$ is also a solution to (7.51) (see Exercise 5). Thus if we find a complex solution to (7.51), we obtain another solution for free.

Finally, it is important to stress the point that the task of finding all of the roots of an arbitrary polynomial is difficult and should be avoided if possible. Methods that are easy to understand and simple to program are of limited effectiveness. The good methods that are known to be reliable and accurate are very complicated and require a major programming effort to implement. In addition, polynomials that arise in applications often tend to have very ill-conditioned roots. Perhaps for these reasons, many problems that used to be solved via polynomials are now solved by other means.

7.5.2 Perturbation Results for Polynomials

Perturbation analysis of polynomial roots has been a subject of mathematical study for many years. The general estimates (7.28) hold, of course, for polynomials. As a consequence, we are faced with the fact that multiple roots or clusters of close roots are ill conditioned. Polynomials that occur in applications often do have multiple roots that, because of rounding error, appear as clusters of close roots. Also, small roots tend to be more sensitive to changes in the coefficients than do large roots. An example of a result that makes sense only for polynomials is the following. Suppose r is a simple root of the polynomial p, defined by (7.50). If a particular coefficient, say a_k, is changed, then r will change. Thus we may think of r as being a function of a_k. It can be shown that r is, in fact, a differentiable function of a_k, and that

$$\partial r / \partial a_k = r^k / p'(r). \tag{7.69}$$

Thus we can conclude, for example, that if $|r| > 1$, then r is more sensitive to changes in a_{n-1}, and a_{n-2} than to changes in, say, a_0 and a_1. If $|r| < 1$, then changes in a_0 and a_1 have a stronger effect than do changes in a_{n-1} and a_{n-2}. Whenever $p'(r)$ is close to zero, changes in *any* coefficient can cause a drastic change in r.

To illustrate these results, consider the polynomial of degree 20 whose roots are the integers 1, 2, 3, ..., 20; that is,

$$p(x) = (x - 1)(x - 2) \cdots (x - 20) = x^{20} - 210x^{19} + \cdots + 20! \quad (7.70)$$

TABLE 7.7 *Roots of the Polynomial*
$p(x) - 2^{-23}x^{19}$, *p Given by* (7.53) *Rounded to Five Decimals*

1.00000	2.00000	$10.09527 \pm 0.64350i$
3.00000	4.00000	$11.79363 \pm 1.65233i$
5.00000	6.00001	$13.99236 \pm 2.51883i$
6.99970	8.00727	$16.73074 \pm 2.81262i$
8.91725	20.84691	$19.50244 \pm 1.94033i$

If the coefficient of x^{19} is changed by the amount 2^{-23}, then the roots (correct to six figures) become those listed in Table 7.7. The formula (7.69) indicates that a change in the coefficient of x^{19} will cause the most drastic change in the roots. Also, the larger roots will be affected most. The estimates (7.28) show how bad the perturbation might be. For example,

$$p'(16) = 4! \, 15! \approx 10^{13}$$

so that, with $g(x) = x^{19}$, $\mu = 2^{-23}$, and $x^* = 16$, (7.28a) gives

$$x^* - x^*(\mu) \approx (16^{19}/10^{13})2^{-23} \approx 10^{10}.$$

With $x^* = 1$, (7.28a) implies that

$$x^* - x^*(\mu) \approx (1/19!)2^{-23} \approx 10^{-17}$$

so that the root at 1 is changed very little.

The above example shows that it is important to interpret "clusters of close roots" properly. The roots of the polynomial (7.70) do not appear to be close together, even though they are very ill-conditioned. Notice, however, that the coefficients of p are as large as

$$20! = 1 \cdot 2 \cdot 3 \cdots 20 \approx 2 \times 10^{18}.$$

Compared to numbers of this size, the roots $\{1, 2, ..., 20\}$ are *very* close together.

A special perturbation of p, which will be important in Section 7.5.4, is when only the constant term a_0 is changed. That is, let $\hat{p}(x)$ be the polynomial defined by

$$\hat{p}(x) = p(x) + \varepsilon \tag{7.71}$$

and suppose x^* is a root of p. Let $\hat{x}^*$ be the root of $\hat{p}$ that is closest to x^*. Then, by the mean value theorem (see the Appendix),

$$p(x^*) - p(\hat{x}^*) = p'(\zeta)(x^* - \hat{x}^*) \tag{7.72}$$

for some ζ between x^* and $\hat{x}^*$. Now, $p(x^*) = 0$ and $p(\hat{x}^*) = \hat{p}(\hat{x}^*) - \varepsilon = -\varepsilon$, so that (7.72) gives the result

$$x^* - \hat{x}^* = \varepsilon/p'(\zeta),$$

provided $p'(\zeta) \neq 0$. If $|p'(x)| \geq \alpha > 0$ for x between x^* and $\hat{x}^*$, then a relative bound is

Bounds on perturbation caused by change in constant term	$$|x^* - \hat{x}^*|/|x^*| \leq |\varepsilon|/(\alpha|x^*|). \tag{7.73}$$

Thus the root x^* will be well-conditioned with respect to changes in the constant term of p whenever $p'(x)$ is not small near x^* and x^* is not small. On the other hand, if either $p'(x)$ is near to zero or x^* is close to zero, then x^* will be ill-conditioned.

7.5.3   Some Root-Finding Techniques

All of the methods discussed in Sections 7.1.2–7.1.4 can be used to find a single real or complex root of a polynomial. Clearly, however, in order to converge to a complex root, the iterates must be complex. Thus it is necessary to use a complex number for the starting value. Algorithm (7.64) should be used to evaluate the iteration function at complex values. In the case of Newton's method, Algorithm (7.66) can be used to evaluate both p and p' at complex values. Notice that this algorithm allows us to compute a complex Newton iterate in approximately $4nA + 4nM$ operations. This is compared with $2nA + 2nM$ operations for real iterates. However, as noted earlier, if x^* is a complex root, then so is $\bar{x}^*$. Hence, if a complex iterate x_k is accepted as a root, then $\bar{x}_k$ may be taken as an approximation to another root. Thus the work required to find two complex roots is roughly the same as the work involved in finding two real roots.

The convergence results derived in Sections 7.2.1 and 7.2.2 carry over directly to the complex case. In particular, the local convergence of Newton's method is still valid. Also, the rate of convergence discussion holds just as before. Newton's method, for example, converges quadratically to a simple root and linearly to a multiple root.

A major difficulty with using iterative methods to find complex roots is the choice of starting values. As seen earlier, it is often necessary to choose x_0 fairly close to a root for the iterates to converge. For this we must have some means of finding a region which contains the root. In the case of a real root, we can evaluate p at several points and examine the signs of these values. Such a process will not work for complex roots. That is, even though p has real coefficients, if x is complex, so is $p(x)$, and the *sign* of $p(x)$ is not even defined. There are, however, some very effective but rather complicated search techniques for locating complex roots. These methods can be thought of as generalizations of the interval methods discussed in Section 7.1.5. Now, however, instead of *intervals*, we look for *circles* in the complex plane that contain roots. An example of such a result is the following. Let z_0 be a complex number for which $p'(z_0) \neq 0$. Then the circle in the complex plane with center z_0 and radius

$$r = n|p(z_0)/p'(z_0)| \qquad (7.74)$$

contains at least one root of p (see Fig. 7.12). Notice that r is just n times the "*Newton correction*" to z_0. The proof of results of this type are beyond the scope of this book. As a simple application consider the equation

$$x^4 - 2x^3 + 6x^2 - 8x + 8 = 0.$$

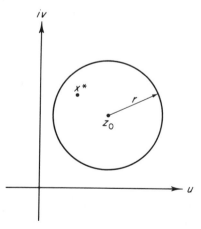

Fig. 7.12 At least one root x^* lies in the circle at z_0, radius $r = n|p(z_0)/p'(z_0)|$.

With $z_0 = 0$, we have $r = 4$ so that the circle centered at zero with radius 4 must contain a root. In fact, this polynomial has roots at $1 \pm i$ and at $\pm 2i$. Hence, we have indeed found a reasonable circle in which to begin searching for roots. Notice, also, that (7.74) gives an excellent **stopping condition for Newton's method**. That is, if the Newton's iterates $x_1, x_2, \ldots$ are computed until $n|x_N - x_{N+1}| \le \varepsilon$, then by applying (7.74) with $z_0 = x_N$, it follows that there is a root x^* with $|x^* - x_N| \le \varepsilon$.

The problem of finding more than one root with an iterative method is made difficult by the fact that the iterates may just keep converging to the same root over and over again. To avoid this, the polynomial must be **deflated** by factoring out the computed roots. This can be done by using Algorithm (7.56) for real roots or (7.64) for complex roots. More precisely, suppose an iterative method has given a real number $\hat{x}$ as an approximate root of p. Algorithm (7.56) will give a polynomial q of degree $n - 1$ that satisfies $p(x) = (x - \hat{x})q(x) + p(\hat{x})$. If $p(\hat{x})$ is small, then the roots of q will be close to the roots of p. If we now apply the iterative method to q, we shall find another (approximate) root of p. The polynomial q can now be deflated and the process repeated until all roots have been found.

If $\hat{x}$ is a complex approximation to a complex root, then we should also accept $\bar{\hat{x}}$ as an approximate root. Both of these approximations can be factored out by using Algorithm (7.64). The result is a real polynomial q of degree $n - 2$. Further (approximate) roots of p can be found by applying the iterative method to q.

The deflated polynomial q will not, of course, have the same roots as p unless $p(\hat{x})$ is actually zero. The fact that generally $p(\hat{x})$ will be small but nonzero means that, by continuing the root-finding process on q, we have introduced an error into the remaining roots. This error will be discussed next.

7.5.4   Numerical Aspects of Deflation

Consider the deflation process defined by (7.52) and computed by Algorithm (7.56). If we continue our root-finding with the polynomial q, then we are, in fact, finding roots of the polynomial

$$\hat{p}(x) = (x - \hat{x})q(x). \tag{7.75}$$

But according to (7.57) and (7.54),

$$\hat{p}(x) = p(x) - p(\hat{x}).$$

Thus roots of q are roots of a polynomial $\hat{p}$ obtained from p by perturbing the constant term by the amount $p(\hat{x})$. The estimate (7.73) shows the effect of this perturbation. We summarize the effect as follows:

> Let $\hat{x}$ be a computed approximation to a root of p, and let q be the polynomial of degree $n - 1$ that satisfies (7.52). If y is any root of q, then there exists a root y^* of p that satisfies
>
> $$|y - y^*|/|y^*| \le |p(\hat{x})|/(|y^*|\alpha), \qquad (7.76)$$
>
> where $0 < \alpha \le |p'(x)|$ for x between y and y^*.

Thus if our root-finding procedure gives a value $\hat{x}$ at which $|p(\hat{x})|$ is small, then well-conditioned roots of p can be found accurately by working with the deflated polynomial q. Of course, ill-conditioned roots of p cannot be found accurately from q because they are not determined accurately by p—the deflation procedure just magnifies this inaccuracy.

The deflation process for two complex roots, as given by Algorithm (7.64), is equivalent to a perturbation of the constant term and the linear term of p. The size of the perturbation is, according to (7.60), given by

$$s\hat{x} + t = p(\hat{x}).$$

Again, we can conclude that if $\hat{x}$ makes $|p(\hat{x})|$ small, then this perturbation will not effect the well-conditioned roots.

In the above discussion, we have ignored the rounding errors during the computation of the coefficients of the deflated polynomial. That is, the Algorithm (7.56) actually computes coefficients $\math'b_0, \math'b_1, \ldots, \math'b_{n-1}$, where

$$\math'b_{k-1} = \mathrm{fl}[a_k + \hat{x}\math'b_k], \qquad k = n - 1, n - 2, \ldots, 1. \qquad (7.77)$$

Here, for ease of notation, we have set $\math'b_{n-1} = 1$. The root-finding process will be continued on the polynomial

$$\hat{q}(x) = x^{n-1} + \math'b_{n-2}x^{n-2} + \cdots + \math'b_0. \qquad (7.78)$$

In order to compare the roots of $\hat{q}$ to the roots of p, define the nth degree polynomial $\hat{p}(x) = x^n + \hat{a}_{n-1}x^{n-1} + \cdots + \hat{a}_1 x + \hat{a}_0$ by

$$\hat{p}(x) = (x - \hat{x})\hat{q}(x). \qquad (7.79)$$

A comparison of the roots of $\hat{p}$ and p, which are different from $\hat{x}$, will be equivalent to a comparison of the roots of $\hat{q}$ and p. Now, by expanding $\hat{p}$, we have

$$\hat{p}(x) = (x - \hat{x})(x^{n-1} + \math'b_{n-2}x^{n-2} + \cdots + \math'b_0)$$
$$= x^n + (\math'b_{n-2} - \hat{x})x^{n-1}$$
$$\quad + (\math'b_{n-3} - \hat{x}\math'b_{n-2})x^{n-2}$$
$$\quad + \cdots + (\math'b_0 - \hat{x}\math'b_1)x - \hat{x}\math'b_0.$$

But, with the notation of Section 2.5.3,

$$\hat{b}_k = \text{fl}[a_{k+1} + \hat{x}\hat{b}_{k+1}]$$
$$= a_{k+1}\langle 1\rangle + \hat{x}\hat{b}_{k+1}\langle 2\rangle$$

so the coefficients $\hat{a}_0, \ldots, \hat{a}_n$ of $\hat{p}$ are, for $k = 0, 1, 2, \ldots, n - 2$,

$$\hat{a}_{k+1} = \hat{b}_k - \hat{x}\hat{b}_{k+1} = a_{k+1}\langle 1\rangle + \hat{x}\hat{b}_{k+1}\langle 2\rangle - \hat{x}\hat{b}_{k+1}$$
$$= (a_{k+1}\langle 1\rangle_0 + \hat{x}\hat{b}_{k+1}\langle 2\rangle_0) + a_{k+1}.$$

Thus we have the estimates

$$|\hat{a}_{k+1} - a_{k+1}| \leq (|a_{k+1}| + 2|\hat{x}\hat{b}_{k+1}|)\mu', \qquad k = 0, 1, \ldots, n - 2.$$
$$(7.80a)$$

For the constant term we note that, from (7.55b, last equation) and (7.54),

$$\text{fl}[p(\hat{x})] = \text{fl}[a_0 + \hat{x}b_0].$$
$$= a_0\langle 1\rangle + \hat{x}b_0\langle 2\rangle$$
$$= a_0 - \hat{a}_0 + a_0\langle 1\rangle_0 + \hat{x}b_0\langle 2\rangle_0.$$

Hence,

$$|a_0 - \hat{a}_0| \leq (|a_0| + 2|\hat{x}b_0|)\mu' + |\text{fl}[p(\hat{x})]|. \qquad (7.80b)$$

Thus the rounding error during the deflation process introduces errors into the coefficients of q that are *equivalent* to perturbations (7.80) of the coefficients of p. These perturbations will be small, provided $\text{fl}[p(\hat{x})]$ is small and $|\hat{x}|$ is not large. Because of the sensitivity of small roots to changes in the coefficients, we state the following rule:

> When polynomial roots are computed one at a time with deflation in between, the roots should be found in *increasing* order of magnitude and with sufficient accuracy that the polynomial evaluated at the approximate roots is small compared to the constant term.

The importance of this rule cannot be overemphasized when p has some roots of small modulus and some with large modulus. If the roots are found in increasing order, then the deflation process preserves the roots very nicely. If large roots are found first, however, then the small roots may be completely destroyed.

To illustrate these points, consider the polynomial

$$p(x) = x^4 - 11.101x^3 + 11.1111x^2 - 1.0111x + .001,$$

which has roots at .001, .1, 1.0, 10.0. Suppose some numerical method has found the approximate root $\hat{x}^* = 10.000005$. The deflation process, with eight-decimal-digit arithmetic, gives

Deflation of a large root

$$b_4 = 1$$
$$b_3 = -11.101 + (10.000005) = -1.100995$$
$$b_2 = 11.1111 + (10.000005)(-1.100995) = .10114400$$
$$b_1 = -1.0111 + (10.000005)(.101144) = .3405000.$$

The deflated polynomial

$$\hat{q}(x) = x^3 - 1.100995x^2 + .101144x + .3405$$

has *no* root close to $x = .001$. In fact, $\hat{q}(.001) = .34051 \ldots$. On the other hand, if the approximate root $\hat{x}^* = .0010000005$ is used to deflate p, the resulting deflated polynomial is

Deflation of a small root $$\hat{q}(x) = x^3 - 11.1x^2 + 11.1x - 1.$$

The polynomial $\hat{p}$ defined by (7.71) is

$$\hat{p}(x) = (x - \hat{x}^*)\hat{q}(x)$$
$$= x^4 - 11.101000\ldots x^3 + 11.111100x^2 - 1.0111000x + .001$$

and this agrees with p to more than eight figures. $\hat{q}$ has roots *very* close to .1, 1.0, and 10.0.

EXERCISES

1   Use Algorithm (7.58b) to evaluate $p(2)$ and $p'(2)$ for the polynomials

   (a) $p(x) = x^4 + 3x^3 - 4x^2 + x - 5$ (b) $p(x) = x^5 + 2x^3 - 5x^2 + 3$.

2   Consider the Horner evaluation Algorithms (3.18) or (7.56) for complex x.
   (a)   Show that if all arithmetic is fully complex (that is, consider a complex multiplication as $2A + 4M$ real operations, and complex addition as $2A$ real operations), then these algorithms require $(4n - 2)A + (4n - 4)M$ real operations. [Assume $a_n = 1$ in Algorithm (3.18).]
   (b)   Suppose the complex arithmetic is coded to take advantage of the fact that a_k are real. (Thus operations of the form $a_k + \hat{x}b_k$ take only one real addition.) Show that the algorithms now require only $(3n - 2)A + (4n - 4)M$ operations.
   (c)   Verify that Algorithm (7.64) requires only $2nA + (2n + 1)M$ real operations.

3   Use Algorithm (7.64) to evaluate $p(1 + i)$ and $p(1 - i)$ for

   (a) $p(x) = x^4 - 7x^3 + 2x^2 - x - 5$ (b) $p(x) = x^6 - 5x^4 - 2x^2 + 5$

4   Write a computer program to implement Algorithm (7.66). Test the program on several

polynomials that you can evaluate by hand. What happens if you give as input a *real* number $\hat{x}$ (that is, a value of 0 for v)?

5 Prove that, if $\hat{x}$ is a complex root of p, then $\bar{\hat{x}}$ is also a root. Does this fact still hold if p has *complex* coefficients?
   [*Hint*: Use the properties: $\bar{a} + \bar{b} = \overline{a + b}$, $\bar{a}\bar{b} = \overline{ab}$, for complex a, b, and $\bar{a} = a$ if a is real.)

6 Let $p(x) = x^n + a_{n-1} x^{n-1} + \cdots + a_1 x + a_0$, where now a_{n-1}, a_{n-2}, $\ldots$, a_1, a_0 are *complex*. Let $\bar{p}(x)$ denote the polynomial with coefficients $\bar{a}_{n-1}, \bar{a}_{n-2}, \ldots, \bar{a}_1, \bar{a}_0$. Show that $P(x) = p(x)\bar{p}(x)$ is a polynomial of degree $2n$ that has *real* coefficients. What is the relation between the roots of P and the roots of p? That is, suppose you find some or all of the roots of P. How are these related to the roots of p?

7 The polynomial $p(x) = x^4 - 6.001x^3 + 12.0004x^2 - 10.0003x + 3$ has roots at $x = 1$ and $x = 3$. Use (7.69) to study the conditioning of these roots. Which root is most sensitive to changes in the constant term? Which is most sensitive to changes in the coefficient of x^3?

8 Write a program to use Newton's method to find roots of a polynomial. Use Algorithm (7.66) to evaluate $p(x_k)$ and $p'(x_k)$, and use (7.74) as a stopping criteria. Once a root has been accepted, deflate it and its conjugate by proceeding with the polynomial whose coefficient are the b_k computed by the Algorithm (7.66).

9 Modify the algorithms (7.56), (7.58), and (7.64) to handle polynomials whose first coefficient a_n is *not* 1.0.

10 Show that Algorithm (7.58b) can also be written:

   1 $p \leftarrow a_{n-1} + \hat{x}$, $pp \leftarrow 1$
   2 For $k = n - 2, n - 3, \ldots, 0$
      2.1 $pp \leftarrow p + \hat{x}pp$
      2.2 $p \leftarrow a_k + \hat{x}p$

11 For the polynomial in Exercise 7, use the deflation Algorithm (7.56) to factor out an approximate root $\hat{x} = 3.00001$. Show that the resulting polynomial has no root close to 1.0.

12 Prove that the number of times the factor $(x - x_i)$ appears in (7.67) is equal to the multiplicity of x_i as a solution to $p(x) = 0$.

Suggestions for Further Reading

General discussions of iterative methods for solving nonlinear equations can be found in Traub (1964) and Ostrowski (1966). The method referred to here as the Dekker–Brent algorithm can be found in Brent (1972). The original version of this algorithm is in the paper by Dekker in the collection edited by Dejon and Henrici (1969). Wilkinson (1963) studies the effects of rounding errors on the solution of nonlinear equations. Methods for solving systems of nonlinear equations can be found in Ortega and Rheinboldt (1970).

Efficient evaluation of polynomials is covered in Knuth (1969). Householder (1970) and Henrici (1977) study methods for finding roots of polynomials.

Chapter 8

ORDINARY DIFFERENTIAL EQUATIONS

8.1 Differential and Difference Equations

Numerical methods for solving differential equations first convert the differen*tial* equation into a differen*ce* equation. In this section we review some basic facts about these two classes of equations. The notation and terminology introduced here will be used throughout the chapter.

8.1.1 Differential Equations

A differential equation is simply an equation that involves an unknown function and some of its derivatives. An equation that involves derivatives of nth order and lower is called an **nth order differential equation**. Almost all nth order equations that occur in applications can be written in the form

$$y^{(n)}(x) = f(x, y(x), y'(x), \ldots, y^{(n-1)}(x)), \qquad a \le x \le b. \qquad (8.1)$$

That is, the nth derivative of the unknown function y can be written explicitly as a function of the independent variable, the function y itself, and lower order derivatives. From now on, when we speak of an nth order equation, we shall mean an equation of the form (8.1).

A **solution to (8.1)** is any function y that is defined on $[a, b]$, has n derivatives on that interval, and satisfies (8.1). It can be shown that under rather weak conditions on f, (8.1) actually has an n-parameter family of solutions. A unique solution can be specified by imposing certain additional conditions on the function y. Conditions of the form

$$y(a) = \eta_1, \qquad y'(a) = \eta_2, \ldots, y^{(n-1)}(a) = \eta_n \qquad (8.2)$$

are called *initial conditions*, and (8.1) with conditions (8.2) is called an *nth order initial value problem*. If some conditions are imposed at $x = a$ and some at $x = b$, then the problem is called a *boundary value problem*.

Thus a general **first order initial value problem** is written as

$$y'(x) = f(x, y(x)), \qquad a \le x \le b$$
$$y(a) = \eta, \tag{8.3}$$

where η is a specified value. An **nth order initial value problem** has the form

$$y^{(n)}(x) = f(x, y, y', y'', \ldots, y^{(n-1)}), \qquad a \le x \le b \tag{8.4a}$$

$$y(a) = \eta_1, \, y'(a) = \eta_2, \ldots, y^{(n-1)}(a) = \eta_n. \tag{8.4b}$$

A second order boundary value problem is given by

$$y''(x) = f(x, y, y'), \qquad a \le x \le b \tag{8.5a}$$

$$y(a) = \eta_1, \qquad y(b) = \eta_2. \tag{8.5b}$$

Most of this chapter will be devoted to solving first order initial value problems (8.3). In Section 8.1.2 we shall briefly discuss nth order initial value problems such as (8.4). Some techniques for solving second-order boundary value problems (8.5) will be described in Section 8.4. It is important to realize that even though equation (8.4) with $n = 2$ and equations (8.5) appear to be very similar, they arise in quite different applications and require *totally* different solution techniques.

An example of the kind of condition that guarantees a unique solution to (8.3) is that f has first partial derivatives with respect to both x and y. In the sequel we shall always assume that f satisfies a condition of this type. In fact, for many of the numerical methods, it is necessary to assume that the solution function has several continuous derivatives. Such assumptions can hold only if f is differentiable. That is, if y satisfies (8.3), then the chain rule for differentiation gives

$$y''(x) = df(x, y(x))/dx$$
$$= \partial f(x, y(x))/\partial x + (\partial f(x, y(x))/\partial y)(dy/dx)$$
$$= f_x(x, y(x)) + f_y(x, y(x))f(x, y(x)). \tag{8.6}$$

In the last line of (8.6) we have used the fact that $dy/dx = f(x, y(x))$ and, of course, f_x, f_y denote the partial derivatives of f. Equation (8.6) shows that y

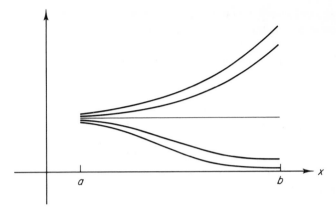

Fig. 8.1   Example of an unstable solution.

has a continuous second derivative only if f has continuous first partial derivatives.

The fact that (8.3a) defines a one-parameter family of functions has important numerical consequences. This family of functions is called the **solution field** for the equation. The initial condition (8.3b) picks out a particular member of this solution field. Suppose, however, that a small change in the initial condition η were to cause a drastic change in the solution, as in Fig. 8.1. In this case the *problem* (8.3) would certainly be considered ill conditioned. If, for example, η is not a floating point number, then the best that any numerical method can do is to solve (8.3), but with η replaced by $\hat{\eta} = \mathrm{fl}[\eta]$. If this replacement changes the solution, then the problem is ill conditioned. An example of an ill-conditioned initial value problem was given in Section 1.2.

One should not be misled into thinking that ill-conditioning is only caused by sensitivity of the solution to the value given at $x = a$. Suppose that the equation has been solved up to a point c between a and b. Then one can think of solving the equation $y'(x) = f(x, y(x)), c \leq x \leq b$, with $y(c)$ equal to the value just determined. If the rest of the solution is very sensitive to the value at $x = c$, then the equation is again ill conditioned.

8.1.2   Systems of Differential Equations

A system of first order differential equations has the general form

$$
\left.
\begin{aligned}
y_1'(x) &= f_1(x,\, y_1,\, y_2,\, \ldots,\, y_n) \\
y_2'(x) &= f_2(x,\, y_1,\, y_2,\, \ldots,\, y_n) \\
&\ \ \vdots \\
y_n'(x) &= f_n(x,\, y_1,\, y_2,\, \ldots,\, y_n)
\end{aligned}
\right\}
\quad a \leq x \leq b.
\tag{8.7a}
$$

Here $f_1, f_2, \ldots, f_n$ are given functions of $n + 1$ variables. If (8.7a) has a solution, then generally there will be an n-parameter *family* of solutions. A unique solution is determined by specifying n auxiliary conditions. As was the case for a single equation, if the conditions have the form

$$y_k(a) = \eta_k, \qquad k = 1, 2, \ldots, n. \tag{8.7b}$$

then the problem is called an **initial value problem**.

Initial value problems for systems of equations arise in many applications. The methods that will be derived for solving a single equation can be easily modified to solve a system of the form (8.7). An important consequence of this fact is that we can then solve nth order initial value problems of the form (8.4). To see how this is done, define a set of n functions $y_1, y_2, \ldots, y_n$ by

$$
\begin{aligned}
y_1(x) &= y(x) \\
y_2(x) &= y'(x) \\
&\;\vdots \\
y_n(x) &= y^{(n-1)}(x).
\end{aligned}
\tag{8.8}
$$

Then (8.4a) becomes

$$y_n'(x) = f(x, y_1, y_2, \ldots, y_n). \tag{8.9a}$$

By differentiating (8.8), we also obtain the equations

$$
\begin{aligned}
y_1'(x) &= y_2(x) \\
y_2'(x) &= y_3(x) \\
&\;\vdots \\
y_{n-1}'(x) &= y_n(x).
\end{aligned}
\tag{8.9b}
$$

Thus (8.9a) and (8.9b) together give a system of n first order equations. The initial conditions (8.4b) become initial conditions (8.7b) for the system (8.9).

To illustrate these ideas, consider the **third order equation**

$$
\begin{aligned}
y''' &= xy'(x) + e^x y(x) + x^2 + 1, \qquad 0 \le x \le 1 \\
y(0) &= 1, \qquad y'(0) = 0, \qquad y''(0) = -1.
\end{aligned}
\tag{8.10a}
$$

An **equivalent system** of first order equations is

$$
\left.
\begin{aligned}
y_1'(x) &= y_2(x) \\
y_2'(x) &= y_3(x) \\
y_3'(x) &= xy_2(x) + e^x y_1(x) + x^2 + 1
\end{aligned}
\right\} \; 0 \le x \le 1 \tag{8.10b}
$$

$$y_1(0) = 1, \qquad y_2(0) = 0, \qquad y_3(0) = -1.$$

The methods of Section 8.2 can be applied to systems of the form (8.10b). The approximations to $y_1(x)$ so obtained will approximate the solution to (8.10a).

8.1.3 Difference Equations

An **nth order difference equation** is a sequence of equations of the form

$$g_k(y_{k+n}, y_{k+n-1}, \ldots, y_k) = 0, \qquad k = 0, 1, 2, \ldots$$

$$y_i = \eta_i, \qquad i = 0, 1, \ldots, n-1. \tag{8.11}$$

Here g_k are given functions of $n+1$ variables and the values $\eta_0, \eta_1, \ldots, \eta_{n-1}$ are specified. A *solution* to such an equation is a sequence $\{y_0, y_1, \ldots, y_{n-1}, y_n, y_{n+1}, \ldots\}$ that satisfies (8.11). A special form of (8.11) that will be of interest is

$$\alpha_n y_{k+n} + \alpha_{n-1} y_{k+n-1} + \cdots + \alpha_0 y_k = 0, \qquad k = 0, 1, \ldots \tag{8.12a}$$

$$y_i = \eta_i, \qquad i = 0, 1, \ldots, n-1. \tag{8.12b}$$

Here the g_k in (8.11) are independent of k and are linear homogeneous functions of all variables. For this reason (8.12) is called a **linear homogeneous difference equation with constant coefficients.** We shall always assume, when writing (8.12a), that $\alpha_n \neq 0$, so that (8.12) is an nth order equation.

Some examples of equations of the form (8.12) are

$$y_{k+1} - y_k = 0$$
$$y_0 = 1 \tag{8.13a}$$

$$y_{k+3} - 2y_{k+2} - y_{k+1} + 2y_k = 0$$
$$y_0 = 0, \qquad y_1 = -3, \qquad y_2 = 1 \tag{8.13b}$$

$$2y_{k+2} - 17y_{k+1} + 8y_k = 0$$
$$y_0 = 2, \qquad y_1 = 1. \tag{8.13c}$$

Since we always assume that $\alpha_n \neq 0$ in (8.12), the equation can be solved for y_{k+n}:

$$y_{k+n} = (-1/\alpha_n)[\alpha_{n-1} y_{k+n-1} + \cdots + \alpha_1 y_{k+1} + \alpha_0 y_k], \qquad k = 0, 1, \ldots. \tag{8.14}$$

This gives a **recursive method** for finding the solution. That is, since y_0,

$y_1, \ldots, y_{n-1}$ are given as part of the equation, (8.14) with $k = 0$ can be used to compute y_n. Then, by using this value in (8.14) with $k = 1$, we determine y_{n+1}, and so forth. For example, if (8.13c) is written in the form (8.14), it becomes

$$y_{k+2} = 8.5y_{k+1} - 4y_k.$$

The solution to (8.13c) can now be computed recursively as

$$y_0 = 2, \qquad y_1 = 1$$

> **Recursive solution to (8.13c)**

$$y_2 = 8.5y_1 - 4y_0 = .5$$

$$y_3 = 8.5y_2 - 4y_1 = .25$$

$$y_4 = 8.5y_3 - 4y_2 = .125.$$

By writing the linear difference equation in the form (8.14), it is clear that one can choose *any* set of values $\{\eta_0, \eta_1, \ldots, \eta_{n-1}\}$ to determine the solution to (8.12). Thus (8.12a) actually defines an n-parameter family of solutions. A particular solution is specified by giving values for $\eta_0, \eta_1, \ldots, \eta_{n-1}$. Just as in the case of differential equations, it is reasonable to consider the sensitivity of the solution to changes in the initial values $\eta_0, \ldots, \eta_{n-1}$. That is, if small changes in these values produce large changes in, say, y_{100}, then the difference equation would be called unstable or ill-conditioned. For example, consider again the difference equation (8.13c). With the given starting values (that is, with $y_0 = 2$, $y_1 = 1$), the value of y_k for $k = 10$ is $y_{10} = .001953125$. However, if the starting values are changed to $y_0 = 2.000001$, $y_1 = 1.000008$, then the value at $k = 10$ becomes $y_{10} = 1073.742\ldots$. Thus, a change of 10^{-6} in the data has caused a change of more than 10^6 in the solution at $k = 10$. The difference equation (8.13c) is extremely ill conditioned.

As was also the situation with differential equations, this ill-conditioning is not limited to the sensitivity to changes in the starting values $\eta_0, \ldots, \eta_{n-1}$ only. In (8.13c), for example, even if accurate values for y_{10} and y_{11} have been found, the problem is then to solve the equation

$$2y_{k+2} - 17y_{k+1} + 8y_k = 0, \qquad k = 10, 11, 12, \ldots$$

$$y_{10}, y_{11} \quad \text{given.}$$

Small perturbations in the values of y_{10} and y_{11} will cause large changes in, say, y_{20}.

As mentioned in the introduction to this section, numerical methods for solving differential equations first replace the differential equation by a difference equation. Then the difference equation is solved, usually by the recursion method illustrated by (8.14). If the original differential equation is unstable, nothing much can be done. But if a *stable differential equation* is replaced by an *ill-conditioned difference equation*, something has clearly gone awry. Much of the theory of numerical solution of differential equations is directed toward obtaining difference equations that are good approximations to the corresponding differential equation but are at least as well conditioned as was the differential equation itself.

8.1.4   Solutions to Difference Equations

The recursive technique described above can and is used to solve linear difference equations. However, in order to analyze the stability of the difference equation, it is necessary to have a clearer relation between the solution and the starting values η_0, η_1, ..., η_{n-1} than can be obtained from the recursion formula (8.14). In fact, what is really needed is a functional relationship between the starting values and the solution; that is, we need a formula such as

$$y_k = F(\eta_0, \eta_1, \ldots, \eta_{n-1}, k), \qquad k = 0, 1, \ldots .$$

For equations of the form (8.12), such a formula is easily obtained. For this purpose suppose we try a solution of the form $y_k = \lambda^k$, $k = 0, 1, \ldots$, where λ is some constant to be determined. If this is substituted into (8.12a), then we require

$$0 = \alpha_n \lambda^{k+n} + \alpha_{n-1} \lambda^{k+n-1} + \cdots + \alpha_0 \lambda^k = \lambda^k (\alpha_n \lambda^n + \cdots + \alpha_0).$$

For $\lambda \neq 0$ this is equivalent to $p(\lambda) = 0$, where $p(\lambda)$ is the polynomial

$$p(\lambda) = \alpha_n \lambda^n + \alpha_{n-1} \lambda^{n-1} + \cdots + \alpha_1 \lambda + \alpha_0 . \tag{8.15}$$

This says that if λ is a root of p, then $y_k = \lambda^k$ satisfies (8.12a). The polynomial p in (8.15) is called the **characteristic polynomial** of the difference equation (8.12). Suppose next that λ_1 and λ_2 are two roots of (8.15). Then one can easily verify that, for any constants a_1, a_2, the sequence $y_k = a_1 \lambda_1^k + a_2 \lambda_2^k$, $k = 0, 1, \ldots$ also satisfies the difference equation. We leave this as an exercise.

To illustrate these ideas, consider (8.13c). The characteristic polynomial is

$$p(\lambda) = 2\lambda^2 - 17\lambda + 8.$$

The roots of this polynomial are $\lambda_1 = .5$ and $\lambda_2 = 8$. Thus, for any a_1, a_2,

$$y_k = a_1 (.5)^k + a_2 (8)^k$$

satisfies the difference equation. Furthermore, the initial values can be satisfied by proper choice of a_1 and a_2. That is, we shall have $y_0 = 2$, provided

$$y_0 = a_1 (.5)^0 + a_2 (8)^0 = a_1 + a_2 = 2$$

and $y_1 = 1$ if

$$y_1 = a_1(.5)^1 + a_2(8)^1 = .5a_1 + 8a_2 = 1.$$

This gives two equations for the parameters a_1, a_2. The solution $a_1 = 2$, $a_2 = 0$ is easily found so that the solution to the difference equation is

$$y_k = 2(.5)^k, \qquad k = 0, 1, 2, \dots.$$

The procedure just used to solve (8.13c) can be used in general to solve difference equations of the form (8.12). If the characteristic polynomial has roots $\lambda_1, \lambda_2, \dots, \lambda_n$, then for any choice of parameters $a_1, a_2, \dots, a_n$, the values y_k, $k = 0, 1, 2, \dots$, given by

> **General solution**
$$y_k = a_1 \lambda_1^k + a_2 \lambda_2^k + \cdots + a_n \lambda_n^k \tag{8.16}$$

satisfy (8.12a). To obtain the proper starting values, that is, to satisfy (8.12b), we choose $a_1, a_2, \dots, a_n$ according to the equations

$$y_0 = a_1 + a_2 + \cdots + a_n = \eta_0$$
$$y_1 = a_1 \lambda_1 + a_2 \lambda_2 + \cdots + a_n \lambda_n = \eta_1$$
$$\vdots$$
$$y_{n-1} = a_1 \lambda_1^{n-1} + a_2 \lambda_2^{n-1} + \cdots + a_n \lambda_n^{n-1} = \eta_{n-1}.$$

This system will always have a unique solution, provided only that the roots $\lambda_1, \dots, \lambda_n$ are all distinct. The case where the characteristic polynomial has less than n distinct roots will not arise in our applications of this theory; hence, we shall not consider this possibility. (See, however, Exercises 5 and 6 for more details.)

8.1.5 Stability of Difference Equations

Consider again the difference equation (8.13c). Its solution was shown to be

$$y_k = a_1(.5)^k + a_2 8^k \tag{8.17a}$$

with $a_1 = 2$, $a_2 = 0$; that is

> **Solution to (8.13c)**
$$y_k = 2(.5)^k. \tag{8.17b}$$

The values of a_1 and a_2 were determined by the initial conditions $y_0 = 2$,

$y_1 = 1$. If, instead of these initial values, we had been given, say, $y_0 = 2.000001$ and $y_1 = 1.000008$, then we would have $a_1 = 2$, $a_2 = 10^{-6}$. That is,

$$2.000001 = y_0 = a_1(.5)^0 + a_2(8)^0$$

$$1.000008 = y_1 = a_1(.5)^1 + a_2(8)^1$$

has the unique solution $a_1 = 2$, $a_2 = 10^{-6}$, so that the solution is now

Perturbed solution

$$y_k = 2(.5)^k + 10^{-6}8^k. \tag{8.18}$$

By comparing (8.18) with (8.17b) we see that the small change in the initial values has caused a change of $10^{-6}\, 8^k$ in the values of y_k. Even for moderate sizes of k, this change is very large. This explains the ill-conditioning of this equation as discussed in Section 8.1.3.

More generally, consider the formula (8.16) and suppose that the initial values $\eta_0, \eta_1, \ldots, \eta_{n-1}$ lead to a value $a_1 = 0$. Now small changes in these initial values may cause a_1 to become nonzero, thus introducing the term $a_1 \lambda_1^k$ into the solution. If $|\lambda_1| < 1$, then this change will be insignificant; in fact, it will go to zero as k increases. On the other hand, if $|\lambda_1| > 1$, then changing a_1 from zero to nonzero causes a perturbation in the solution by the quantity $a_1 \lambda_1^k$ that goes to infinity as k increases—*no matter how small* a_1 *is*. If $|\lambda_1| = 1$, then the perturbation will not grow, nor will it die out. If, however, the solution *itself* is decreasing in magnitude, then the *relative* perturbation will increase.

Generally, we say that a solution to a difference equation is *stable* if it is well conditioned in the numerical sense. That is, small changes in the data should produce only small changes in the solution. If the solution is decreasing, however, then the meaning of "small changes" in this statement may have to be interpreted in the sense of *absolute error*. A precise definition of stability for difference equations is rather complicated. For our purposes, the following will suffice.

Definition 8.1 A solution to the difference equation (8.12) that involves only the roots $\lambda_1, \lambda_2, \ldots, \lambda_m$ of the characteristic polynomial is **stable** if all of the other roots $\lambda_{m+1}, \ldots, \lambda_n$ satisfy $|\lambda_k| < 1$ for $k = m + 1, \ldots, n$.

This definition corresponds to the notion of *strong stability* as used in more advanced books on numerical solutions of differential equations. This terminology is used in order to distinguish a concept of *weak* stability, in which certain of the other roots $\lambda_{m+1}, \ldots, \lambda_n$ are allowed to have modulus equal to 1.

8.1.6   Approximating Differential Equations

All of the numerical methods to be discussed in this chapter begin by first approximating the differential equation by a difference equation. A numerical solution to the difference equation is then determined. The first part of this process introduces a truncation error, while the second stage involves rounding errors. In order to illustrate how a differential equation can be approximated by a difference equation, consider the equation (8.3) and the approximation

$$y'(x) \cong [y(x + h) - y(x)]/h, \tag{8.19}$$

where h is some small but fixed parameter. We introduce the notation

$$x_k = a + kh, \qquad k = 0, 1, \ldots,$$

so that $a = x_0 < x_1 < x_2 < \cdots$. If we now let y_k, $k = 0, 1, \ldots,$ represent approximations to $y(x_k)$, $k = 0, 1, \ldots,$ where $y(x)$ is the solution to (8.3), then (8.19) suggests that

$$y'(x_k) \cong (y_{k+1} - y_k)/h.$$

Thus the differential equation (8.3) becomes

$$(y_{k+1} - y_k)/h = f(x_k, y_k), \qquad k = 0, 1, \ldots,$$

with $y_0 \cong y(x_0) = y(a)$ suggesting that $y_0 = \eta$. In other words, we obtain the difference equation

$$y_{k+1} = y_k + hf(x_k, y_k), \qquad k = 0, 1, \ldots \tag{8.20}$$

$$y_0 = \eta.$$

This is called **Euler's method** for solving the differential equation (8.3). Note that (8.20) is a first order difference equation of the form (8.11) with $n = 1$. It is not, however, a linear equation unless $f(x, y)$ is linear in y. For any f, however, the solution can be easily computed recursively. This will be done in Algorithm (8.22) below.

We can compare (8.20) with (8.3) by using Taylor's theorem. That is, in (8.20) y_{k+1} is an approximation to $y(x_{k+1})$, where $y(x)$ satisfies (8.3). Suppose we expand $y(x_{k+1})$ in a Taylor series:

$$y(x_{k+1}) = y(x_k + h) = y(x_k) + hy'(x_k) + \tfrac{1}{2}h^2 y''(\xi_k). \tag{8.21a}$$

Here ξ_k is some value between x_k and x_{k+1}. Now y satisfies (8.3) so that $y'(x_k) = f(x_k, y(x_k))$, and (8.21a) can be written as

$$y(x_{k+1}) = y(x_k) + hf(x_k, y(x_k)) + \tfrac{1}{2}h^2 y''(\xi_k). \tag{8.21b}$$

By comparing (8.21b) with (8.20), we see that Euler's method is obtained by omitting the term $(\frac{1}{2}h^2)y''(\xi_k)$ in the expansion of $y(x_{k+1})$ at the point x_k.

The methods presented in this chapter can all be thought of as having been obtained by expanding the solution function in a Taylor series at a particular point, and then omitting certain terms in the expansion. The omitted terms account for the truncation error. The size of the truncation error will be shown in Section 8.3 to be related to the accuracy of the computed approximations y_k. Hence it is important to be able to estimate this truncation error. Algorithms for solving differential equations should use an estimate for the truncation error to adjust the size of the steplength h. This will be discussed more fully in Section 8.4.2. For now, we illustrate method (8.20) by the following partial algorithm:

Euler's Method Algorithm

$$\left[\begin{array}{l}\text{Algorithm for Euler's method to solve (8.3). Input should include}\\ a, b, \eta, \text{ an initial value for } h, \text{ and a subprogram to evaluate } f(x, y).\end{array}\right]$$

1	Input $\{a, b, \eta, h\}$
2	$x \leftarrow a, y \leftarrow \eta$
3	While $x \le b$
	3.1 $y \leftarrow y + hf(x, y)$
	3.2 $x \leftarrow x + h$
	3.3 Output $\{x, y\}$
	3.4 Estimate the truncation error and adjust h
4	Halt

(8.22)

The reason for adjusting h in step 3.4 is to control the accuracy of the computed approximation. Thus the input to the algorithm may also include an indication of the desired accuracy.

EXERCISES

1   Verify that the following differential equations have the indicated solutions:

(a)
$$y'(x) = x^{-1/3}, \qquad -1 \le x \le 1$$
$$y(-1) = \tfrac{3}{2}$$
Solution: $y(x) = \tfrac{3}{2}x^{2/3}$

(b)
$$y'(x) = -y \tan x, \qquad 0 \le x \le \pi$$
$$y(0) = 1$$
Solution: $y(x) = \cos x$

(c)
$$y'(x) = y^2, \qquad 0 \le x \le 2$$
$$y(0) = 1$$
Solution: $y(x) = 1/(1 - x)$.

[Note that the *equations* (a) and (b) have discontinuities in the interval, but the *solution* is well-behaved. Equation (c) is continuous but the solution has a singularity at $x = 1$.]

2   Write the following initial value problems as systems of first order equations:

(a) $y''(x) = 2y'(x) + y(x)^2 + 2x$, $0 \leq x \leq 1$
      $y(0) = 1$, $y'(0) = 0$

(b) $xy''(x) + x^2 y'(x) + x^3 y(x) = 3$, $1 \leq x \leq 2$
      $y(1) = 1$, $y'(1) = 1$.

3   Write out the characteristic polynomials for the difference equations (8.13a) and (8.13b). Find the roots of these polynomials.
    [*Hint*: $\lambda = 1$ is a root of both polynomials.]
    What is the solution that satisfies the initial conditions?

4   Let λ_1 and λ_2 be two roots of the characteristic polynomial (8.15). Show that for any constants a_1, a_2 the sequence $y_k = a_1 \lambda_1^k + a_2 \lambda_2^k$, $k = 0, 1, 2, \ldots$, satisfies the difference equation (8.12a).

5   Let $p(\lambda)$ be the characteristic polynomial for the difference equation (8.12a) and suppose that $p(\lambda_1) = p'(\lambda_1) = 0$. That is, suppose λ_1 is a root of multiplicity 2. Show that $y_k = \lambda_1^k$ and $\hat{y}_k = k\lambda_1^k$ are solutions to the difference equation.

6   Let $p(\lambda)$ be the characteristic polynomial for the difference equation (8.12a). Suppose that λ_1, $\lambda_2, \ldots, \lambda_m$ are the *distinct* roots of p, with λ_i a root of multiplicity m_i, $i = 1, 2, \ldots, m$. That is, $p(\lambda_i) = p'(\lambda_i) = \cdots = p^{(m_i - 1)}(\lambda_i) = 0$, but $p^{(m_i)}(\lambda_i) \neq 0$. Show that

$$\lambda_i^k, \; k\lambda_i^k, \; k^2\lambda_i^k, \; \ldots, \; k^{m_i - 1}\lambda_i^k, \qquad i = 1, 2, \ldots, m,$$

give n distinct solutions to the difference equation.

7   Use Algorithm (8.22) with $h = .1$ fixed (i.e., ignore step 3.4) to compute an approximate solution to the equation $y'(x) = y(x)$, $0 \leq x \leq .5$, $y(0) = 1$. Compare your results with the exact solution $y(x) = e^x$.

8.2   Basic Methods

In this section we derive the most popular methods for solving equations of the form (8.3). These methods differ in the size of their truncation error, the way in which they are affected by rounding errors, and the manner in which they are implemented as computer programs.

8.2.1   Runge–Kutta Methods

We begin by discussing a class of methods that approximate the differential equation (8.3) by a first order difference equation. The simplest of these so-called *Runge–Kutta methods* is Euler's method (8.20). Since this method was obtained by omitting terms in the expansion of $y(x_{k+1})$ that contained powers of h greater than one, it is called a *first order method*.

In order to obtain higher order methods, we must expand $y(x_{k+1})$ up to terms involving h^p for $p > 1$. In order to simplify these derivations, it is convenient to introduce a special notation for such terms. Specifically, we shall use

$$O(h^p)$$

to represent *any* function $Q(h)$ that has the property that $|Q(h)/h^p|$ is bounded as h goes to zero. For example, we can

$$\begin{array}{llll}
\text{denote} & 2h^3 & \text{by} & O(h^3) \\
\text{denote} & 3h^2 + 5h^4 & \text{by} & O(h^2) \\
\text{denote} & -287h^4 e^{-h} & \text{by} & O(h^4) \\
\text{denote} & h^2 O(h) & \text{by} & O(h^3) \\
\text{denote} & O(h^2) + O(h^2) & \text{by} & O(h^2) \quad \text{and} \\
\text{denote} & O(h^2)O(h^3) & \text{by} & O(h^5).
\end{array} \qquad (8.23)$$

The justification of these statements will be left as an exercise (Exercise 13).

If we assume that $y(x)$ is three times continuously differentiable, then Taylor's theorem gives

$$y(x_{k+1}) = y(x_k) + hy'(x_k) + \tfrac{1}{2}h^2 y''(x_k) + \tfrac{1}{6}h^3 y'''(\xi_k) \qquad (8.24a)$$

for some ξ_k between x_{k+1} and x_k. By using the notation just introduced, together with $y'(x_k) = f(x_k, y(x_k))$, we see that

$$y(x_{k+1}) = y(x_k) + hf(x_k, y(x_k)) + \tfrac{1}{2}h^2 \, df(x, y(x))/dx \Big|_{x=x_k} + O(h^3).$$
$$(8.24b)$$

The derivative of $f(x, y(x))$ with respect to x is given by formula (8.6). However, in order to avoid partial derivatives of $f(x, y)$, we use the approximation

$$df(x, y(x))/dx = [f(x + h, y(x + h)) - f(x, y(x))]/h + O(h). \quad (8.25)$$

This is just formula (5.3) with $O(h)$ representing the truncation error (5.4) when $h = x_1 - x_2$. If (8.25) with $x = x_k$ is substituted into (8.24b), we find

$$y(x_{k+1}) = y(x_k) + hf(x_k, y(x_k))$$
$$+ \frac{h^2}{2}\left[\frac{f(x_{k+1}, y(x_{k+1})) - f(x_k, y(x_k))}{h}\right] + O(h^3). \quad (8.26)$$

Here the term represented by $O(h^3)$ includes the $O(h^3)$ term in (8.24b) as well as the term $(h^2/2)O(h)$, which comes from (8.25). If we ignore the $O(h^3)$ term in (8.26), we are left with the difference equation

$$y_{k+1} = y_k + hf(x_k, y_k) + \frac{h^2}{2}\left(\frac{f(x_{k+1}, y_{k+1}) - f(x_k, y_k)}{h}\right),$$

which simplifies to

$$y_{k+1} = y_k + \tfrac{1}{2}h(f(x_{k+1}, y_{k+1}) + f(x_k, y_k)). \tag{8.27}$$

This is a *second order* approximation to the differential equation (8.3). For reasons that will be explained in Section 8.2.2, it is called the **trapezoidal method**. Unfortunately (8.27) cannot be used easily to compute y_1, y_2, ... recursively because y_{k+1} occurs on both sides of the formula. If $f(x, y)$ is nonlinear in y, then (8.27) is a nonlinear equation for y_{k+1}. Formulas of this type are called **implicit methods** and will be discussed in more detail in Section 8.2.3.

For now we will show how an **explicit** expression for y_{k+1} in terms of y_k can be obtained. Assume that $f(x, y)$ is continuously differentiable with respect to y and consider the following expansion

$$f(x_{k+1}, y(x_{k+1})) = f(x_{k+1}, y(x_k) + hy'(x_k) + O(h^2))$$
$$= f(x_{k+1}, y(x_k) + hy'(x_k)) + O(h^2). \tag{8.28}$$

Here we have first expanded $y(x_{k+1})$ in a Taylor series and then have expanded f at the point $(x_{k+1}, y(x_k) + hy'(x_k))$. If next we substitute (8.28) into (8.26) we find, after rearranging terms, that

$$y(x_{k+1}) = y(x_k) + \tfrac{1}{2}h[f(x_k, y(x_k))$$
$$+ f(x_{k+1}, y(x_k) + hf(x_k, y(x_k))] + O(h^3). \tag{8.29}$$

Now by dropping the $O(h^3)$ term, we obtain an *explicit* formula for y_{k+1}:

$$y_{k+1} = y_k + \tfrac{1}{2}h[f(x_k, y_k) + f(x_{k+1}, y_k + hf(x_k, y_k))]. \tag{8.30}$$

This is called the **Runge–Kutta formula of second order**.

Higher order Runge–Kutta formulas can be derived in exactly this same manner. There are many ways to expand the derivatives of $f(x, y(x))$ and hence it is possible to derive a great many Runge–Kutta methods. One of the most popular of these, because of its accuracy and simplicity, is defined by the formulas:

$$y_{k+1} = y_k + \tfrac{1}{6}[a_1 + 2a_2 + 2a_3 + a_4], \qquad k = 0, 1, \dots$$

Fourth order Runge–Kutta method	$a_1 = hf(x_k, y_k)$

$$a_1 = hf(x_k, y_k)$$
$$a_2 = hf(x_k + \tfrac{1}{2}h, y_k + \tfrac{1}{2}a_1)$$
$$a_3 = hf(x_k + \tfrac{1}{2}h, y_k + \tfrac{1}{2}a_2) \tag{8.31a}$$
$$a_4 = hf(x_k + h, y_k + a_3).$$

It can be shown (Exercise 2), by expanding f as it appears in the definition of a_2, a_3, and a_4 in Taylor's series around the point (x_k, y_k), that (8.31a) is obtained by omitting $O(h^5)$ terms. Hence (8.31a) is a *fourth order* method.

This method is used rather frequently because of its good accuracy and the ease with which it can be programmed. To illustrate this latter fact, we give the following algorithm:

Algorithm for Fourth Order Runge–Kutta Method

$\bigg[$ Algorithm to solve (8.3) by the fourth order Runge–Kutta method (8.31a). Input should include a, b, η, an initial value for h, and a subprogram to evaluate $f(x, y)$. $\bigg]$

1 Input $\{a, b, \eta, h\}$
2 $x \leftarrow a, y \leftarrow \eta$
3 While $x + h \leq b$
  3.1 $a_1 \leftarrow hf(x, y)$
  3.2 $a_2 \leftarrow hf(x + .5h, y + .5a_1)$
  3.3 $a_3 \leftarrow hf(x + .5h, y + .5a_2)$ (8.31b)
  3.4 $a_4 \leftarrow hf(x + h, y + a_3)$
  3.5 $y \leftarrow y + \frac{1}{6}(a_1 + 2a_2 + 2a_3 + a_4)$
  3.6 $x \leftarrow x + h$
  3.7 Output $\{x, y\}$
  3.8 Estimate the truncation error and adjust h
4 Halt

The most troublesome part of this algorithm is step 3.8, as we shall see in Section 8.3.2. Another serious drawback is that steps 3.1, 3.2, 3.3, and 3.4 require $f(x, y)$ to be evaluated *four* times in order to compute the new value of y. This multiple evaluation of f is avoided by the methods that we discuss next.

8.2.2 Multistep Methods

The Runge–Kutta methods use an approximation to $y(x_k)$ to obtain an approximation to $y(x_{k+1})$. However, if we have already obtained several (approximate) values for $y(x)$, say $y(x_k)$, $y(x_{k-1})$, $y(x_{k-2})$, ..., then it is reasonable to use more than just $y(x_k)$ to determine $y(x_{k+1})$. A method that uses n approximate values of $y(x)$ to compute the next value is called an *n*-**step method**. Thus the Runge–Kutta methods are all *one-step* methods. For $n > 1$ we sometimes call the method a *multistep* method.

A one-step method uses the value $y_0 = \eta$ to compute y_1, which is used to find y_2, and so on. A two-step method, however, requires values of y_0 *and* y_1 in order to determine y_2. Clearly, a two-step method cannot be used to find

y_1 because not enough information is available. Generally, an n-step method must have values for $y_0, y_1, \ldots, y_{n-1}$ in order to get started. These **starting values** must be computed by one-step methods, or by combinations of methods that can use the values that are currently available. In the following discussion, we shall assume that some starting method, such as that of Runge–Kutta, has been used to get the necessary initial values.

The easiest way to derive multistep methods is to write the differential equation (8.3) as an **integral equation**:

$$y(x + t) = y(x) + \int_x^{x+t} f(s, y(s)) \, ds, \qquad a \le x < x + t \le b. \quad (8.32)$$

We now replace the integral by a numerical integration formula. For example, if we let $t = h$ and then apply the **rectangle rule** (see Exercise 6, Section 5.2), we have

$$\int_x^{x+h} f(s, y(s)) \, ds = hf(x, y(x)) + \frac{h^2}{2} f'(\xi, y(\xi)), \quad (8.33)$$

where ξ is some point between x and $x + h$. Thus (8.32) becomes

$$y(x + h) = y(x) + hf(x, y(x)) + \tfrac{1}{2}h^2 f'(\xi, h(\xi)). \quad (8.34)$$

With $x = x_k$ so that $x + h = x_{k+1}$, this leads to the difference equation

$$y_{k+1} = y_k + hf(x_k, y_k). \quad (8.35)$$

This is, in fact, just Euler's method (8.20) again. If in (8.32) the **trapezoidal rule**

$$\int_x^{x+h} f(t) \, dt = \frac{h}{2}(f(x) + f(x + h)) - \frac{f''(\xi)}{12} h^3$$

is used [see (5.19), (5.21)], we also obtain a formula derived earlier, namely, (8.27). Other formulas can be derived by using more complicated numerical itegration formulas. For example, **Simpson's rule** [(5.26) with $a = x$, $b = x + 2h$, $c = x + h$, $b - a = 2h$] is

$$\int_x^{x+2h} f(s, y(s)) \, ds = \frac{h}{3}[f(x, y(x)) + 4f(x + h, y(x + h))$$

$$+ f(x + 2h, y(x + 2h))]$$

$$- \tfrac{1}{90} f^{(4)}(\xi, y(\xi))h^5 \quad (8.36)$$

and leads to the equation

$$y(x + 2h) = y(x) + \tfrac{1}{3}h[f(x, y(x)) + 4f(x + h, y(x + h))$$

$$+ f(x + 2h, y(x + 2h))] - \tfrac{1}{90} f^{(4)}(\xi, y(\xi)h^5.$$

By setting $x = x_k$ and dropping the truncation error term, we obtain the difference equation

$$y_{k+2} = y_k + \tfrac{1}{3}h[f(x_k, y_k) + 4f(x_{k+1}, y_{k+1}) + f(x_{k+2}, y_{k+2})]. \quad (8.37)$$

This formula, called **Simpson's method** or sometimes **Milne's method**, is implicit in the same sense as the trapezoidal method (8.27). That is, the new point y_{k+2} occurs on both sides of the equation.

It is easy to see that *any* numerical integration formula that uses the value of the integrand at the upper limit of integration will lead to an implicit method. All of the numerical integration formulas that were derived in Chapter 5 are of this type. However, by using the techniques of Chapter 5, it is possible to derive integration formulas such as

$$\int_x^{x+h} f(t)\, dt = \frac{h}{2}[3f(x) - f(x - h)] + \frac{5}{12} h^3 f''(\xi) \quad (8.38a)$$

$$\int_x^{x+h} f(t)\, dt = \frac{h}{24}[55f(x) - 59f(x - h) + 37f(x - 2h) - 9f(x - 3h)]$$

$$+ \frac{251}{720} h^5 f^{(4)}(\xi). \quad (8.38b)$$

These so-called **open integration formulas** do not use the value of the integrand at the upper limit of integration. They are not generally used for numerical integration because they are not as accurate as comparable **closed formulas**; that is, formulas that do use the value of the integrand at the endpoints [see Exercise 5 for comparison of open and closed formulas; Exercise 6 indicates how formulas (8.38) are derived].

Now suppose that (8.38a) is used in (8.32) with $t = h$:

$$y(x + h) = y(x) + \tfrac{1}{2}h[3f(x, y(x)) - f(x - h, y(x - h))] + \tfrac{5}{12}h^3 f''(\xi).$$

By setting $x = x_k$, so that $x + h = x_{k+1}$, and dropping the truncation error term, we obtain the difference formula

Adams–Bashforth method of order 2	$y_{k+1} = y_k + \tfrac{1}{2}h[3f(x_k, y_k) - f(x_{k-1}, y_{k-1})]. \quad (8.39a)$

This gives values for $y_2, y_3, \ldots$, provided that values for y_0 and y_1 are given. y_0 is, of course, just η, and y_1 can be computed with a one-step method.

A similar derivation with (8.38b) gives the difference equation

Adams–Bashforth method of order 4	$y_{k+1} = y_k + \frac{1}{24}h[55f(x_k, y_k) - 59f(x_{k-1}, y_{k-1})$ $+ 37f(x_{k-2}, y_{k-2}) - 9f(x_{k-3}, y_{k-3})].$ (8.39b)

Here we must have values for y_0, y_1, y_2, and y_3 in order to apply the formula to get $y_4, y_5, \ldots$. The methods (8.39) are called *Adams–Bashforth methods*. The latter formula, when used in conjunction with the implicit Adams–Moulton formula (8.45b) as described in Section 8.2.3, is perhaps the most widely used multistep method. We illustrate the use of (8.39a) by the following algorithm:

Algorithm for the Adams–Bashforth Method

$$\begin{bmatrix} \text{Algorithm for solving (8.3) by using the Adams–Bashforth for-} \\ \text{mula (8.39a). Input must include } a, b, \eta, \text{ a value for } h, \text{ and a} \\ \text{function subprogram to evaluate } f(x, y). \end{bmatrix}$$

1	Input $\{a, b, \eta, h\}$
2	$x \leftarrow a, y_0 \leftarrow \eta$
3	$f_0 \leftarrow f(x, y_0)$
4	$f_1 \leftarrow f(x + h, y_0 + hf_0)$
5	$y_1 \leftarrow y_0 + .5h(f_0 + f_1)$
6	$x \leftarrow x + h$
7	While $x + h \le b$
	7.1 $f_1 \leftarrow f(x, y_1)$
	7.2 $y_1 \leftarrow y_1 + .5h(3f_1 - f_0)$
	7.3 $x \leftarrow x + h$
	7.4 Output $\{x, y_1\}$
	7.5 $f_0 \leftarrow f_1$
8	Halt

(8.40)

Steps 4 and 5 use the second order Runge–Kutta method (8.30) to compute y_1. Steps 7.1 and 7.2 then determine approximations to $y(x_k)$ according to formula (8.39a). A program based on this algorithm should be used with care since no estimate of the error is included. One possible check on the error is to run the program twice with different values for h. A comparison of the results will give an indication of their accuracy (see Section 8.3 for more details).

In a similar manner formula (8.39b) can be implemented. The starting values y_1, y_2, y_3 should be computed by the Runge–Kutta Algorithm (8.31b).

8.2.3 Predictor–Corrector Methods

In the preceding paragraph we derived several *implicit* methods for solving (8.3). It will be shown in Section 8.3 that implicit methods are generally more accurate than comparable explicit methods. Moreover, they often lead to difference equations that are better conditioned than those produced by explicit methods. The problem with using implicit methods is, of course, the fact that the new value y_{k+1} is only defined implicitly by the formula. As already noted, if $f(x, y)$ is nonlinear in y, then an implicit method gives y_{k+1} as a solution to a nonlinear equation. The nonlinear equations that occur here have a special form, however; namely, the fixed point form (7.5). Consider, for example, the trapezoidal method (8.27). For fixed k the value of y_{k+1} as defined by (8.27) is the solution to the equation

| Fixed |
| point |
| equation |

$$y = g(y), \tag{8.41a}$$

where

$$g(y) = y_k + \tfrac{1}{2}h(f(x_{k+1}, y) + f(x_k, y_k)). \tag{8.41b}$$

As described in Section 7.1, (8.41a) suggests an iteration of the form

| Iteration |
| to solve |
| (8.41a) |

$$y^{(i+1)} = g(y^{(i)}), \qquad i = 0, 1, 2, \ldots, \tag{8.41c}$$

where $y^{(0)}$ is some starting value. Here we use superscripts to denote the *iterates* that we hope will converge to the value y_{k+1} that satisfies (8.27). The discussion in Section 7.3 can be used to show that if h is sufficiently small, then (8.41c) will, in fact, converge to a solution of (8.41a) (Exercise 5). The question now arises of when iteration (8.41c) should be stopped. That is, which $y^{(i)}$ should we accept as the value to be used as y_{k+1}? The answer to this question is simple, provided that the starting value $y^{(0)}$ has been properly chosen. Specifically, if $y^{(0)}$ is determined by a suitable *explicit* formula, then we need to do only *one* iteration of (8.41c). That is, $y^{(1)}$ can be accepted as the value of y_{k+1}. To illustrate this point, suppose we use Euler's method (8.20) to compute $y^{(0)}$; that is,

$$y^{(0)} = y_k + hf(x_k, y_k). \tag{8.42a}$$

Then the value of $y^{(1)}$ as given by (8.41b,c) is

$$y^{(1)} = y_k + \tfrac{1}{2}h(f(x_{k+1}, y^{(0)}) + f(x_k, y_k)). \tag{8.42b}$$

If these formulas are applied to the differential equation

$$y'(x) = y, \tag{8.43}$$

the result is

$$y^{(0)} = y_k + hy_k = (1 + h)y_k$$

$$y^{(1)} = y_k + \tfrac{1}{2}h(y^{(0)} + y_k) = y_k + \tfrac{1}{2}h((1 + h)y_k + y_k)$$

so that

$$y^{(1)} = (1 + h + \tfrac{1}{2}h^2)y_k. \tag{8.44a}$$

Now the *exact* value y_{k+1}, as determined by the trapezoidal rule (8.27) is

$$y_{k+1} = y_k + \tfrac{1}{2}h(y_{k+1} + y_k),$$

which gives

$$\boxed{\begin{array}{l} \text{Exact} \\ \text{value} \\ \text{of } y_{k+1} \end{array}} \quad y_{k+1} = \left(\frac{1 + h/2}{1 - h/2}\right)y_k = \left(1 + h + \frac{h^2}{2} + \frac{h^3}{4} + \frac{h^4}{8} + \cdots\right)y_k. \tag{8.44b}$$

But the solution to the differential equation (8.43) that equals y_k at x_k is just $y(x) = e^{x - x_k}y_k$. Thus we have

$$\boxed{\begin{array}{l} \text{Exact} \\ \text{solution} \end{array}} \quad\quad\quad y(x_{k+1}) = e^h y_k. \tag{8.44c}$$

Hence the factors in (8.44a) and (8.44b) that multiply y_k should be approximations to e^h, where we recall that

$$e^h = 1 + h + (h^2/2) + (h^3/3!) + \cdots. \tag{8.44d}$$

By comparing (8.44c) and (8.44d) with (8.44a) and (8.44b), we see that $y^{(1)}$ is just as good an approximation to $y(x_{k+1})$ as y_{k+1} is. The point here is that the iteration (8.41c) may converge to y_{k+1}; but we are not interested in the value of y_{k+1}. What we really want is an approximation to $y(x_{k+1})$, and with $y^{(0)}$ given by Euler's method $y^{(1)}$ is a good approximation.

The value of $y^{(1)}$ given by (8.44a) can be thought of as having been obtained as follows: First an explicit formula (Euler's method) was used to **predict** a value for y_{k+1}. This predicted value was then substituted into the right-hand side of an implicit formula (trapezoidal method) to obtain a **corrected** value for y_{k+1}. This corrected value could be corrected again, in keeping with the iteration (8.41c). However, this will usually not improve y_{k+1} as an approximation to the solution $y(x)$ of the differential equation.

Generally a **predictor–corrector method** consists of a pair of formulas, one of which is explicit, the other implicit. At each value of k for which both formulas are defined, the explicit formula is used to predict a value for y_{k+1}.

This value is then corrected by substituting it into the right-hand side of the implicit formula. The most popular predictor–corrector methods use the *Adams–Bashforth formulas* (8.39a) and (8.39b) as predictors. The corresponding corrector formulas are the (implicit) **Adams–Moulton formulas**:

$$y_{k+1} = y_k + \tfrac{1}{2}h[f(x_{k+1}, y_{k+1}) + f(x_k, y_k)] \qquad (8.45a)$$

and

$$y_{k+1} = y_k + \tfrac{1}{24}h[9f(x_{k+1}, y_{k+1}) + 19f(x_k, y_k)$$
$$- 5f(x_{k-1}, y_{k-1}) + f(x_{k-2}, y_{k-2})]. \qquad (8.45b)$$

[Note that (8.45a) is just the trapezoidal formula. The derivation of (8.45b) is described in Exercise 14.]

It will be shown in Section 8.3.2 that there is an important reason, besides accuracy and conditioning, for using predictor–corrector methods. Namely, the difference between the predicted and corrected values gives a simple estimate of the accuracy. We conclude this discussion with an algorithm based on the Adams pair (8.39a), (8.45a) of predictor–corrector formulas. The quantity "est" in this algorithm is an error estimate. The proper interpretation of it will be explained in the next section.

Adams Predictor–Corrector Algorithm

$$\left[\begin{array}{l}\text{Algorithm to solve (8.3) by using the predictor–corrector for-}\\ \text{mulas (8.39a) and (8.45a). The input consists of values for } a, b, \eta,\\ \qquad h, \text{ and a subprogram to evaluate } f(x, y).\end{array}\right]$$

1 Input $\{a, b, \eta, h\}$
2 $x \leftarrow a,\ y_0 \leftarrow \eta$
3 $f_0 \leftarrow f(x, y_0)$
4 $f_1 \leftarrow f(x + h, y_0 + hf_0)$
5 $y_1 \leftarrow y_0 + .5h(f_0 + f_1)$
6 $x \leftarrow x + h$
7 While $x + h \le b$
   7.1 $f_1 \leftarrow f(x, y_1)$ (8.46)
   7.2 $y_1^p \leftarrow y_1 + .5h(3f_1 - f_0)$
   7.3 $x \leftarrow x + h$
   7.4 $f_1^p \leftarrow f(x, y_1^p)$
   7.5 $y_1 \leftarrow y_1 + .5h(f_1^p + f_1)$
   7.6 est $\leftarrow |y_1^p - y_1|/6.0$
   7.7 Output $\{x, y_1, \text{est}\}$
   7.8 $f_0 \leftarrow f_1$
8 Halt

As in Algorithm (8.40), we use the second order Runge–Kutta formula (8.30) to compute the starting value y_1. The predicted value is computed at step 7.2 and is corrected in step 7.5. Note that step 7.4 is an extra evaluation of f. Thus Algorithm (8.46) is more expensive to use than Algorithm (8.40), especially so if f is difficult to evaluate. We do, however, get a more accurate answer plus an error estimate as a result of this extra evaluation.

8.2.4 Methods for Systems of Equations

The methods derived above can be applied easily to systems of differential equations. For simplicity, consider the case of two equations

$$y'(x) = f(x, y(x), z(x))$$
$$z'(x) = g(x, y(x), z(x)), \qquad a \le x \le b \qquad (8.47)$$
$$y(a) = \eta, \qquad z(a) = \xi.$$

Euler's method (8.20) can be applied to this system in the form

Euler's method for (8.47)

$$y_{k+1} = y_k + hf(x_k, y_k, z_k)$$
$$z_{k+1} = z_k + hg(x_k, y_k, z_k), \qquad k = 0, 1, 2, \ldots \qquad (8.48)$$
$$y_0 = \eta, \qquad z_0 = \xi.$$

Similarly, the second order Runge–Kutta method (8.30), when applied to (8.47), becomes

Runge–Kutta method for (8.47)

$$y_{k+1} = y_k + \tfrac{1}{2}h(f(x_k, y_k, z_k)$$
$$+ f(x_{k+1}, y_k + hf(x_k, y_k, z_k), z_k + hg(x_k, y_k, z_k)))$$
$$z_{k+1} = z_k + \tfrac{1}{2}h(g(x_k, y_k, z_k) \qquad (8.49)$$
$$+ g(x_{k+1}, y_k + hf(x_k, y_k, z_k), z_k + hg(x_k, y_k, z_k)))$$
$$y_0 = \eta, \qquad z_0 = \xi.$$

The other Runge–Kutta formulas generalize in exactly the same way as do the multistep methods. For example, the Adams–Bashforth method (8.39b) becomes

Adams–Bashforth method for (8.47)

$$y_{k+1} = y_k + \tfrac{1}{24}h[55f(x_k, y_k, z_k) - 59f(x_{k-1}, y_{k-1}, z_{k-1})$$
$$+ 37f(x_{k-2}, y_{k-2}, z_{k-2}) - 9f(x_{k-3}, y_{k-3}, z_{k-3})]$$
$$z_{k+1} = z_k + \tfrac{1}{24}h[55g(x_k, y_k, z_k) - 59g(x_{k-1}, y_{k-1}, z_{k-1})$$
$$+ 37g(x_{k-2}, y_{k-2}, z_{k-2}) - 9g(x_{k-3}, y_{k-3}, z_{k-3})]$$

$$(8.50)$$

with $y_0 = \eta$, $z_0 = \xi$. Values for y_1, y_2, y_3 and z_1, z_2, z_3 must be found by using, say, a Runge–Kutta method.

Just as for a single equation, it can be shown that all of these methods for systems can be obtained by expanding the pair of functions $y(x)$, $z(x)$, in Taylor series and then truncating the series.

EXERCISES

1 Verify the expansion (8.28) by showing that, if $y(x)$ is twice continuously differentiable, and $f(x, y)$ is differentiable with respect to y, then $f(x + h, \; y(x + h)) = f(x + h, \; y(x) + hy'(x)) + 0(h^2)$.

2 By expanding f as it appears in the definition of a_2, a_3, a_4 in (8.31), show that this method is obtained by omitting $O(h^5)$ terms.

3 Write an algorithm to compute y_1, y_2, ... according to method (8.30).

4 Write a program to implement Algorithm (8.31b). Omit step 3.8, but for each equation you solve, run the program twice; first with what you think is a reasonable value for h, and then with half that value. Compare the results of these two runs. [See also Exercise 8, Section 8.3.] Test your program with the equation $y'(x) = wy$, $0 \le x \le 1$, $y(0) = 1$, for values of $w = -10$, -5, -1, 1, 2, 5. [The exact solution is $y(x) = e^{wx}$.]

5 Compare the integration formula (8.38a) with the simple trapezoidal formula. Which has the smaller truncation error? Which method would you prefer for computing an approximation to $\int_0^{.3} \cos x \, dx$?

6 Derive formula (8.38a) by integrating the polynomial that interpolates f at the points x and $x - h$. Derive formula (8.38b) by using the polynomial that interpolates f at x, $x - h$, and $x - 2h$.

7 Write out the details to show how (8.39b) is obtained.

8 Write out an algorithm for implementing formula (8.39b). Use the Runge–Kutta Algorithm (8.31b) to obtain the starting values y_1, y_2, and y_3.

9 Program Algorithm (8.40). As suggested in the discussion following the algorithm, check the accuracy of your results by running the program with two values of h. Test the program on the equations in Exercise 4.

10 Use Euler's method to solve the system of equations: $y'(x) = z(x)$, $z'(x) = y(x)$, $0 \le x \le 1$, $y(0) = z(0) = 1$.

11 By using the technique described in Section 8.1.2, transform the equation

$$y''(x) = xy'(x) + x^2 y + 1, \qquad 0 \le x \le 1$$

$$y(0) = 1, \qquad y'(0) = 1$$

into a system of first order equations. Then solve the system by using Euler's method.

12 Write out an algorithm, similar to (8.40), to solve *systems* of differential equations by formula (8.39a).

13 Prove that each of the expressions in (8.23), when divided by the appropriate power of h, remains bounded as $h \to 0$.

14 Derive a closed integration formula by integrating from x to $x + h$ the polynomial that interpolates the integrand at the points $x - 2h$, $x - h$, x, and $x + h$. Use this integration formula to obtain (8.45b).

15 Prove that, for h sufficiently small, the iteration (8.41c) is locally convergent, as defined in Section 7.2.

[*Note*: How small h must be will depend on the size of the derivatives of f.]

8.3 Error Estimates

The greatest difficulty in solving differential equations numerically is to estimate the errors in the computed solution. In this section we consider various aspects of this problem.

8.3.1 Local and Global Errors

As we have repeatedly observed, the difference equations that are used to approximate the differential equations can be obtained by truncating a series expansion of the solution function. The size of these truncated terms provides a first measure of the accuracy of the method. Consider, for example, Euler's method (8.20). It was derived by omitting the last term from the expansion

$$y(x_{k+1}) = y(x_k) + hf(x_k, y(x_k)) + \tfrac{1}{2}h^2 y''(\xi). \tag{8.51}$$

Now suppose we use Euler's method to compute y_{k+1}, *with* $y_k = y(x_k)$. That is, we assume that y_k is exact and then use the method to compute y_{k+1} as follows:

$$y_{k+1} = y_k + hf(x_k, y_k) = y(x_k) + hf(x_k, y(x_k)). \tag{8.52}$$

If we subtract (8.52) from (8.51), we find

$$y(x_{k+1}) - y_{k+1} = \tfrac{1}{2}h^2 y''(\xi). \tag{8.53}$$

Thus the truncated term measures the error in y_{k+1} *when y_k is exact.*

Generally, if y_{k+1} is determined by the numerical method, but *with exact values* for y_k, y_{k-1}, ..., then the error in y_{k+1} is called the **local error** for the method. As illustrated above with Euler's method, the truncated term equals the local error.

The **global error** for a particular method is the error in y_{k+1} when all previous values y_k, y_{k-1}, ... are *also* determined by the method. That is, the global error takes into account the accumulation of errors. To see the difference between these two errors, consider again Euler's method, this time applied to the equation

$$y'(x) = y(x), \qquad 0 \le x \le 1$$
$$y(0) = 1.$$

(8.54)

This equation has the solution $y(x) = e^x$. The approximations $y_1, y_2, \ldots$ are determined by the difference equation

$$y_{k+1} = y_k + hf(x_k, y_k) = y_k + hy_k$$
$$= (1 + h)y_k, \qquad k = 0, 1, \ldots$$

(8.55)

with $y_0 = 1$. Equation (8.55) is a linear difference equation. It is easily verified that the solution is $y_k = (1 + h)^k$, $k = 0, 1, \ldots$. Thus the *global error* is

$$E_G = |y(x_k) - y_k| = |e^{x_k} - (1 + h)^k| = |e^{kh} - (1 + h)^k|.$$

A complicated but elementary argument (see Exercise 1) shows that this can be estimated as

$$E_G \cong \tfrac{1}{2}h(e^{x_k} - 1) \le \tfrac{1}{2}h(e - 1) \qquad \text{for} \quad 0 \le x_k \le 1.$$

(8.56)

On the other hand, the local error was shown above to be $(\tfrac{1}{2}h^2)y''(\xi)$. For this equation, $y''(\xi) = e^\xi$, so that the local error is

$$E_L = \tfrac{1}{2}h^2 |y''(\xi)| = \tfrac{1}{2}h^2 e^\xi, \qquad x_k \le \xi \le x_{k+1}.$$

Hence we have the result

$$E_L \le \tfrac{1}{2}h^2 e.$$

(8.57)

By comparing (8.57) with (8.56) we see that the local error is approximately $2h$ times smaller than the global error.

In the above example, it was possible to analyze the global error because the solution to the differential equation was known. In general, this will not be the case. In fact, it is usually not possible even to estimate the global error with much precision. The best that can be done is to look for *qualitative* relations between the local and global errors and then use the local error estimates derived in the next paragraph as a guide to the size of the global error.

8.3.2 Estimating the Local Error

Since the local error in a method is just the term that was deleted in the derivation of the method, it is sometimes a simple matter to estimate this error. For example, consider again Euler's method for which the local error in y_{k+1} is

$$E_L = \tfrac{1}{2}h^2 y''(\xi), \tag{8.58a}$$

where ξ is some value between x_k and x_{k+1}. Now since $y'(x) = f(x, y(x))$, we have, according to (8.6),

Local error formula

$$E_L = \tfrac{1}{2}h^2 (f_x + ff_y)_{x=\xi}. \tag{8.58b}$$

Thus if bounds on f, f_x, and f_y are available, then E_L can be bounded. For example, suppose that Euler's method is used to solve the equation

$$y'(x) = y \cos x + \sin x, \qquad 0 \le x \le \pi$$

$$y(0) = 1.$$

Then $f(x, y) = y \cos x + \sin x$, so that

$$|f(x, y)| \le |y| + 1$$

$$|f_x(x, y)| \le |-y \sin x + \cos x| \le |y| + 1$$

$$|f_y(x, y)| \le |\cos x| \le 1.$$

Thus by (8.58b) the local error is bounded by

$$|E_L| \le \tfrac{1}{2}h^2 (|f_x| + |f||f_y|)_{x=\xi}$$

$$\le \tfrac{1}{2}h^2 [(|y(\xi)| + 1 + (|y(\xi)| + 1)]$$

$$= h^2 (|y(\xi)| + 1).$$

The estimate

Local error estimate

$$|E_L| \cong h^2 (|y_{k+1}| + 1)$$

follows immediately.

The problem with applying this technique to more complicated methods is that the expression for the local error is simply too complex to estimate accurately. For example, the method (8.39a) has local error

$$\tfrac{5}{12}h^3 f''(\xi, y(\xi)). \tag{8.59}$$

Here f'' denotes the second derivative of $f(x, y(x))$ *with respect to* x. If this differentiation is carried out, the result is

$$d^2 f(x, y(x))/dx^2 = f_{xx} + 2ff_{xy} + f_{yy} f^2 + f_x f_y + f_y^2 f.$$

Thus to estimate (8.59), one would have to estimate f and all of its first and second partial derivatives.

The Runge–Kutta formulas generally have extremely complicated local error terms. It can be shown, for example, (see Exercise 3) that the second order Runge–Kutta method (8.30) has local error

Local error for Runge–Kutta method	$\frac{1}{6}h^3[\frac{3}{2}(f_{yy}f^2 + 2f_{xy}f + f_{xx}) - f_x - ff_y].$	(8.60)

In practice, these expressions for the local error are seldom used directly to estimate the local error. Instead we settle for the following estimates that are less accurate but more easily computed. Suppose we have two methods for solving the equation, one with local error $O(h^p)$ and the other with local error $O(h^q)$, where $q \geq p$. Let y_{k+1} and $\bar{y}_{k+1}$ denote the approximations produced by these methods. It is not hard to see (Exercise 4) that an $O(h^p)$ local error, for example, can be written as

$$\tau_k h^p + O(h^{p+1}),$$

where τ_k is some constant. Thus if we assume that $y_k, y_{k-1}, \ldots$ are exact, then we can write

$$y_{k+1} = y(x_{k+1}) + \tau_k h^p + O(h^{p+1})$$
$$\bar{y}_{k+1} = y(x_{k+1}) + \bar{\tau}_k h^q + O(h^{q+1})$$

(8.61)

for some constants τ_k and $\bar{\tau}_k$.

We now subtract these equations to obtain

$$y_{k+1} - \bar{y}_{k+1} = \tau_k h^p - \bar{\tau}_k h^q + O(h^{p+1}) = (\tau_k - \bar{\tau}_k h^{q-p})h^p + O(h^{p+1}).$$

(8.62)

If h is small and $q > p$, then $\tau_k - \bar{\tau}_k h^{q-p} \cong \tau_k$, and we can write

$$y_{k+1} - \bar{y}_{k+1} \cong \tau_k h^p + O(h^{p+1}).$$

(8.63)

But the expression on the right of (8.63) is (approximately) the local error in y_{k+1}. Thus we have

Local error estimate when $p < q$	local error in $y_{k+1} \cong y_{k+1} - \bar{y}_{k+1}$.	(8.64)

If $p = q$, then we can obtain a more precise estimate. From (8.62) we have

$$y_{k+1} - \bar{y}_{k+1} = (\tau_k - \bar{\tau}_k)h^p + O(h^{p+1})$$
$$= [(\tau_k - \bar{\tau}_k)/\bar{\tau}_k]\bar{\tau}_k h^p + O(h^{p+1})$$
$$= [(\tau_k - \bar{\tau}_k)/\bar{\tau}_k](\bar{y}_{k+1} - y(x_{k+1})) + O(h^{p+1}).$$

This last equality follows from (8.61). Thus if $\tau_k \neq \bar{\tau}_k$, then we have

| Local error |
| estimate | local error in $\bar{y}_{k+1} \cong [\bar{\tau}_k/(\tau_k - \bar{\tau}_k)](y_{k+1} - \bar{y}_{k+1})$.   (8.65) |
| when $p = q$ |

Hence the difference in the computed values y_{k+1} and $\bar{y}_{k+1}$ can be used to estimate the local error in $\bar{y}_{k+1}$.

An important application of (8.65) is in connection with the predictor-corrector methods of Section 8.2.3. Consider, for example, the formulas (8.39a) and (8.45a) that were used in Algorithm (8.46). The local errors for these formulas are $\frac{5}{12}y'''(\eta)h^3$ and $-\frac{1}{12}y'''(\xi)h^3$, respectively. By using the method of Exercise 4, we can write these local errors as

$$\tfrac{5}{12}y'''(x_k)h^3 + O(h^4) \qquad\qquad (8.66a)$$

and

$$-\tfrac{1}{12}y'''(x_k) + O(h^4). \qquad\qquad (8.66b)$$

If we let $y_{k+1}^{(p)}$ denote the predicted value and $y_{k+1}^{(c)}$ the corrected value, then, under the assumption that $y_{k+1}^{(c)}$ satisfies the corrector formula (8.45a), we can write

$$y_{k+1}^{(p)} = y(x_{k+1}) + \tau_k^{(p)}h^3 + O(h^4)$$

$$y_{k+1}^{(c)} = y(x_{k+1}) + \tau_k^{(c)}h^3 + O(h^4),$$

where

$$\tau_k^{(p)} = \tfrac{5}{12}y'''(x_k), \qquad \tau_k^{(c)} = -\tfrac{1}{12}y'''(x_k).$$

Now the estimate (8.65) can be applied to give the result:

| Local error |
| estimate |
| for formulas | local error in $y_{k+1}^{(c)} \cong -\tfrac{1}{6}(y_{k+1}^{(p)} - y_{k+1}^{(c)})$.   (8.67) |
| (8.39a), |
| (8.45a) |

The quantity on the right of (8.67) is, in fact, just the value *est* computed in Algorithm (8.46).

Similarly, the fourth order formulas (8.39b), (8.45b) can be used as a predictor-corrector pair. The local errors for these formulas are

$$\tfrac{251}{720}y^{(5)}(\eta)h^5$$

and

$$-\tfrac{19}{720}y^{(5)}(\xi)h^5,$$

respectively. These can be combined, according to (8.65), to give the estimate

Local error estimate for formulas (8.39b), (8.45b)	local error in $y_{k+1}^{(c)} \cong -\tfrac{19}{270}(y_{k+1}^{(p)} - y_{k+1}^{(c)})$.	(8.68)

The multistep methods require only one function evaluation at each step while the predictor–corrector methods need an additional evaluation but also provide an estimate of the local error. The Runge–Kutta methods, however, require *several* evaluations of f at each step. If we try to use two Runge–Kutta methods to compute y_{k+1} *and* a local error estimate, the number of function evaluations can easily get out of hand. An idea, due to Fehlberg, is to use a pair of Runge–Kutta formulas in which some of the evaluations of f that are needed by one of the formulas can also be used in the other formula. It turns out that such formula pairs must be of different orders. The relation (8.64), however, can be used to estimate the local error in the lower order approximation. The most widely used **Runge–Kutta–Fehlberg formulas** are

$$y_{k+1} = y_k + h[\tfrac{25}{216}a_1 + \tfrac{1408}{2565}a_3 + \tfrac{2197}{4104}a_4 - \tfrac{1}{5}a_5] \tag{8.69a}$$

$$\bar{y}_{k+1} = y_k + h[\tfrac{16}{135}a_1 + \tfrac{6656}{12825}a_3 + \tfrac{28561}{56430}a_4 - \tfrac{9}{50}a_5 + \tfrac{2}{55}a_6], \tag{8.69b}$$

where

$$a_1 = f(x_k, y_k), \qquad a_2 = f(x_k + \tfrac{1}{4}h, y_k + \tfrac{1}{4}ha_1)$$

$$a_3 = f(x_k + \tfrac{3}{8}h, y_k + h(\tfrac{3}{32}a_1 + \tfrac{9}{32}a_2))$$

$$a_4 = f(x_k + \tfrac{12}{13}h, y_k + h(\tfrac{1932}{2197}a_1 - \tfrac{7200}{2197}a_2 + \tfrac{7296}{2197}a_3)) \tag{8.69c}$$

$$a_5 = f(x_k + h, y_k + h(\tfrac{439}{216}a_1 - 8a_2 + \tfrac{3680}{513}a_3 - \tfrac{845}{4104}a_4))$$

$$a_6 = f(x_k + \tfrac{1}{2}h, y_k + h(-\tfrac{8}{27}a_1 + 2a_2 - \tfrac{3544}{2565}a_3 + \tfrac{1859}{4104}a_4 - \tfrac{11}{40}a_5)).$$

The **estimate (8.64) for the local error** in y_{k+1} as given by (8.69a) is just

$$y_{k+1} - y(x_{k+1}) \cong -h(\tfrac{1}{360}a_1 - \tfrac{128}{4275}a_3 - \tfrac{2197}{75240}a_4$$
$$+ \tfrac{1}{50}a_5 + \tfrac{2}{55}a_6). \tag{8.70}$$

Thus to compute y_{k+1} by (8.69a) requires four evaluations of f. With two additional evaluations, we can compute the local error estimate (8.70). The ability to monitor the local error and adjust h as described in Section 8.4.2 is well worth the extra work.

Finally, a simple way to make use of (8.64) is to use a single method to compute y_{k+1} and $\bar{y}_{k+1}$ but with different values of h. For example, suppose that we use a certain method, with local error $O(h^p)$ and step size h, to compute y_{k+1}. Now use the same method, but with step size $\frac{1}{2}h$, to compute $\bar{y}_{k+1/2} \cong y(x_k + \frac{1}{2}h)$ and then $\bar{y}_{k+1}$. Equation (8.61) becomes

$$y_{k+1} = y(x_{k+1}) + \tau_k h^p + O(h^{p+1})$$
$$\bar{y}_{k+1} = y(x_{k+1}) + \tau_k(\tfrac{1}{2}h)^p + O(h^{p+1})$$
$$= y(x_{k+1}) + \tau_k(1/2^p)h^p + O(h^{p+1}).$$

Thus in (8.61) we have $\bar{\tau}_k = \tau_k/2^p$, so (8.65) gives

$$\text{local error in } \bar{y}_{k+1} \cong 1/(2^p - 1)(y_{k+1} - \bar{y}_{k+1}). \tag{8.71}$$

These estimates are extremely important. They give a practical means for estimating the local error and hence for estimating the global error in the approximate solution.

8.3.3 Global Error—Stability

Suppose that in the computation of y_k a local error of size e_k occurs; then one might expect that the global error in y_k would be just the sum of the local errors up to this point; that is $e_1 + e_2 + \cdots + e_k$. Unfortunately, the situation is not so simple. The problem arises from the fact that we are actually computing a solution to a difference equation. As was pointed out in Section 8.1.5, solutions to difference equations may be ill conditioned. The effect of this ill-conditioning is to magnify the local errors $e_1, e_2, \ldots$ so that they appear as very large errors in y_k. Consider, for example, the multistep method

$$y_{k+1} = 4y_k - 3y_{k-1} - 2hf(x_{k-2}, y_{k-2}). \tag{8.72}$$

By expanding $y(x_{k-1})$ and $y'(x_{k-2})$ in a Taylor series at x_k, it can be shown (see Exercise 5) that the local error for this method is

$$-2h^2 f''(\xi_k). \tag{8.73}$$

However, suppose that we use this method to solve the (trivial) differential equation

$$y'(x) = 0, \qquad 0 \le x$$
$$y(0) = 0 \tag{8.74}$$

TABLE 8.1 *Solution to (8.76) as Determined by (8.72)*

x_k	y_k	Global error $\|y_k - y(x_k)\|$
0	1.000	0
.1	1.105...	0
.2	1.221...	0
.3	1.346...	0.004
⋮	⋮	⋮
.7	1.636...	.3774
.8	1.240...	1.102
.9	−.740...	3.200
1.0	−6.559...	9.277

whose solution is, of course, $y(x) = 0$. If we use starting values $y_0 = 0$ and $y_1 = \varepsilon$, so that y_1 has local error ε, then we find

$$y_2 = 4y_1 - 3y_0 = 4\varepsilon$$
$$y_3 = 4y_2 - 3y_1 = 13\varepsilon$$
$$y_4 = 4y_3 - 3y_2 = 40\varepsilon$$
$$y_5 = 4y_4 - 3y_3 = 121\varepsilon \tag{8.75}$$
$$y_6 = 4y_5 - 3y_4 = 364\varepsilon$$
$$y_7 = 4y_6 - 3y_5 = 1093\varepsilon.$$

Thus the local error ε in y_1 is magnified tremendously as further values are computed. In order to illustrate this effect on a nontrivial equation, consider the method (8.72) applied to the equation

$$y'(x) = y(x), \qquad 0 \le x$$
$$y(0) = 1 \tag{8.76}$$

and with $h = .1$. According to the local error (8.74), we should obtain reasonable results. However, with exact (to six figures) values for y_1, y_2 the computed values for y_3, y_4, ..., y_{10} are shown in Table 8.1. Thus, in spite of the fact that the local error is of the size $h^2 = 10^{-2}$, the global error increases steadily until the computed approximations y_k are totally wrong.

The reason why (8.72) gives such bad results when applied to the simple equations (8.74) and (8.76) is that we are trying to compute an ill-conditioned solution to a difference equation. That is, when (8.72) is applied to (8.76), we obtain the **difference equation**

$$y_{k+1} = 4y_k - 3y_k - 2hy_{k-2}$$
$$y_0 = 1. \tag{8.77}$$

The characteristic polynomial for (8.77) is

$$\lambda^3 - 4\lambda^2 + 3\lambda + 2h. \tag{8.78}$$

When $h = 0$, (8.78) has roots 1, 0, 3; thus, for small but nonzero h, we can expect the roots to be

$$\lambda_1 \cong 1, \qquad \lambda_2 \cong 0, \qquad \lambda_3 \cong 3.$$

As shown in Section 8.1.4, any solution to (8.77) must have the form

$$y_k = a\lambda_1^k + b\lambda_2^k + c\lambda_3^k. \tag{8.79}$$

The coefficients a, b, c are determined by the starting values y_0, y_1, y_2. The exact solution to (8.76) is $y(x) = e^x$; hence the solution (8.79) should *not* contain the term $c\lambda_3^k$ because it grows too rapidly. Even if the starting values y_0, y_1, y_2 are chosen so that $c = 0$ in (8.79), rounding errors and local errors will eventually cause $c \neq 0$, and this term will gradually dominate. In fact, the last few entries in Table 8.1 clearly illustrate this fact. That is, the errors here are increasing as powers of three.

The technique that we have used to show the instability of (8.72) can be used to analyze any method. The trick in this analysis is to apply the numerical method to simple test equations of the form (8.74) and (8.76). More generally, consider the family of equations

| Test |
| equation |

$$y'(x) = wy(x), \qquad x \geq 0$$
$$y(0) = 1 \tag{8.80}$$

Here w is a parameter that can be chosen to obtain different types of solutions. That is, the solution is just $y(x) = e^{wx}$, so that if

$$w > 0 \quad \text{then} \quad y(x) \to \infty \quad \text{as} \quad x \to \infty$$
$$w = 0 \quad \text{then} \quad y(x) = 1 \quad \text{for all} \quad x$$
$$w < 0 \quad \text{then} \quad y(x) \to 0 \quad \text{as} \quad x \to \infty$$
$$w \quad \text{is complex, then} \quad y(x) \quad \text{oscillates.}$$

Thus the *general* behavior of a wide class of differential equations can be modeled by this simple test equation. Moreover, when any of the methods of Section 8.2 are applied to (8.80), the result is a linear difference equation of the type (8.12). Hence the theory of Sections 8.1.4 and 8.1.5 can be used to analyze the difference equation.

For example, if the Adams–Bashforth method (8.39b) is applied to the test equation (8.80), the result is the difference equation

$$y_{k+1} = (1 + \tfrac{55}{24}hw)y_k - \tfrac{59}{24}hwy_{k-1}$$
$$+ \tfrac{37}{24}hwy_{k-2} - \tfrac{9}{24}hwy_{k-3}, \qquad k = 3, 4, \ldots . \tag{8.81}$$

Here we let $y_0 = 1$, but the rest of the initial values y_1, y_2, y_3 must be found by some other method.

We shall draw several conclusions about the methods of Section 8.2 by looking at their performance when applied to the test equation (8.80). The proof that these conclusions are also valid for a quite general class of equations is beyond our scope.

We would like to say that a method of Section 8.2 is *stable* if, when applied to the test equation (8.80), the resulting difference equation is stable as specified by Definition 8.1. Unfortunately the situation is complicated by the fact that the difference equation depends on h and w. For some values of these parameters, the difference equation may be stable, for others unstable. Furthermore, Definition 8.1 refers to the stability of a *particular* solution to the difference equations. Some solutions may be stable, others unstable. To make this more precise, we first note that the difference equations actually depend on the product hw. See, for example, (8.81). Thus the **roots of the characteristic equation** can be written as

$$\lambda_1(hw), \lambda_2(hw), \ldots, \lambda_n(hw). \tag{8.82}$$

Next consider the three cases $w = 0$, $w > 0$, and $w < 0$. Since $h > 0$, we can express these equivalently as $hw = 0$, $hw > 0$, and $hw < 0$. The case when w is complex can be treated in a similar manner. When $hw = 0$, all of the methods derived in Section 8.2, *except for Milne's method* (8.37), reduce to the simple difference equation

$$y_{k+1} = y_k, \qquad k = 0, 1, \ldots . \tag{8.83}$$

We shall return to Milne's method later. For now we note that (8.83) implies that the roots (8.82) satisfy

$$\lambda_1(0) = 1, \qquad \lambda_i(0) = 0 \qquad \text{for} \quad i = 2, 3, \ldots, n. \tag{8.84}$$

Furthermore, when $w = 0$, the solution to the test equation is just $y(x) = 1$. Since the local error involves a *derivative* of y, it follows that the local error for any of these methods, when applied to (8.80) with $w = 0$, is zero. Thus the method should give the exact solution; that is, we should have $y_k = 1, k = 0, 1, 2, \ldots$. But (8.84) ensures that this will indeed be the case since the solution is just $y_k = \lambda_1(0)^k = 1$ for $k = 0, 1, 2, \ldots$. We shall assume that the roots in (8.82) are numbered so that (8.84) holds. Moreover, keep in mind that the root $\lambda_1(hw)$ for small hw, is the important root that will give our desired approximation.

For the cases where $hw \neq 0$, the difference equation is much more complicated than (8.83); in fact, it is of the type (8.81). However, the roots (8.82) of the characteristic polynomial are continuous functions of the parameter hw. Hence, for hw small but nonzero, the roots may be expected to be as

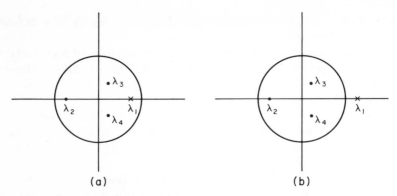

Fig. 8.2   Roots of the characteristic polynomial with hw small but nonzero, in relation to the unit circle $\{z : |z| = 1\}$ in the complex plane.

shown in Fig. 8.2. In Fig. 8.2a the root $\lambda_1(hw)$ has moved inside the unit circle, whereas in Fig. 8.2b this root has moved outside the circle. If $w > 0$, then the exact solution $y(x) = e^{wx}$ is increasing, so the approximate solution y_k should also increase. This can only happen if $|\lambda_1(hw)| > 1$; that is, the situation in Fig. 8.2b should hold. If $w < 0$, then the solution is decreasing, so Fig. 8.2a should apply. In either case we see by this simple continuity argument that if hw is sufficiently small, then the approximate solution y_k, $y_{k+1}, \ldots$, as given by the difference equation, is stable.

If hw is so large that the roots $\lambda_i(hw)$, $i > 1$, are outside the unit circle, then the desired solution to the difference equation may be unstable. That is, in the situation of Fig. 8.3a, the desired solution decreases, whereas the solution λ_2^k increases and alternates in sign. Figure 8.3b indicates the possibility

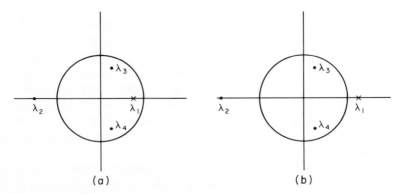

Fig. 8.3   Roots of the characteristic polynomial for large value of hw, relative to the unit circle. (a) hw large and negative. (b) hw large and positive.

where $|\lambda_2(hw)| > |\lambda_1(hw)|$, so that the unwanted solution λ_2^k eventually dominates the desired part.

To illustrate these ideas, consider the Adams–Bashforth method (8.39a). If this method is applied to the test equation (8.80), the resulting difference equation is

$$y_{k+1} = (1 + \tfrac{3}{2}hw)y_k - \tfrac{1}{2}hwy_{k-1}, \qquad k = 1, 2, \ldots . \tag{8.85a}$$

The **roots of the characteristic polynomial** are

$$\lambda_1(hw) = \tfrac{1}{2}(1 + \tfrac{3}{2}hw + \sqrt{1 + hw + \tfrac{9}{4}h^2w^2})$$

$$\lambda_2(hw) = \tfrac{1}{2}(1 + \tfrac{3}{2}hw - \sqrt{1 + hw + \tfrac{9}{4}h^2w^2}).$$

For hw small, $\lambda_1(hw) \cong 1 + hw$, and $\lambda_2(hw) \cong \tfrac{1}{2}hw$. Hence, $|\lambda_2(hw)| < 1$, and $\lambda_1(hw) > 1$ whenever $w > 0$, so that the approximate solution $y_k = \lambda_1(hw)^k$ will increase just as does the exact solution $y(x) = e^{wx}$. If $w < 0$, then $\lambda_1(hw) < 1$ so the approximate solution decreases to zero. Suppose, however, that hw is not small. In fact, let $w = -200$ and $h = .01$. Then we have $\lambda_1(hw) = .4142\ldots$ and $\lambda_2(hw) = -2.4142\ldots$ so that our **approximate solution** will be

$$y_k = \alpha(.4142\ldots)^k + \beta(-2.4142\ldots)^k. \tag{8.85b}$$

Accurate starting values will make β small. However, for large k, the second part of the solution (8.85b) will dominate, and the approximate solution will bear no resemblance to the exact solution $y(x) = e^{-200x}$. We summarize this discussion as follows:

For any of the methods of Section 8.2, except for Milne's method (8.37), the approximate solution $y_k, y_{k+1}, \ldots$ will be stable provided h is sufficiently small. In terms of the test equation (8.80), the roots $\lambda_i(hw)$ of the characteristic polynomial must satisfy $|\lambda_i(hw)| < 1$, except perhaps for the root $\lambda_1(hw)$, for which $\lambda_1(0) = 1$.

In Section 8.4.2 we shall discuss how one can change the step size h in order to control the local error. Fortunately, for many equations a value of h that is small enough to guarantee an acceptable local error is also small enough so that the approximate solution is stable. In cases where this is not true, the instability in the approximate solution eventually leads to large local errors. Hence a large increase in the local error estimates serves as a warning that the solution is unstable and h should be reduced.

The above analysis holds for the Adams multistep methods. Note that there is no stability problem with the Runge–Kutta methods because there is only one solution to the difference equation. Milne's method must be analyzed in a manner similar to that used for (8.72). That is, if Milne's method is applied to the test equation, we obtain the difference equation

| Milne's method applied to test equation |

$$(1 - \tfrac{1}{3}hw)y_{k+2} = \tfrac{4}{3}hwy_{k+1} + (1 + \tfrac{1}{3}hw)y_k. \tag{8.86}$$

The characteristic polynomial for this equation has roots

$$\lambda_1 = [2hw + \sqrt{9 + 3h^2w^2}]/(3 - hw)$$

$$\lambda_2 = [2hw - \sqrt{9 + 3h^2w^2}]/(3 - hw).$$

If $hw > 0$, then $|\lambda_1| > 1$ and $|\lambda_2| < 1$, hence the desired solution $y_k = \alpha\lambda_1^k$ is stable. However, when $hw < 0$, then $|\lambda_2| > 1$ and the solution is unstable. Milne's method is said to be *weakly* unstable because it is actually stable for certain equations (see Exercise 7 for another such method). The small local error in Milne's method makes it a useful method whenever the weak instability does not occur.

8.3.4 Effect of Rounding Errors

So far we have considered only the effect of truncation error on the approximate solution. Suppose now that the values $\hat{y}_k, \hat{y}_{k+1}, \ldots$ are a computed solution to the difference equation. That is, they do not satisfy the difference equation exactly. We should like to examine the global errors in these computed values. By introducing some notation, we can give a result that nicely summarizes this entire discussion of errors, and at the same time shows how rounding errors affect the approximations. For this, consider an n-step method that has local errors $O(h^{p+1})$ and let $\hat{y}_0, \hat{y}_1, \ldots, \hat{y}_{n-1}, \hat{y}_n, \ldots,$ $\hat{y}_k, \ldots$ be the computed solution to a particular differential equation. Assume that the following conditions hold:

(1) The exact solution $y(x)$ to the differential equation has $p + 1$ continuous derivatives.
(2) h is sufficiently small that the computed solution is stable.
(3) The (absolute) rounding errors that occur when $\hat{y}_{k+1}$ is computed are bounded by some number δ.
(4) The initial values $\hat{y}_0, \ldots, \hat{y}_{n-1}$ are computed by some appropriate starting method and have global error bounded by σ; that is

$$|\hat{y}_i - y(x_i)| \leq \sigma, \qquad i = 0, 1, \ldots, n - 1.$$

Then there are positive numbers A, B, C, D, which are independent of h so that

Global error bound

$$|\hat{y}_k - y(x_k)| \leq (A\sigma + B(\delta x_k/h) + Ch^p)e^{Dx_k}. \qquad (8.87)$$

Thus the global error at x_k depends on three things: The accuracy of the starting values as measured by $A\sigma$, the local error as given by Ch^p, and the rounding errors $B\delta/h$. As often happens in numerical methods, we can reduce the truncation error (local error) by reducing h, but doing so will increase the rounding error. The factor e^{Dx_k} in (8.87) will generally be small. It indicates, however, that if x_k is *very* large, then the global error may eventually tend to grow.

An important consequence of this estimate is the relation between the global error and the starting values. In particular, according to (8.87) it is not sensible to use a low-order starting method for a high-order multistep method. Ideally the terms $A\sigma$ and Ch^p should be comparable. Thus, for example, the fourth order Runge–Kutta method should be used to determine starting values for the fourth order Adams method.

EXERCISES

1 Complete the proof of estimate (8.56) by showing that $|e^{kh} - (1 + h)^k| \leq \frac{1}{2}h(e^{kh} - 1)$.
[*Hint*: Let $\alpha_k = e^{kh} - (1 + h)^k$. Prove that $\alpha_{k+1} = (1 + h)\alpha_k + \frac{1}{2}h^2 e^{\xi_k}$, where ξ_k is some value between 0 and h. From this conclude that $|\alpha_{k+1}| \leq (1 + h)|\alpha_k| + \frac{1}{2}h^2 e$ whenever $|h| \leq 1$. Now use induction to show that $|\alpha_k| \leq \beta_k$, where β_0, β_1, ... is the solution to the difference equation $\beta_{k+1} = (1 + h)\beta_k + \frac{1}{2}h^2 e$, $\beta_0 = 0$. Finally show that this difference equation has the solution $\beta_k = \frac{1}{2}h[(1 + h)^k - 1]$ and that $\beta_k \leq \frac{1}{2}h(e^{kh} - 1)$.)

2 Use Euler's method to solve the equation $y'(x) = ((2/x) - 1)y$, $1 \leq x \leq 5$, $y(1) = 1$. Start with $h = .1$ and use (8.58) to adjust h until the local error (estimate) is approximately 10^{-2}.

3 Prove that the local error for the second order Runge–Kutta method (8.30) is given by (8.60).

4 By expanding $y(x_{k+1})$ out to $O(h^3)$ terms, show that the local error for Euler's method can be written as $\frac{1}{2}y''(x_k)h^2 + O(h^3)$. Similarly, by expanding $y(x_{k-1})$ and $y(x_{k+1})$ out to $O(h^4)$ terms, show that the local error for (8.39a) is $\frac{5}{12}y'''(x_k)h^3 + O(h^4)$.

5 Prove that the method (8.72) has local error (8.73).

6 Prove that the limit, as $k \to \infty$, $h \to 0$, but $hk = x$ is fixed, of $\lambda_1(hw)$ as given below (8.85a) is just e^{wx}. Thus, the approximation $y_k = \lambda_1(hw)^k$ "converges" to the exact solution $y(x) = e^{wx}$.

7 Use the formula $y_{k+1} = y_{k-1} + 2hf(x_k, y_k)$ to solve the equation $y'(x) = -2y(x) + 1$, $y(0) = 2$ for $0 \leq x \leq 5$. Compare your results with the exact solution $y(x) = e^{-2x} + 1$. Now use this method to solve the equation $y'(x) = 2y(x) + 1$, $y(0) = 2$. (This method is weakly unstable.)

8 Write a program to solve (8.3) by using Euler's method. At each step compute y_{k+1} using a value of h and then with $.5h$. Now estimate the local error according to (8.71) and adjust h to keep this error estimate less than 10^{-4}. Test your program on the equation $y'(x) = wy$, $0 \leq x \leq 1$, $y(0) = 1$, with several values of w.

8.4   Practical Aspects of Solving Differential Equations

In this section we summarize the most commonly used methods for solving initial value problems. We also mention some of the difficulties that arise in practical applications.

8.4.1   Summary of the Methods

A variety of methods for solving equation (8.3) were derived in Section 8.2. Section 8.3 was devoted to the study of local error estimates for these methods. It seems worthwhile to gather together the various methods with their appropriate local error estimates.

We begin with the *explicit methods*:

Euler's Method
$$y_{k+1} = y_k + hf(x_k, y_k), \qquad k = 0, 1, \ldots$$
Local error $\frac{1}{2}h^2 y''(\xi_k)$.

Runge–Kutta, order 2
$$y_{k+1} = y_k + \tfrac{1}{2}h[f(x_{k+1}, y_k + hf(x_k, y_k)) + f(x_k, y_k)]$$
Local error $O(h^3)$.

Runge–Kutta, order 4
$$y_{k+1} = y_k + \tfrac{1}{6}(a_1 + a_2 + a_3 + a_4), \qquad k = 0, 1, \ldots$$
with a_1, a_2, a_3, a_4 given by equations (8.31a). Local error $O(h^5)$.

Adams–Bashforth, order 2

$$y_{k+1} = y_k + \tfrac{1}{2}h[3f(x_k, y_k) - f(x_{k-1}, y_{k-1})], \qquad k = 1, 2, \ldots$$

Local error $\frac{5}{12}y'''(\xi_k)h^3$.

Adams–Bashforth, order 4

$$y_{k+1} = y_k + \tfrac{1}{24}h[55f(x_k, y_k) - 59f(x_{k-1}, y_{k-1})$$
$$+ 37f(x_{k-2}, y_{k-2}) - 9f(x_{k-3}, y_{k-3})], \qquad k = 3, 4, \ldots$$

Local error $\frac{251}{720}y^{(5)}(\xi_k)h^5$.

The most commonly used *implicit* methods are the following:

Trapezoidal Rule (also known as *Heun*'s method, or the *Adams–Moulton* method of *order 2*)

$$y_{k+1} = y_k + \tfrac{1}{2}h[f(x_{k+1}, y_{k+1}) + f(x_k, y_k)]$$

Local error $-\frac{1}{12}y'''(\xi_k)h^3$.

Simpson's Rule (also known as *Milne*'s method)

$$y_{k+1} = y_{k-1} + \tfrac{1}{3}h[f(x_{k+1}, y_{k+1}) + 4f(x_k, y_k) + f(x_{k-1}, y_{k-1})],$$
$$k = 1, 2, \ldots$$

Local error $-\frac{1}{90}y^{(5)}(\xi_k)h^5$.
Weakly unstable.

Adams–Moulton Method, order 4

$$y_{k+1} = y_k + \tfrac{1}{24}h[9f(x_{k+1}, y_{k+1}) + 19f(x_k, y_k) - 5f(x_{k-1}, y_{k-1})$$
$$+ f(x_{k-2}, y_{k-2})]$$

Local error $-\frac{19}{720}y^{(5)}(\xi_k)h^5$.

The implicit formulas are generally used as part of a *predictor–corrector* pair. The most common of these pairs are the following:

Second order Adams Predictor–Corrector
Predictor: Adams–Bashforth, order 2.
Corrector: Trapezoidal method.
Local error $\cong \frac{1}{6}(y^{(p)}_{k+1} - y^{(c)}_{k+1})$.

> *Fourth order Adams Predictor–Corrector*
> Predictor: Adams–Bashforth, order 4.
> Corrector: Adams–Moulton, order 4.
>
> Local error $\cong \frac{19}{270}(y_{k+1}^{(p)} - y_{k+1}^{(c)})$
> $\cong \frac{1}{14}(y_{k+1}^{(p)} - y_{k+1}^{(c)})$.

> *Milne's Predictor–Corrector Method*
>
> Predictor: $y_{k+1}^{(p)} = y_{k-3} + \frac{4}{3}h[2f(x_k, y_k) - f(x_{k-1}, y_{k-1})$
> $\qquad\qquad\qquad + 2f(x_{k-2}, y_{k-2})]$
>
> Corrector: Simpson's rule.
> Local error $\cong \frac{1}{29}(y_{k+1}^{(p)} - y_{k+1}^{(c)})$.
> This method is weakly unstable.

Finally we recall the

> *Runge–Kutta–Fehlberg 4/5 Method*
>
> $$y_{k+1} \text{ given by (8.69a)}$$
> $$\bar{y}_{k+1} \text{ given by (8.69b)}$$
>
> Local error estimated by (8.70).

8.4.2 Controlling the Error

As the estimate (8.87) clearly shows, there are two ways to control the global error, namely, to change h or to change p, the order of the method. If our local error estimate for y_{k+1} is too big, then we should recompute y_{k+1} with a smaller value for h, or with a larger value for p. On the other hand, if our local error estimate is extremely small, then efficiency considerations suggest that h should be increased or p decreased before further values y_{k+2}, y_{k+3}, ... are computed. Some techniques related to a change of step size or order will now be discussed.

First, consider how the step size might be adjusted to obtain a reasonable local error; that is, a local error that is not too small nor too large. Suppose the method we are using has local error $E_L(x_k)$ at x_k. If, for a given *error tolerance* ε, we can ensure that

$$|E_L(x_k)| \cong h\varepsilon, \qquad k = 1, 2, \ldots, N, \tag{8.88}$$

then, for a stable method, we can conclude that the global error E_G at x_N satisfies

$$|E_G| \cong Nh\varepsilon. \qquad (8.89)$$

Now suppose we have an estimate $e(h)$ for the local error with step size h. If the method has local error $O(h^{p+1})$, then we must have

$$e(h) \cong \gamma h^{p+1} \qquad (8.90)$$

for some constant γ. In order to adjust h so that (8.89) is satisfied, consider the replacement

$$h \leftarrow \alpha h. \qquad (8.91a)$$

From (8.90) we have

$$e(\alpha h) \cong \gamma \alpha^{p+1} h^{p+1}$$

so that (8.88) holds for the next step if

$$\alpha h\varepsilon \cong |E_L(x_{k+1})| \cong e(\alpha h) \cong \gamma \alpha^{p+1} h^{p+1} \cong \alpha^{p+1} e(h).$$

By "solving" this approximate equation for α, we find

$$\alpha \cong [h\varepsilon/e(h)]^{1/p}. \qquad (8.91b)$$

Thus (8.91) gives a new value for h that will keep the local error at x_{k+1} to a reasonable size. In order to avoid underestimates of the local error, (8.91b) is often modified to

$$\alpha = .9[h\varepsilon/e(h)]^{1/p}. \qquad (8.92)$$

The estimate $e(h)$ can be any of the local error estimates as derived in Section 8.3.2. The following algorithm uses this process in connection with the Runge–Kutta–Fehlberg method (8.69).

Algorithm to Solve (8.3) with the Runge–Kutta–Fehlberg Method

> Input should include a, b, η, an initial value for h, an absolute error tolerance ε, and a subprogram to evaluate $f(x, y)$. The algorithm adjusts h to satisfy (8.85) and (8.86).

1 Input $\{a, b, \eta, h\}$
2 $c_1 \leftarrow \frac{25}{216}$, $c_2 = \frac{1408}{2565}$, $c_3 \leftarrow \frac{2197}{4104}$, $c_4 \leftarrow -.20$
3 $c_5 \leftarrow \frac{1}{360}$, $c_6 \leftarrow \frac{128}{4275}$, $c_7 \leftarrow -\frac{2197}{75240}$, $c_8 \leftarrow \frac{1}{50}$, $c_9 \leftarrow \frac{2}{55}$
4 $c_{10} \leftarrow \frac{3}{8}$, $c_{11} \leftarrow \frac{3}{32}$, $c_{12} \leftarrow \frac{9}{32}$, $c_{13} \leftarrow \frac{12}{13}$, $c_{14} \leftarrow \frac{1932}{2197}$
5 $c_{15} \leftarrow -\frac{7200}{2197}$, $c_{16} \leftarrow \frac{7296}{2197}$
6 $c_{17} \leftarrow \frac{439}{216}$, $c_{18} \leftarrow \frac{3680}{513}$, $c_{19} \leftarrow -\frac{845}{4104}$
7 $c_{20} \leftarrow -\frac{8}{27}$, $c_{21} \leftarrow -\frac{3544}{2565}$, $c_{22} \leftarrow \frac{1859}{4104}$, $c_{23} \leftarrow -\frac{11}{40}$
8 $x \leftarrow a$, $y \leftarrow \eta$

(*continued*)

9   While $x \leq b$
    9.1 $a_1 \leftarrow f(x, y)$
    9.2 $a_2 \leftarrow f(x + .25h, y + .25a_1 h)$
    9.3 $a_3 \leftarrow f(x + c_{10}h, y + h(c_{11}a_1 + c_{12}a_2))$ (8.93)
    9.4 $a_4 \leftarrow f(x + c_{13}h, y + h(c_{14}a_1 + c_{15}a_2 + c_{16}a_3))$
    9.5 $a_5 \leftarrow f(x + h, y + h(c_{17}a_1 - 8a_2 + c_{18}a_3 + c_{19}a_4))$
    9.6 $a_6 \leftarrow f(x + .5h, y + h(c_{20}a_1 + 2a_2 + c_{21}a_3 + c_{22}a_4 + c_{25}a_5))$
    9.7   est $\leftarrow -h(c_5 a_1 + c_6 a_3 + c_7 a_4 + c_8 a_5 + c_9 a_6)$
    9.8 $\alpha \leftarrow .9(h\varepsilon/\text{est})^{1/4}$
    9.9   If $\alpha \geq 1$ Then
         9.9.1 $y \leftarrow y + h(c_1 a_1 + c_2 a_3 + c_3 a_4 + c_4 a_5)$
         9.9.2   Output $\{x, y\}$
         9.9.3 $h \leftarrow \alpha h$
         9.9.4 $x \leftarrow x + h$
         Else
         9.9.5 $h \leftarrow \alpha h$
    9.10 Continue

Step 9.8 computes the value of α according to (8.92) with $e(h)$ given by (8.70). Note that the value of p to be used in (8.70) is 4 since y_{k+1} is determined by a method that has local error $O(h^5)$. If $\alpha \geq 1$, then the local error in y is acceptable, h and x are updated in steps 9.9.3 and 9.9.4, and the next value is computed. If $\alpha < 1$, then the local error in y is too big, so we reduce h in step 9.9.5 and go back to compute a new value for y with this smaller h.

The process for changing h when using a multistep method is much more complicated. The problem is that new initial values must be determined. Consider, for example, a *three-step* method in which an approximation to $y(x_k)$ has been computed. Suppose our step size controller has suggested that h should be reduced. Then the values of y_{k-1} and y_{k-2} must be changed since these no longer correspond to $y(x_k - h)$ and $y(x_k - 2h)$ (see Fig. 8.4).

Fig. 8.4   When h is changed, x_{k-1} and x_{k-2} have new values.

Thus, new values of y_{k-1} and y_{k-2} must be computed. These values can be determined from previous values by using a Runge–Kutta method or by interpolation. That is, if, as in Fig. 8.4, the new values of x_{k-1} and x_{k-2} lie between the old x_{k-2} and x_k, then the new values of y_{k-1}, y_{k-2} can be obtained by evaluating the polynomial that interpolates the approximations y_{k-2}, y_{k-1}, y_k that have just been computed (see Exercise 4.)

The process of restarting a multistep method is simplified somewhat if the algorithm includes methods of several *orders*; that is, several values of p. The reason for this is that lower order methods also require fewer starting values. For example, an algorithm that can choose between Euler's method, the second order Adams–Bashforth and the fourth order Adams–Bashforth methods has at its disposal a one-step, a two-step, and a three-step method. Thus to restart the three-step method, simply use the most recent accurate value of y_k with the one-step method to get y_{k+1}. Then use these values with the two-step method to determine y_{k+2}, and the three-step method can be applied to compute y_{k+3}. Such algorithms are obviously quite complicated. When accuracy as well as efficiency are important, this complexity is entirely justified. Except for the question of how best to determine p, the essential parts of a general-purpose differential equation solver are covered by the previous discussion. The choice of an optimal value for p is beyond our scope.

8.4.3   Stiff Equations

The technique described in the previous paragraph for adjusting h works quite well except for a class of equations called **stiff equations**. To illustrate this type of equation and why our step size controller fails, consider the equation

$$y'(x) = -100 + 100x + 1, \qquad 0 \le x \le 10$$
$$y(0) = 1. \tag{8.94}$$

The solution to this equation is $y(x) = e^{-100x} + x$. When $x_k \ge 1$, say, the exact solution is *very* close to $y(x) = x$, hence the local error in *any* method will be quite small. The estimates for the local error, as described in Section 8.3.2, together with the technique of Section 8.4.2, will allow a rather large value of h to be used. However, (8.94) is similar to the test equation (8.80), but with $w = -100$. For stability we must have $|hw|$ small. Thus it is likely that the step size estimator (8.92) will produce a value of h that destroys the stability of the method.

For example, if Euler's method is used, then the local error for $x \ge 1$ is

$$\tfrac{1}{2}h^2 y''(\xi_k) \cong \tfrac{1}{2}h^2 10^4 e^{-100x} \cong \tfrac{1}{2}h^2 10^{-39}.$$

Thus the local error is sufficiently small for *any* value of $h \le 1$. However, Euler's method applied to (8.94) gives the difference equation

$$y_{k+1} = (1 - 100h)y_k + 100hk + 1, \qquad k = 0, 1, 2, \ldots$$
$$y_0 = 1. \tag{8.95}$$

If $|1 - 100h| > 1$, then this difference equation is unstable (see Exercise 5).

Generally a differential equation is called *stiff* if the solution has a rapidly decaying part and another part that changes slowly. The rapidly decaying part corresponds to the test equation with w large and negative. Thus, in order for $|hw|$ to be small, h will have to be *very* small. On the other hand, as soon as the slowly changing part of the solution begins to dominate, then the local error estimates will indicate that a large value of h is acceptable.

In practice, when we try to solve a stiff equation, the instability caused by too large a value for h also creates a large local error. Thus a rapid growth in local error should be viewed with suspicion (see Exercise 5). Special methods called "stiffly stable" methods should be used in this case. These methods are less efficient than those discussed above and should not be used for nonstiff equations.

EXERCISES

1   Derive the local error estimate

$$E_{\mathrm{L}} \cong \tfrac{1}{29}|y_{k+1}^{(p)} - y_{k+1}^{(c)}|$$

for Milne's predictor–corrector method.

2   Write an algorithm for the Adams–Bashforth/Adams–Moulton predictor–corrector method. Use the fourth order Runge–Kutta formula to compute starting values. Each time the Adams formulas are applied, compute the local error estimate and determine α according to (8.89) with $p = 4$. If $.5 \le \alpha \le 1.5$, do not change h and just proceed to the next point. If $\alpha > 1.5$, double h, whereas if $\alpha < .5$, h should be halved. In either case restart the solution with the Runge–Kutta formula. Test your program on the test equation with several values for w.

3   Write a program to implement Algorithm (8.93). Test your program on the test equation (8.80) with several values for w. Also, try solving the equations given in Exercise 1 of Section 8.1. Observe what happens when the solution becomes singular.

4   Write a step-size-changer program for a three-step method based on the idea of interpolation. That is, with input y_{k-2}, y_{k-1}, y_k, x_k, h, and α, your subprogram should determine the polynomial of degree two that takes the values y_{k-2}, y_{k-1}, y_k at the points $x_k - 2h$, $x_k - h$, and x_k, respectively. Now evaluate the polynomial at the points x_k, $x_k - \alpha h$, $x_k - 2\alpha h$, and call the results $\hat{y}_k$, $\hat{y}_{k-1}$, $\hat{y}_{k-2}$. Output these values together with $\hat{h} = \alpha h$.

5   Write a program to solve (8.94) with Euler's method. At each step estimate the local error by using (8.58b) and adjust h according to (8.92) with $p = 1$. Compare your computed approximations with the exact solution to the equation.

8.5 Boundary Value Problems

The methods discussed so far in this chapter are not directly applicable to solving boundary value problems. In this section we consider two techniques for solving such problems.

8.5.1 Basic Concepts

In this section we shall briefly consider two techniques that are used to solve equations of the form (8.5), that is,

Boundary value problem

$$y''(x) = f(x, y, y'), \qquad a \le x \le b \tag{8.96a}$$

$$y(a) = \eta_1, \qquad y(b) = \eta_2. \tag{8.96b}$$

Recall that the feature that distinguishes (8.96) from a second order initial value problem (8.4) is that (8.96b) involves conditions at *both* end points a and b. The conditions (8.96b) are called boundary conditions for the equation. In practice, a variety of boundary conditions occur, such as

Other boundary conditions

$$y'(a) = \tau, \qquad y'(b) = \sigma \qquad \text{or} \tag{8.97a}$$

$$y(a) + 2y'(a) = 5, \qquad y(b) + 8y'(b) = 6. \tag{8.97b}$$

Equations with boundary conditions such as (8.97a) or (8.97b) can be treated in much the same manner as the simpler equation (8.96). Hence we shall restrict our discussion to boundary conditions of the form (8.96b). See Exercise 1 for some ideas concerning more general conditions.

Under rather weak conditions on f, it can be shown that (8.96) has a unique solution. [For the more general boundary conditions (8.97b), it is necessary to impose some additional conditions.] In the sequel we always assume that a unique solution exists.

The important point to observe about (8.96) is that the methods of Section 8.2 cannot be applied directly. The reason is that all of those methods started at one endpoint $x = a$ and computed the solution step by step. In the case of (8.96) not enough information is given at *either* endpoint to allow a step-by-step solution.

8.5.2 Finite Difference Method

The first method we describe for solving (8.96) gives the approximate solution "all at once." That is, we obtain an approximate solution y_0, $y_1, \ldots, y_k, \ldots, y_N$ as a solution to a set of algebraic equations, rather than as a solution to a difference equation. To illustrate this idea, consider first the **special case** where the right-hand side depends only on x and y. That is, suppose (8.96a) has the form

$$y''(x) = f(x, y), \qquad a \le x \le b.$$

Now replace the second derivative by a *finite difference*:

$$y''(x) \cong [y(x - h) - 2y(x) + y(x + h)]/h^2, \qquad (8.98)$$

where $h > 0$ is some small parameter. According to Exercise 7, this approximation involves a trunction error

$$-\tfrac{1}{12}h^2 y^{(4)}(\xi) \qquad \text{for some } \xi \text{ in } (x - h, x + h).$$

We again use the standard notation

$$x_k = a + kh, \qquad y_k \cong y(x_k), \qquad k = 0, 1, \ldots, N. \qquad (8.99a)$$

Now, however, it is necessary for the last point x_N to be exactly equal to b; hence we must have

$$h = (b - a)/N. \qquad (8.99b)$$

With this notation, we can write (8.98) as

$$y''(x_k) \cong (y(x_{k-1}) - 2y(x_k) + y(x_{k+1}))/h^2. \qquad (8.100)$$

Thus equation (8.96a) can be approximated by

$$[y(x_{k-1}) - 2y(x_k) + y(x_{k+1})]/h^2 \cong f(x_k, y(x_k)).$$

This suggests the approximate equation

$$y_{k-1} - 2y_k + y_{k+1} = h^2 f(x_k, y_k). \qquad (8.101a)$$

Equation (8.101a) makes sense for $k = 1, 2, \ldots, N - 1$. Thus we have $N - 1$ equations for the approximations $y_0, y_1, \ldots, y_N$. But we also have the boundary conditions (8.96b) that give

$$y_0 = \eta_1, \qquad y_N = \eta_2. \qquad (8.101b)$$

Altogether then (8.101) gives $N + 1$ equations for the $N + 1$ unknowns y_0, $y_1, \ldots, y_N$. The equations (8.101), written as a **system of equations**, has the form

$$
\begin{aligned}
-2y_1 + y_2 &= -\eta + h^2 f(x_1, y_1) \\
y_1 - 2y_2 + y_3 &= h^2 f(x_2, y_2) \\
y_2 - 2y_3 + y_4 &= h^2 f(x_3, y_3) \qquad (8.102) \\
&\;\; \ddots \qquad \ddots \qquad \vdots \\
y_{N-2} - 2y_{N-1} &= -\xi + h^2 f(x_{N-1}, y_{N-1}).
\end{aligned}
$$

Here we have substituted (8.101b) directly into (8.101a).

If $f(x, y)$ is *linear* in y, then (8.102) is a system of linear equations for the unknowns $y_1, y_2, \ldots, y_{N-1}$. This system is, in fact, *tridiagonal*, and can be solved by the method of Section 6.2.6. If f is nonlinear in y, then (8.102) is a *system of nonlinear equations*. Such systems can be solved by methods related to those of Chapter 7. The study of such methods is beyond our scope.

The most general kind of equation for which the above process leads to a linear system is

| More general linear equation | $y''(x) = p(x)y'(x) + q(x)y(x) + r(x), \qquad a \le x \le b,$
 $y(a) = \eta, \qquad y(b) = \xi.$ | (8.103) |

That is, the function $f(x, y, y')$ is linear in y and y'. The functions $p(x)$, $q(x)$, and $r(x)$ in (8.103) are given functions that are usually continuous and sometimes differentiable. By using the approximation

$$y'(x) \cong [y(x + h) - y(x - h)]/2h \tag{8.104}$$

together with (8.98), we obtain from (8.103) the **approximate equation**

$$\frac{y(x - h) - 2y(x) + y(x + h)}{h^2}$$

$$\cong p(x)\left(\frac{y(x + h) - y(x - h)}{2h}\right) + q(x)y(x) + r(x).$$

With the usual notation (8.99), and with p_k, q_k, r_k denoting the values $p(x_k)$, $q(x_k)$, $r(x_k)$, respectively, we have the equation

$$\frac{y_{k-1} - 2y_k + y_{k+1}}{h^2} = p_k\left(\frac{y_{k+1} - y_{k-1}}{2h}\right) + q_k y_k + r_k.$$

A simple rearrangement of this gives

$$(1 + \tfrac{1}{2}hp_k)y_{k-1} - (2 + h^2 q_k)y_k + (1 - \tfrac{1}{2}hp_k)y_{k+1} = h^2 r_k. \tag{8.105}$$

This equation is defined for $k = 1, 2, \ldots, N - 1$, and, of course, we also have $y_0 = \eta$, $y_N = \xi$. Thus (8.105) is a sytem of $N - 1$ linear equations for the unknowns $y_1, y_2, \ldots, y_{N-1}$. The sytem (8.105) is tridiagonal and can be solved by the method of Section 6.2.6.

To illustrate this technique, consider the equation

$$y''(x) = -y(x), \qquad 0 \le x \le \pi/2$$
$$y(0) = 1, \qquad y(\pi/2) = 0. \tag{8.106}$$

The linear system (8.105) [or (8.101)] is

$$-(2 - h^2)y_1 + \qquad y_2 \qquad\qquad\qquad\qquad\qquad = -1$$

$$y_1 - (2 - h^2)y_2 + \qquad y_3 \qquad\qquad\qquad = 0$$

$$y_2 - (2 - h^2)y_3 + y_4 \qquad\qquad = 0 \qquad (8.107)$$

$$\ddots \qquad\qquad \ddots \qquad\qquad \vdots$$

$$y_{N-2} - (2 - h^2)y_{N-1} = 0.$$

Some values of the solution to this system with $N = 40$, $h = \pi/80$ are given in Table 8.2. The exact solution is $y(x) = \cos x$.

TABLE 8.2 *Approximate Solution to (8.103), Determined by the Divided Difference Method, with* $N = 40$

k	x_k	y_k	$y(x_k)$
1	0	1.0	0
5	.19634950	.98080120	.98078529
10	.39269900	.92390639	.92387956
15	.58904850	.83150218	.83146968
20	.78539800	.70713996	.70710690
25	.98174749	.55559950	.55557041
30	1.1780970	.38270511	.38268366
35	1.3744465	.19510184	.19509061
40	1.5707963	0	0

The algebraic system (8.105) is obtained from the differential equation (8.96) by using the approximations (8.98) and (8.104). These approximations have truncation errors $O(h^2)$. Thus we should expect that the local errors in the solution y_k to (8.105) will be $O(h^2)$. Now, however, there is no accumulation of errors so the global error is also $O(h^2)$. That is, it can be shown that the solution $y_1, y_2, \ldots, y_{N-1}$ to (8.105) satisfies

$$y_k - y(x_k) = O(h^2).$$

8.5.3 Shooting Methods

In order to solve (8.96) when f is a *nonlinear* function of y or y', it is often necessary to use a kind of trial and error technique called **shooting**. The idea

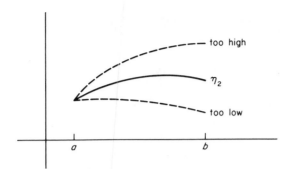

Fig. 8.5 Shooting to find correct $y'(a)$.

here is to try and guess a value α for $y'(a)$, so that the solution $y_\alpha(x)$ of the **initial value problem**

$$y''(x) = f(x, y, y'), \qquad a \le x \le b$$
$$y(a) = \eta_1 \tag{8.108}$$
$$y'(a) = \alpha$$

satisfies $y_\alpha(b) = \eta_2$. That is, we try to guess what value $y'(a)$ must have so that the solution takes the correct value at the other endpoint. (See Fig. 8.5). The initial value problem (8.108) can, of course, be solved by the methods of Section 8.2. That is, (8.108) is reduced via the technique of Section 8.1.2 to a system of two first order equations. This system is then solved as described in Section 8.2.4. Thus, a rough description of a **shooting method** for solving (8.96) is

(1)   Choose a value α.
(2)   Solve (8.108) by a method in Section 8.2.
(3)   If the approximate solution is not suitably close to η_2 at $x = b$, adjust α and go back to step 2.

Many strategies have been proposed for choosing values for α. Most of these are based on the observation that we are really attempting to solve the **nonlinear equation**

$$y_\alpha(b) = \eta_2. \tag{8.109}$$

That is, think of α as an independent variable. For a particular value of α, the value of $y_\alpha(b)$ is found by solving (8.108). We want to find the value of α that satisfies (8.109). To solve (8.109), we could try any of the methods of Chapter 7. The following algorithm uses the secant method 7.12 to try to solve (8.109).

Algorithm to Solve (8.96) by a Secant Shooting Method

> Input consists of values for a, b, η_1, η_2, two starting values α_0, α_1, and a stopping parameter ε. Also a subprogram for solving the initial value problem (8.108) is required.

$$
\begin{aligned}
&1 \quad \alpha \leftarrow \alpha_0 \\
&2 \quad \text{Solve (8.108), set } z_0 = y_N \\
&3 \quad \text{If } |z_0 - \eta_2| \leq \varepsilon, \text{halt.} \\
&4 \quad \alpha \leftarrow \alpha_1 \\
&5 \quad \text{Solve (8.108), set } z_1 = y_N \\
&6 \quad \text{While } |z_1 - \eta_2| > \varepsilon \\
&\quad 6.1 \quad \alpha \leftarrow [\alpha_1(z_0 - \eta_2) - \alpha_0(z_1 - \eta_2)]/(z_0 - z_1) \\
&\quad 6.2 \quad z_0 \leftarrow z_1 \\
&\quad 6.3 \quad \text{Solve (8.108), set } z_1 = y_N \\
&\quad 6.4 \quad \alpha_0 \leftarrow \alpha_1, \alpha_1 \leftarrow \alpha \\
&7 \quad \text{Output } \{y_0, y_1, \ldots, y_N\}
\end{aligned}
$$

(8.110)

In order to simplify this algorithm, we have not included a test for non-convergence. Since each iteration requires the solution of a second order initial value problem, it is unlikely that one would want to do more than a few iterations. That is, actual implementations of this algorithm would probably limit the number of iterations to three or four, at most.

Similar algorithms based on Newton's method or interval techniques can also be constructed. Interactive methods with graphic display terminals are also used for finding an appropriate value of α.

Shooting methods, unlike the finite difference methods, are applicable to nonlinear equations. They are, however, rather expensive to use and can be unstable. For example, if the solution to (8.108) is very sensitive to α, then the iteration used to solve (8.109) may not converge. Sometimes such instability can be avoided by starting at $x = b$ and going *backward*. That is, replace (8.108) with the initial value problem

$$
\begin{aligned}
y''(x) &= f(x, y, y'), \qquad b \geq x \geq a \\
y(b) &= \eta_2 \\
y'(b) &= \alpha.
\end{aligned}
$$

(8.111)

This equation can be solved with the methods of Section 8.2 but with negative values for h.

EXERCISES

1 Consider the equation $y''(x) = f(x, y)$, $a \leq x \leq b$, with boundary conditions (8.97a). Approximate the boundary conditions by the equations

$$[y(a + h) - y(a)]/h = \tau, \qquad [y(b) - y(b - h)]/h = \sigma$$

and use the approximation (8.98). Derive an algebraic system similar to (8.102) for solving the equation with these boundary conditions.

2 For each of the following equations and with $N = 5$, write out the equations (four of them) corresponding to (8.105).

(a) $y''(x) = x^2 y'(x) + y(x)$, $1 \leq x \leq 2$
$y(x) = 1$, $y(2) = 1$

(b) $y''(x) = (\cos x)y'(x) + (\sin x)y(x) + \pi$, $0 \leq x \leq \pi$
$y(0) = 1$, $y(\pi) = 0$

3 Write a program to solve the linear system (8.107). Let N be an input parameter and solve the system for several values of N ($N = 10, 20, 50, \ldots$). Compare your results with the exact solution $y(x) = \cos x$.

4 Write a subprogram to use Euler's method to solve the initial value problem (8.108). Input should include values of a, b, η_1, and α. Now use this subprogram to solve the boundary value problem

$$y''(x) = \tfrac{1}{2}y - [2(y')^2/y], \quad 0 \leq x \leq 1$$

$$y(0) = 1, \quad y(1) = 1.5.$$

Start by using values of $y'(0)$ near 0.

5 Describe a bisection-type algorithm similar to (8.110) for solving (8.109).

6 Write a program to implement Algorithm (8.110). Use any method you like to solve the initial value problem (8.108). Limit the program to a maximum of five iterations. Test the program on the equations

$$y''(x) = -y, \quad 0 \leq x \leq \pi$$
$$y(0) = 1, \quad y(\pi) = -1$$
$$\text{exact solution:} \quad y(x) = \cos x$$

$$y''(x) = y, \quad 0 \leq x \leq 1$$
$$y(0) = 1, \quad y(1) = e$$
$$\text{(exact solution:} \quad y(x) = e^x).$$

7 By expanding $y(x - h)$ and $y(x + h)$ in a Taylor series out to terms involving $y^{(4)}$, show that the truncation error in the approximation (8.98) is given by $\frac{1}{12}h^2 y^{(4)}(\xi)$ for some $\xi \in [x - h, x + h]$.

8 Prove that if $p(x) \geq 0$, $q(x) > 0$ for $a \leq x \leq b$, then the system (8.105) can be solved by the Jacobi or Gauss–Seidel iterative methods, as described in Section 6.8.

Suggestions for Further Reading

A thorough discussion of numerical methods for solving initial value problems is found in Gear (1971). An analysis of the Adams-type methods with FORTRAN programs is given in Shampine and Gordon (1975). Numerical aspects of these techniques are covered by Babuska et al. (1966).

Two-point boundary value problems are studied by Keller (1968). More general boundary value problems (equations in *two* variables) are analyzed in Varga (1962).

APPENDIX

In this appendix we summarize some mathematical results that are used throughout the text. Most of these theorems are proven in a normal first-year calculus course. The reader is referred to any standard calculus text for proofs and further discussions.

Mean Value Theorem   Let $f(x)$ be any function that is continuous on the interval $a \leq x \leq b$ and is differentiable for $a < x < b$. Then, for any two distinct points x, y in the interval $[a, b]$, there is a value ξ between x and y such that

$$[f(x) - f(y)]/(x - y) = f'(\xi).$$

Taylor's Theorem   Let $f(x)$ be a function that has n continuous derivatives for $a \leq x \leq b$. For any values c and x between a and b, there is a value ξ between x and c such that

$$f(x) = f(c) + f'(c)(x - c) + \frac{f''(c)}{2}(x - c)^2$$

$$+ \cdots + \frac{f^{(k)}(c)}{k!}(x - c)^k + \cdots + \frac{f^{(n-1)}(c)}{(n-1)!}(x - c)^{n-1} + E,$$

where

$$E = [f^{(n)}(\xi)/n!](x - c)^n.$$

The above is called the *Taylor series expansion of f around the point c.*

334

Rolle's Theorem Let $f(x)$ be continuous for $a \le x \le b$ and differentiable for $a < x < b$. If $f(x) = f(y) = 0$ for two values x and y between a and b, then there is a value z between x and y for which $f'(z) = 0$.

Intermediate Value Theorem Let $f(x)$ be continuous for $a \le x \le b$. For any x and y between a and b, if u satisfies

$$f(x) \le u \le f(y),$$

then there is a z between x and y with $f(z) = u$.

Mean Value Theorem for Integrals Let $f(x)$ and $g(x)$ be continuous functions for $a \le x \le b$. If $g(x)$ does not change sign in this interval, then there is a value ξ between a and b such that

$$\int_a^b f(x)g(x)\,dx = f(\xi)\int_a^b g(x)\,dx.$$

Implicit Function Theorem Let $f(x, y)$ be continuous and have continuous partial derivatives with respect to x and y for all x and y in some region. Suppose (x_0, y_0) is a point at which $f(x_0, y_0) = 0$. If $(\partial f(x_0, y_0)/\partial x) \ne 0$, then for all y sufficiently close to y_0, $f(x, y) = 0$ has a unique solution $x = g(y)$, where g is also continuously differentiable; in fact,

$$\frac{dg}{dy}(y) = -\left(\frac{\partial f}{\partial x}(g(y), y)\right)^{-1}\frac{\partial f}{\partial y}(g(y), y).$$

BIBLIOGRAPHY

Abramowitz, M., and Stegun, I. eds. (1964). "Handbook of Mathematical Functions with Formulas, Graphs, and Mathematical Tables." National Bureau of Standards, Applied Mathematics Series 55, U.S. Government Printing Office, Washington, D.C.

Ahlberg, J. H., Nilson, E. N., and Walsh, J. L. (1967). "The Theory of Splines and Their Applications." Academic Press, New York.

Babuška, I., Prager, M., and Vitasek, E. (1966). "Numerical Processes in Differential Equations." Wiley (Interscience), New York.

Brent, R. P. (1972). "Algorithms for Minimization without Derivatives." Prentice-Hall, Englewood Cliffs, New Jersey.

Conte, S. D., and deBoor, C. (1972). "Elementary Numerical Analysis: An Algorithmic Approach," 2nd ed. McGraw-Hill, New York.

Dahlquist, G., Björck, A., and Anderson, N. (1974). "Numerical Methods." Prentice-Hall, Englewood Cliffs, New Jersey.

Davis, P. J., and Rabinowitz, P. (1975). "Methods of Numerical Integration," 2nd ed. Academic Press, New York.

Dejon, B., and Henrici, P., eds. (1969). "Constructive Aspects of the Fundamental Theorems of Algebra." Wiley (Interscience), New York.

Dorn, W. S., and McCracken, D. D. (1972). "Numerical Methods with FORTRAN IV Case Studies." Wiley, New York.

Fike, C. T. (1968). "Computer Evaluation of Mathematical Functions." Prentice-Hall, Englewood Cliffs, New Jersey.

Forsythe, G. E., and Moler, C. B. (1967). "Computer Solution of Linear Algebraic Systems." Prentice-Hall, Englewood Cliffs, New Jersey.

Forsythe, G. E., Malcolm, M. A., and Moler, C. B. (1977). "Computer Methods for Mathematical Computations." Prentice-Hall, Englewood Cliffs, New Jersey.

Gear, C. W. (1971). "Numerical Initial Value Problems in Ordinary Differential Equations." Prentice-Hall, Englewood Cliffs, New Jersey.

Hart, J. F., et al. (1968). "Handbook of Computer Approximations." Wiley, New York.

Henrici, P. (1974). "Applied and Computational Complex Analysis," Vol. 1. Wiley, New York.

Henrici, P. (1964). "Elements of Numerical Analysis." Wiley, New York.

Hildebrand, F. B. (1974). "Introduction to Numerical Analysis," 2nd ed. McGraw-Hill, New York.

Householder, A. S. (1970). "The Numerical Treatment of a Single Nonlinear Equation." McGraw-Hill, New York.

Isaacson, E., and Keller, H. B. (1966). "Analysis of Numerical Methods." Wiley, New York.

Keller, H. B. (1968). "Numerical Methods for Two-point Boundary Value Problems." Ginn/Blaisdell, Waltham, Massachusetts.

Knuth, D. C. (1969). "The Art of Computer Programming, Vol. 2: Seminumerical Algorithms." Addison-Wesley, Reading, Massachusetts.

Krylov, V. J. (1967). "Approximate Calculation of Integrals." Macmillan, New York.

Lawson, C. L., and Hanson, R. J. (1974). "Solving Least Squares Problems." Prentice-Hall, Englewood Cliffs, New Jersey.

Ortega, J. M., and Rheinboldt, W. C. (1970). "Iterative Solution of Nonlinear Equations in Several Variables." Academic Press, New York.

Ostrowski, A. (1966). "Solution of Equations and Systems of Equations," 2nd ed. Academic Press, New York.

Ralston, A. (1965). "A First Course in Numerical Analysis." McGraw-Hill, New York.

Schultz, M. H. (1973). "Spline Analysis." Prentice-Hall, Englewood Cliffs, New Jersey.

Shampine, L. F., and Gordon, M. K. (1975). "Computer Solution of Ordinary Differential Equations." W. H. Freeman, San Francisco.

Sterbenz, P. H. (1974). "Floating Point Computation." Prentice-Hall, Englewood Cliffs, New Jersey.

Stewart, G. W. (1973). "Introduction to Matrix Computations." Academic Press, New York.

Stroud, A. H., and Secrest, D. (1966). "Gaussian Quadrature Formulas." Prentice-Hall, Englewood Cliffs, New Jersey.

Traub, J. (1964). "Iterative Methods for the Solution of Equations." Prentice-Hall, Englewood Cliffs, New Jersey.

Varga, R. S. (1962). "Matrix Iterative Analysis." Prentice-Hall, Englewood Cliffs, New Jersey.

Wilkinson, J. H. (1963). "Rounding Errors in Algebraic Processes." Prentice-Hall, Englewood Cliffs, New Jersey.

Young, D. M. (1971). "Iterative Solution of Large Linear Systems." Academic Press, New York.

INDEX

Computer Science and Applied Mathematics

A SERIES OF MONOGRAPHS AND TEXTBOOKS

Editor

Werner Rheinboldt

University of Maryland

HANS P. KÜNZI, H. G. TZSCHACH, and C. A. ZEHNDER. Numerical Methods of Mathematical Optimization: With ALGOL and FORTRAN Programs, Corrected and Augmented Edition

AZRIEL ROSENFELD. Picture Processing by Computer

JAMES ORTEGA AND WERNER RHEINBOLDT. Iterative Solution of Nonlinear Equations in Several Variables

AZARIA PAZ. Introduction to Probabilistic Automata

DAVID YOUNG. Iterative Solution of Large Linear Systems

ANN YASUHARA. Recursive Function Theory and Logic

JAMES M. ORTEGA. Numerical Analysis: A Second Course

G. W. STEWART. Introduction to Matrix Computations

CHIN-LIANG CHANG AND RICHARD CHAR-TUNG LEE. Symbolic Logic and Mechanical Theorem Proving

C. C. GOTLIEB AND A. BORODIN. Social Issues in Computing

ERWIN ENGELER. Introduction to the Theory of Computation

F. W. J. OLVER. Asymptotics and Special Functions

DIONYSIOS C. TSICHRITZIS AND PHILIP A. BERNSTEIN. Operating Systems

ROBERT R. KORFHAGE. Discrete Computational Structures

PHILIP J. DAVIS AND PHILIP RABINOWITZ. Methods of Numerical Integration

A. T. BERZTISS. Data Structures: Theory and Practice, Second Edition

N. CHRISTOPHIDES. Graph Theory: An Algorithmic Approach

ALBERT NIJENHUIS AND HERBERT S. WILF. Combinatorial Algorithms

AZRIEL ROSENFELD AND AVINASH C. KAK. Digital Picture Processing

SAKTI P. GHOSH. Data Base Organization for Data Management

DIONYSIOS C. TSICHRITZIS AND FREDERICK H. LOCHOVSKY. Data Base Management Systems

WILLIAM F. AMES. Numerical Methods for Partial Differential Equations, Second Edition

ARNOLD O. ALLEN. Probability, Statistics, and Queueing Theory: With Computer Science Applications

ALBERT NIJENHUIS AND HERBERT S. WILF. Combinatorial Algorithms. Second Edition.

JAMES S. VANDERGRAFT. Introduction to Numerical Computations

In preparation

AZRIEL ROSENFELD. Picture Languages, Formal Models for Picture Recognition

ISAAC FRIED. Numerical Solution of Differential Equations